D0847641

Elements of Early Modern Physics

ELEMENTS OF EARLY MODERN PHYSICS

J. L. Heilbron

125548

UNIVERSITY OF CALIFORNIA PRESS
Berkeley • Los Angeles • London

University of California Press
Berkeley and Los Angeles, California

University of California Press, Ltd.
London, England

© 1982 by
The Regents of the University of California

Printed in the United States of America

1 2 3 4 5 6 7 8 9

Library of Congress Cataloging in Publication Data

Heilbron, J. L.
Elements of early modern physics.

Chapter 1 and 2 reprinted with minor corrections from the author's Electricity in the 17th and 18th centuries. Chapter 3 condenses and updates original account of electricity.
Bibliography: p.
Includes index.
1. Electricity—History. 2. Physics—History. I. Title.
QC507.H482 537′.09032 81-40327
ISBN 0-520-04554-8 AACR2
ISBN 0-520-04555-6 (pbk.)

QC
507
.H482
1982

Contents

Preface

There is no synthetic history of early modern science that meets contemporary standards of scope and scholarship. It would be good to have one, not only for itself but also as a buttress and correction to the rapidly growing historiography of modern science. The up-to-date synthesizer must attend to institutions as well as to ideas, to the context as well as to the content of science. A few elements of such a synthesis make up the two introductory chapters of my book *Electricity in the 17th and 18th Centuries: A Study of Early Modern Physics* (1979). Some reviewers have suggested that these chapters be issued separately for the use of students, or, to put the matter in a fairer light, that they be made available as an inexpensive stopgap until a closer approximation to a proper synthesis arrives. The chapters are reprinted here except for the correction of a few misprints.

I have not been able to stop there. Because of the structure of the original book, examples from the history of electricity were not often included in the introductory chapters. No description of early modern physics that omits electricity could qualify even as a stopgap. I have accordingly rewritten and condensed the original account of electricity into a third chapter to complete these *Elements*. The rewriting enabled me to tie the history of electricity closer to general themes than the larger format allowed and to incorporate new material about the study of electricity at the Royal Society of London during Newton's presidency.

The first of the book's three chapters presents the general principles to which physical theory at different times conformed or that otherwise mediated its development: peripatetic philosophy, corpuscularism, Newton's attractions, Newtonian forces and fluids, the impulse towards quantification. Where the ground has been tilled before, I have emphasized application rather than analysis of principles. The chapter opens with an account of the changing meaning and scope of 'physics' and closes with examples of the successful mathematizing of its newer branches. These sections break new ground.

The second chapter describes the institutional frameworks in which physics was cultivated in the seventeenth and eighteenth centuries. I had two purposes in mind when preparing it. The first was to show the opportunities offered and the constraints imposed by organized learning; historians of science often qualify a person as a member of this or that society, academy, or religious order, or as a professor here or there, without explaining the relevance of the affiliation (if any) to the matter at hand. The second purpose was to provide the beginnings of a demography of physicists, their numbers, salaries, and career goals. These factors conditioned the pace and extent of study of natural phenomena at a level of support and urgency quite different from what physicists enjoy today or had a century ago.

The single most important contributor to the support of the study of physics in the seventeenth century was the Catholic Church and, within it, the Society of Jesus. From about 1670 to about 1750, private lecturers played an important part in keeping up 'experimental philosophy;' while throughout most of the eighteenth century universities and academies dominated the investigation of physical phenomena. I consider each group in turn, Jesuits, academicians, professors, and private lecturers. Cross-national comparisons are made where useful and practicable.

Chapter III presents the case of electricity. I chose it for several reasons. Firstly, the magnitude of its advance. The subject came into existence about 1600, with an inventory of bodies able to perform electrical attraction and a misleading, qualitative theory of its true cause. By 1800 electricians had abandoned the search for true causes, worked out the principles of electrostatics, established the basis for a mathematical theory, and opened the vast new domain of galvanism. In these particulars electricity was the bellwether of the flock of physical sciences created during the Scientific Revolution. Secondly and thirdly, electricity was unique among branches of Enlightenment physics in amusing the public, who enjoyed seeing others shocked, and in showing, in treatments for paralysis and lightning, that science might be useful. Electricity became the exemplar of physical science during the eighteenth century, whence the propriety of taking its history as illustrative of early modern physics.

The example brings significant new results and interpretations. Much of the historiography of early modern science has centered on the development of terrestrial and celestial mechanics, on the spread of the 'corpuscular philosophy,' and on the grand cosmological disputes, or squabbles, between the sectaries of Descartes and of Newton. I find, however, that despite their disagreement over theory, in practice the Newtonian experimental philosopher thought in much the same terms as his Cartesian counterpart, aether being to the one what subtle matter was to the other; that each side held experiment in high esteem; and that the achievement of quantification confounded the programs of both. Again, the 'Copernican Revolution' does not adequately represent the transition from medieval natural philosophy to classical physics. The bullish personality of Galileo, local jealousies, the post-Tridentine paranoia of the Roman Church, and the apparent bearing of scripture on questions of cosmic geometry combined to introduce into astronomy issues that divided men otherwise able to cooperate in the creation of a new science. Galileo's propagandistic masterpiece, the *Dialogue on the Two Chief Systems of the World*, still hoodwinks historians into believing that peripatetics contributed nothing to the Scientific Revolution but unreasoning opposition. Study of the development of electricity, which was theologically and cosmologically neutral, points the way to a juster estimate of the contributions, the expectations, and the changing composition of the early modern physicists.

The belief, common among historians who concern themselves only with Britain and France, that university professors made only a small and continually declining contribution to natural philosophy during the seventeenth and eighteenth centuries also fails before the facts. Between one-third and one-half of the electricians whose work is noticed in *Electricity* were affiliated with universities. Preliminary study of the early histories of meteorology and thermodynamics gives a similar result. The institute of physics, usually considered an invention of the nineteenth century, may be discerned at a few leading universities at the end of the Ancien Régime.

Some reviewers have found it difficult to accept the finding that not metaphysical commitments but new instruments gave the main impulse to the development of electrical theory. Their resistance is consonant with a pervasive bias in the recent historiography of science: the tendency to make general theory, or world view, or deep principle, the driving force in the growth of scientific ideas. Our case history shows that the metaphysics, of the paradigms and research programs supposed to guide scientists are seldom close enough to experimental work and theory construction to order them in useful ways.

It is a pleasure again to thank P. Forman, G. Freudenthal, R. Hahn, R. Home, T. S. Kuhn, A. Quinn, and S. Weart for valuable suggestions about the original manuscript; the Università Gregoriana (Rome), the Biblioteca Nazionale-Centrale (Florence), the Académie des Sciences (Paris), the Royal Society and the British Library (London), the Royal Observatory (Herstmonceux), the Deutsche Staatsbibliothek (Berlin), the American Philosophical Society (Philadelphia), the Bancroft Library (Berkeley), the Yale University Library, and the Burndy Library (Smithsonian Institution, Washington, D.C.), for permission to quote unpublished material; and the staff of the Office for History of Science and Technology at the University of California, Berkeley, for their intelligence and vigilance in preparing the final typescript.

A Note on the Notes

The notes give abbreviated titles of books and omit those of journal articles; full titles of both and other pertinent information will be found in the Bibliography. Arabic italics are used for volume numbers of journals, roman numerals for individuals of a multi-volume work or of a manuscript collection (excepting the Sloane Mss.). References within this book are given in the form '*infra* (or *supra*), XII.2,' meaning chapter XII, section 2. Small superscripts thus (^{2}) indicate the edition cited. 'x:y (1900)' signifies part or item y of volume x; if the volume number is not given, the form is '1900:y.' The following abbreviations are also used:

ADB	*Allgemeine deutsche Biographie.* 56 vols. Leipzig, 1875–1912.
AKSA	Kungl. Svenska Vetenskapsakademien, Stockholm. *Der Königl. schwedischen Akademie der Wissenschaften, Abhandlungen, aus der Naturlehre, Haushaltungskunst und Mechanik.* Tr. Abraham Gotthelf Kästner. Hamburg and Leipzig, 1749–65, Leipzig, 1766–83.
AS	Académie des Sciences, Paris.
BF	Benjamin Franklin.
BL	British Library.
BFP	Benjamin Franklin. *Papers.* Ed. Leonard W. Labaree et al. New Haven, 1959+.
BFP (Smyth)	Benjamin Franklin. *The Writings of Benjamin Franklin.* Ed. A. H. Smyth. 10 vols. New York, 1905–7.
CAS	Akademii͡a nauk S.S.S.R., Leningrad. *Commentarii academiae scientiarum imperialis petropolitanae.*
CK	Carteggio Kircheriano. Mss. of Athanasius Kircher (mainly incoming letters). 13 vols. Università Gregoriana, Rome.
DBI	*Dizionario biografico degli italiani.* Rome, 1960+.
DM	William Gilbert. *De magnete.* London, 1600. *DM*(Mo) signifies the translation by P. F. Mottelay (New York, 1893); *DM*(Th) that of S. P. Thompson (London, 1900).
DNB	*Dictionary of National Biography.* 22 vols. Oxford, 1921–2.
DSB	*Dictionary of Scientific Biography.* New York, 1970+.
EO	Benjamin Franklin. *Experiments and Observations on Electricity.* *EO* without further qualification will refer to the annotated edition of *EO*^{5} (London, 1774), published by I. B. Cohen as *Benjamin Franklin's Experiments* (Cambridge, 1941). Other editions, when required, will be plainly indicated, e.g., *EO*^{1}, the first English edi-

	tion (London, 1751–4); *EO* (Dal[2]), the second French edition of T. Dalibard (Paris, 1756); *EO* (Wilcke), the German version of J. C. Wilcke (Leipzig, 1758). For titles of these editions see under Franklin in the Bibliography.
FN	I. B. Cohen. *Franklin and Newton*. Philadelphia, 1956.
GGA	*Göttingische gelehrte Anzeigen*.
GM	*Gentleman's Magazine*.
HAS	Académie des Sciences, Paris. *Histoire*.
HAS/Ber	Akademie der Wissenschaften, Berlin. *Histoire*.
JB	Royal Society of London. Journal Book.
JHI	*Journal of the History of Ideas*.
JP	*Journal de physique*.
MAS	Académie des Sciences, Paris. *Mémoires*.
MAS/Ber	Akademie der Wissenschaften, Berlin. *Mémoires*.
Mss. Gal.	Manoscritti Galileiani. Biblioteca Nazionale-Centrale, Florence.
NCAS	Akademii͡a nauk S.S.S.R., Leningrad. *Novi commentarii academiae scientiarum imperialis petropolitanae*.
PT	Royal Society of London. *Philosophical Transactions*.
RHS	*Revue d'histoire des sciences et de leurs applications*.
RS	Royal Society of London.
RS Edin.	Royal Society of Edinburgh.
Sloane Mss.	Sloane Manuscripts. British Library, London.
VE	Alessandro Volta. *Epistolario*. 5 vols. Bologna, 1949–55.
VO	Alessandro Volta. *Le Opere*. 7 vols. Milan, 1918–29.

NOTE ON CONVERSIONS

The basic units of reference are the Paris livre of 1726 (silver) and the louis of 24 livres (gold). The value of any other currency is taken as the ratio of its precious metal content, as given by Martini, *Metrologia* (1883), to that of the livre or the louis. Some frequently used conversions:

Currency	*Symbol*	*Equivalent*
Livre	#	
British pound/guinea	£/gn.	25#
Dutch florin	f.	2#
Reichsthaler	RT	4#
Swedish daler (copper)	Dkmt	⅓#

CHAPTER I

Physical Principles

1. THE SCOPE OF 'PHYSICS'

At the beginning of the seventeenth century 'physics' signified a qualitative, bookish science of natural bodies in general. It was at once wider and narrower than the subject that now has its name: wider in its coverage, which included organic and psychological as well as inorganic phenomena; and narrower in its methods, which recommended neither mathematics nor experiment. The width of coverage and the depreciation of mathematics derived from Aristotle; the indifference to experiment, as opposed to everyday experience, from the authors of peripatetic textbooks.

The *libri naturales*, or physical books of the Aristotelian corpus, begin with a treatise called *Physica*, which sets the categories of analysis of all natural bodies: form, matter, cause, chance, motion, time, place. After this *physica generalis* come the treatises of *physica specialis* or *particularis*, applications of the general principles to the heavens *(De caelo)*, to inorganic nature *(De generatione et corruptione, Meteorologica)*, to organic nature *(De partibus animalium)*, and to man *(De anima)*. The text books of the early seventeenth century offered epitomes of these ancient works, or rather epitomes of sixteenth-century handbooks, of which the most influential were the compendia of J. J. Scaliger and the enlightened commentaries of the Coimbra Jesuits.[1] Typical texts of the early seventeenth century are the *Idea philosophiae naturalis* (1622) of Frank Burgersdijck and the *Physiologia peripatetica* (1600) of Johannes Magirus, both long-lived, widely used and often reprinted, and now food only for the ultimate epitomizer, the historian.[2]

The authors of these texts were not physicists in the modern sense, but either professional philosophers or beginning physicians awaiting preferment in the practice of medicine. Above all the textbook writers were pedagogues, who aimed to supply not new material, but an improved arrangement of the old. One

1. Reif, *J. Hist. Ideas*, 30 (1969), 17–23; Ruestow, *Physics* (1973), 6, 17. The *Commentarii conimbricences* were widely used even in Protestant universities, e.g., Cambridge (H. F. Fletcher, *Milton*, II [1961], 169, 561), Oxford at least until 1678 (Reilly, *Line* [1969], 13), and many German schools. The Coimbra Jesuits define natural philosophy as the study of elements and of the bodies compounded of them. Coll. conim., *Comm. in Phys.*, I (1602), cols. 17–23.

2. Allen, *JHI*, 10 (1949), 225, 240, 243; Costello, *Schol. Curr.* (1958), 83–102; Ruestow, *Physics* (1973), 16–32; Thorndike, *Hist.*, VII (1958), 402–6; and H. F. Fletcher, *Milton*, II (1961), 168–75, including Magirus' table of contents.

went so far as to recommend false doctrines properly ordered over sound ones badly digested.[3] None suggested that physics might be advanced by experiment. The subtitle of Burgersdijck's *Idea* makes his objective clear enough: 'methodus definitionum et controversiarum physicarum,' a handbook of definitions and disputations for students wishing to wrangle over physics.[4] 'There are more things in Heaven and Earth, Horatio, than are dreamt of in your philosophy,' says Prince Hamlet. 'And more things in our compendia of physics,' answers a textbook writer, 'than can be found in earth or heaven.'[5]

For up-to-date general texts and reference works on physics, which describe experiments and the instruments used to perform them, one must look to books on natural magic, to J. C. Sturm's *Collegium curiosum* or the compendia of the Jesuit polymaths. Later we shall examine this literature, which kept the study of electricity alive during the seventeenth century. Now we need only confirm that, like early-modern physics, natural magic included all the sciences, physical and biological. According to Gaspar Schott, S. J., perhaps the best writer on the subject, 'magia universalis naturae et artis' covers vision, light, and everything pertaining to them; sound and hearing, with like accessories; white magic or applied mathematics; and the hidden, rare and uncommon things, the secrets of stones, plants and animals.[6]

The quantified portions of physical science fell not to physics in the seventeenth century but to 'mixed' or 'applied' mathematics, which customarily included astronomy, optics, statics, hydraulics, gnomonics, geography, horology, fortification, navigation and surveying. The association of mathematics with application gave philosophers who did not understand it a colorable cause to despise it; as John Wallis wrote of his experience at Cambridge in the 1630s, 'Mathematicks . . . were scarce looked upon as Academical Studies, but rather Mechanical, as the business of Traders, Seamen, Carpenters, Surveyers of Lands, or the like, and perhaps some Almanack-makers in London.' Neglect of mathematics in the English universities was doubtless linked to its odor of practicality, just as emphasis on it at Gresham College, London, reflected the concerns of City merchants and tradesmen.[7] 'Arithmetic,' says John Webster, in his well-known attack on the universities, '[is] transmitted over to the hands of Merchants and Mechanicks;' geometry is the province of 'Masons, Carpenters, Surveyors;' as for applied mathematics par excellence, 'in all the scholastick learning there is not found any piece . . . so rotten, ruinous, absurd and de-

3. Reif, *JHI*, 30 (1969), 29 quoting Bartholomaus Keckermann.

4. Ruestow, *Physics* (1973), 16.

5. Lichtenberg, *Aphorismen* (1902–8), III: 2, 37–8.

6. Schott, *Magia opt.* (1671), sig. 003v–004r. Cf. Heinrichs in Diemer, ed., *Wissenschaftsbegriff* (1970), 42. The Coimbra Jesuits take natural magic to be applied physics; Coll. conim., *Comm. in Phys.*, I (1602), cols. 18, 25.

7. Allen, *JHI*, 10 (1949), 226, 228, 231, 249; Johnson, *JHI*, 1 (1940), 430–4; Greaves, *Puritan Rev.* (1969), 65–7, 70; C. Hill, *Intell. Orig.* (1965), 122–3; Wallis in Scriba, RS, *Not. Rec.*, 25 (1970), 27.

formed' as Oxbridge astronomy.[8] No doubt Wallis and Webster exaggerated, but even those who defended the universities against the charge of neglect of mathematics conceded its tie to practical application.[9]

In the Jesuit schools, mathematics was taught, and taught well, but only in the vernacular, while the philosophy course spoke and wrote Latin. The mathematicians were so indulged because their technical terms, particularly those relating to fortification, could not be translated conveniently into the language of Cicero.[10] Perhaps the greatest mathematician trained by the Jesuits, Descartes, left school, he says, with the conviction that mathematics was 'useful only in the mechanical arts.'[11] Quantifying physics therefore implied a radical readjustment of the divisions of knowledge, including the downgrading of physics from philosophy to applied mathematics. It would be an uncomfortable process.[12]

'Physics' continued to be understood in its Aristotelian extent throughout the seventeenth century. Molière's bourgeois gentilhomme asks his philosophical tutor what physics is and receives in reply, '[the science] that explains the principles of natural things, and the properties of bodies; that discourses about the nature of the elements, metals, minerals, stones, plants and animals; and [that] teaches us the causes of all the meteors.' Molière's friend, the Cartesian physicist Jacques Rohault, says the same ('the science that teaches us the reasons and causes of all the effects that Nature produces') and he tries to give an account of everything, including human psychology, in a physics text that had a peculiarly long life.[13] John Harris' *Lexicon technicum* (1704) boils down Rohault's definition to 'the Speculative Knowledge of all Natural Bodies,' and adds angelology, on the authority of Locke. Then there is Sturm's important *Physica electiva,* which does not treat the organic world; not because the subject was foreign to physics, 'naturae seu naturalium rerum scientia,' but because Sturm died before he could reach it, only 2200 pages into his work.[14] Meanwhile nothing stopped or replaced Rohault's treatise, which reached a twelfth French edition in the 1720s, and frequently came forth in Latin, fur-

8. J. Webster, *Acad. exam.* (1654), 41–2.

9. Wilkins and Ward, *Vind. acad.* (1654), 36, 58–9; H. F. Fletcher, *Milton,* I (1956), 363–70, II (1961), 310, 314. Cf. Costello, *Schol. Curr.* (1958), 102–4.

10. Dainville, *XVII*[e] *siècle,* no. 30 (1956), 62–8. What mathematics was taught at Oxbridge also used the vernacular, e.g., Blundeville's *Exercises;* H. F. Fletcher, *Milton,* II (1961), 311–21.

11. Descartes, *Discourse* [1637] (1965), 8. Cf. Boutroux, *Isis,* 4 (1921–2), 276–94, for the separation of mechanics (i.e., mathematical theory of simple machines) from physics in the 17th century.

12. See Ruestow, *Physics* (1973), 111–12, for the interesting case of B. de Volder.

13. Molière, *Bourgeois gentilhomme* (1667), Act 2, sc. 6; Rohault, *Traité* (1692[6]), I, 1. Cf. Fontenelle, 'Préface' to *HAS* (1699), vii: 'Ce qui regarde la conservation de la vie appartient particulièrement à la physique.'

14. J. Harris, *Lexicon technicum* (1704), s.v. 'Physicks'; Sturm, *Phys. electiva* (1697–1722). Cf. Schimank in Manegold, *Wissenschaft* (1969), 456–7; and Hartsoeker, *Conjectures physiques* (1706), who promises to complete his text with an account of biology.

nished with the notes of Samuel Clarke, which grew increasingly and belligerently Newtonian. Clarke's last version, translated into English by his brother John in 1723, was still used at Cambridge in the 1740s, long after its generous conception of the scope of physics—not to mention its Cartesian text and strange notes—was outmoded.[15]

Adoption of the modern meaning of 'physics,' like the developments in science it reflected, did not come abruptly. The word continued to be used in its older, broader sense even as it was being qualified and specialized. The lexicons naturally retained the oldest usage: Richelet (1706) gives the science of 'the causes of all natural effects,' and it is the same in the standard dictionaries in the chief European languages throughout the century. An exception is Johnson's *Dictionary* (1755 ff.), which has no entry for 'physics'; for 'physical' it offers a choice among 'relating to material or natural philosophy,' 'medicinal,' and, what some might prefer, 'not moral.'[16] Paulian's *Dictionnaire de physique* includes botany and physiology; that of Monge and his collaborators (1793) rejects them after showing their impropriety in entries for 'abeille' and 'abdomen.'[17] This subtle rejection scarcely ended the use of physics in the old inclusive sense. In the guide to 'Wissenschaftskunde,' as practiced in the Braunschweig gymnasium in 1792, we learn once again that physics is the science of 'all things that make up the Körperwelt,' and properly includes medical subjects as well as natural history.[18] The *Journal de physique,* founded in 1773, calls for papers in natural history; a leading German scientist recommends the study of agriculture as 'such an interesting branch of physics;' and the Paris Academy of Sciences in 1798 offers a prize in 'physics' for the best paper on 'the comparison of the nature, form and uses of the liver in the various classes of animals.'[19]

Yet the *Journal de physique* had, among its subclassifications, one for 'physique' in the modern sense, under which it published papers on mechanics, electricity, magnetism, and geophysics. Since these papers made up less than half the journal, most of the items in a periodical ostensibly devoted to physics were not classed as physics by its editors. Other examples of the simultaneous use of 'physics' in the ancient and modern senses may be found in the class designations of learned societies. Originally the Paris Academy had two

15. Hoskin, *Thomist,* 24 (1951), 353–63; Casini, *L'universo-macchina* (1969), 112–36; Hans, *New Trends* (1951), 51.

16. Richelet, *Dictionnaire françois,* 604; 'physics' had not yet appeared in the 1818 edition of Johnson's *Dictionary,* although *Encycl. Brit.* (1771), III, 478, allows it as a synonym for 'natural philosophy.' Cf. the *Vocabulario* of the Accademia della Crusca; J. C. Adelung, *Grammatisch-kritisches Wörterbuch* (1793[2]); P. C. V. Boiste, *Dictionnaire universel* (1823[6]).

17. Cf. Silliman, *Hist. Stud. Phys. Sci.,* 4 (1974), 140.

18. Eschenberg, *Lehrbuch* (1792), 169, 198–9, 217.

19. *JP,* 1 (1773), vii; Achard to Magellan, 6 Aug. 1784, in Carvalho, *Corresp.* (1952), 107; *MAS,* 1 (1798), ii–ix. Cf. d'Alembert's proposal of 1777 for a 'prix de physique' for questions in anatomy, botany and chemistry. Maindron, *Rev. sci.,* 18 (1880), 1107–17.

classes, one 'mathematical' (geometry, astronomy, mechanics), the other 'physical' (anatomy, biology, chemistry). In 1785 it added two new subclasses, experimental physics and natural history / mineralogy. Experimental 'physics' (new meaning) went into the class of mathematics, and natural history into that of 'physics' (old meaning). A similar juggle occurred in naming the divisions of the Koninklijke Maatschappij der Wetenschappen in 1807. The subgroup 'physics' then fell into the class of 'experimental and mathematical sciences' along with, and distinct from, anatomy, botany, chemistry, etc. In a draft of the organization, however, the class had been called 'physical and mathematical sciences' and the subgroup, 'experimental physics.'[20] The draft employs the old usage and the final version the new.

EXPERIMENTAL PHYSICS

The chief agent in changing the scope of physics was the demonstration experiment. The new instruments of the seventeenth century, and above all the air pump, having been invented, developed, and enjoyed outside the university, began to make their way slowly into the schools at the beginning of the eighteenth century. In discussing, say, the nature of the air, the up-to-date professor of physics not only talked but showed, extinguishing cats and candles *in vacuo* and weighing the atmosphere. Excellent pedagogues, they saw the advantage of similar illustrations of general concepts: the beating of pendula, the composition of forces, the conservation of 'motion' (momentum) in collisions, the principles of geometrical optics, the operation of the lodestone. Virtually the entire repertoire of experiments pertained to physics in the modern sense. There were three chief reasons for this narrowing. First, the biological sciences did not lend themselves readily to demonstration experiments. Second, the established instrument trade, which already made teaching apparatus like globes, telescopes, and surveying gear, could more easily supply the professor of experimental physics the closer his wants to those of his colleagues in applied mathematics. Third, Newton's first English and Dutch disciples, thinking to follow his experimental and mathematical way, radically restricted the purview of natural philosophy.

It is sometimes said that the adjectives in the title of Newton's major work, the *Mathematical Principles of Natural Philosophy* (1687), were intended to emphasize the distance between it and Descartes' *Principles of Philosophy* (1644), which had refashioned traditional physics in a qualitative manner. To Descartes' arrogance, breadth and imprecision Newton opposed caution, narrowness and exactitude: he confined himself to the application of mathematical laws of motion, said to be taken from experiment, to a few problems in mechanics and physical astronomy. Newton's limited mathematical principles

20. Maindron, *Académie* (1888), 50; R. Hahn, *Anatomy* (1971), 99–100; R. J. Forbes, *Marum*, III (1971), 6–7; Guerlac, *Hist. Stud. Phys. Sci.*, 7 (1976), 194–5n.

were immediately advertised as exhaustive in John Keill's *Introductio ad veram physicam* (1702), translated less presumptuously as *Introduction to Natural Philosophy* (1720), which does not pass beyond general mechanics. Keill was perhaps the first lecturer at Oxford to illustrate his course on natural philosophy with experiments; and, as will appear, one of his associates, J. T. Desaguliers, became the leading British exponent of the new experimental physics.[21]

The most influential of the narrowers of physics were the Dutch Newtonians, W. J. 'sGravesande and Pieter van Musschenbroek, whose teaching careers lasted from 1717 to 1761. Both drank in British natural philosophy at its source, 'sGravesande (who began his career as a lawyer) while on a diplomatic mission to London in 1715, Musschenbroek just after graduating M.D. at Leyden the same year. With the help of the Dutch ambassador to England, 'sGravesande, who had kept up a schoolboy interest in geometry, became professor of mathematics and astronomy at Leyden (1717). A few years later he published perhaps the first modern survey of physics, *Physices elementa mathematica experimentis confirmata, sive introductio ad philosophiam newtonianam* (1720–1). It was incontinently translated into English, as *Mathematical Elements of Natural Philosophy,* in two competing editions, one made by Desaguliers, who reached print first by dictating to four copyists at a time, the other overseen by Keill, whose chief help was an old priest ignorant of natural philosophy. And these volumes were only hors d'oeuvres: 'sGravesande's book had two more Latin and four more English versions before he died in 1742.[22]

The French, after attacking 'sGravesande for preferring contrived experiments to 'simple, naive, and easy observations,' and for pretending that there was no physics but Newton's, tried to ignore him. Voltaire did not allow them to do so; he went to Leyden to ask the professor 'whose name begins with an apostrophe' for help in preparing his influential *Eléments de la philosophie de Newton* (1738).[23] When 'sGravesande's book did appear in French, in 1747, it bore the title *Eléments de physique,* etc., suggesting that, by then, 'physique' was understood to mean 'natural philosophy confirmed by experiments.' The inference is confirmed by the enthusiastic review in the *Journal de sçavans* for 1748, which extolled the *Eléments* for its 'very great quantity of curious experiments, which teach about everything now known in physics.' The same journal had earlier praised Musschenbroek's *Essai de physique* (1739) on the same ground.[24] Now both these books, deemed complete, omit the biological and geological sciences, and almost all of chemistry and meteorology.

By the middle of the eighteenth century the British and the French were

21. Schofield, *Mechanism* (1970), 25–8.

22. Brunet, *Physiciens* (1926), 41–2, 51, 75, 96; Allamand in 'sGravesande, *Oeuvres* (1774), II, x–xi, xxi; Torlais, *Rochelais* (1937), 19–20; Ruestow, *Physics* (1973), 117–19.

23. Knappert, *Janus,* 13 (1908), 249–57; 'sGravesande did not think very highly of Voltaire's popularization.

24. Brunet, *Physiciens* (1926), 104–5, 122, 128; Schofield, *Mechanism* (1970), 140–1; Schimank in Manegold, *Wissenschaft* (1969), 468.

composing texts in the Dutch style. Desaguliers wrote an elaborate *Course of Experimental Philosophy* (1734–44) in two volumes quarto that did not cover much more than mechanics.[25] J. A. Nollet issued six volumes of *Leçons de physique* beginning in 1743; except for a short digression on the nature of the senses, in connection with the question of the divisibility of matter, Nollet's lengthy text concerns only mechanics, hydrostatics and hydrodynamics, simple machines, pneumatics and sound, water and fire (from a physical point of view), light, electricity, magnetism, and elementary astronomy. The reviewers were impressed: 'Apart from a few general principles . . . the entire study of physics today reduces to the study of experimental physics.'[26] 'True physics is the science of the Newtons and the Boyles; one marches only with the baton of experiment in one's hands, true physics has become experimental physics.'[27]

In Germany the narrowing of physics was begun independently of the Dutch Newtonians by Christian von Wolff. His *Generally Useful Researches for Attaining to a more Exact Knowledge of Nature and the Arts,* completed in three volumes in 1720/1, describes demonstrations given in his lectures on physics, and every detail, 'to within a hair' as he says, needed to build the instruments to repeat them. 'We must spare no effort and no expense to permit nature to reveal to us what she usually hides from our eyes.' In the event Wolff left her some secrets; he restricted himself to gross mechanics, hydrostatics, pneumatics, meteorology, fire, light, color, sound and magnetism. Only two chapters of the work, some sixty of two thousand pages, concern biology and psychology; the one considers animals chiefly as subjects for investigation *in vacuo,* and the other treats sense organs as examples of optical and mechanical principles. Similarly a representative text of the next generation, J. G. Krüger's *Naturlehere* (1740), esteemed for its 'order, thoroughness and clarity,' gives up less than five percent of its space to plants and animals.[28]

The first important text explicitly to exclude 'the whole theory of plants, animals and man' from its domain was G. E. Hamberger's *Elementa physicae* (1735²), which drew its principles from Wolff's philosophy. Hamberger's book is particularly good evidence of a change in meaning of 'physics' since, as a physician, he might be expected to have advertised biological science where he could. The change in operational meaning was thus explained by the author of an excellent *Institutiones physicae* long used in Austria and Catholic Germany: the etymological meaning of 'physics,' the study of all natural things, 'physics in the largest sense,' is not a practicable subject. He confines himself to

25. Cf. *ibid.*, 463; Schofield, *Mechanism* (1970), 80–8.

26. Desfontaines, *Jugements sur quelques ouvrages nouveaux,* IV (1744), 49, quoted by Brunet, *Physiciens* (1926), 131n. Brunet emphasizes the Dutch ties of Nollet, who visited Leiden and London in 1736. Cf. *ibid.*, 108–9, 113–14, 117, 124–5, 151.

27. Memorandum of 1762, quoted by Anthiaume, *Collège* (1905), I, 221.

28. Wolff, *Allerh. Nützl. Vers.* (1745–7³), I, Vorr., III, Vorr. and pp. 456–515; Börner, *Nachrichten,* I (1749), 75.

'physica stricte talis,' to general principles, astronomy, and the usual branches of experimental physics.[29]

The best German physics text of the eighteenth century, J.C.P. Erxleben's *Anfangsgründe der Naturlehre,* dates from 1772. It covers the material then standard: motion, gravity, elasticity, cohesion, hydrostatics, pneumatics, optics, heat, electricity, magnetism, elementary astronomy, geophysics. Its third edition (1784), brought up to date by Lichtenberg's incisive notes, sold out in eighteen months. More editions were called for, with still more notes; 'because of the fast trot of physics, much became old or useless while the book was in press.' It was translated into Danish; Volta toyed with an Italian version; while everyone, according to Lichtenberg, rushed to learn German 'for the admirable purpose of being able to read the best that is written in physics in Europe.'[30] There was also something passable in English.[31] None of these fine texts so much as hinted at the earlier intimacy between their subject and the biological sciences.

This liberation, or rather the demonstration experiment that effected it, had its dark side for serious savants. Demonstrations became too popular; people, even students, came to physics lectures expecting to be entertained. Kästner says that he gave up teaching from Erxleben's text because most of his students only 'wished to see physics, not to learn anything about it.'[32] A French school teacher at the turn of the century, Antoine Libes, scolds Nollet for serving up hasty, uncritical flim-flam, the 'plaything of childhood and the instrument of charletanism,' under the 'perfidious name of experimental physics.' 'Physique' had come to have a frivolous connotation. Daire, in his *Epithetes françoises* (1759), gives 'agréable' and 'curieuse' among its synonyms. The *Almanach dauphin* for 1777 names four Parisian practitioners under 'physicien.' One of them, Rabiqueau, operated a cabinet of curiosities filled with automata, 'which he makes play and move when asked [and paid] to'; another, Comus, 'known

29. Biwald, *Inst. phys.* (1779²), I, Prol., §§1–5. Similar moves are made in Maximus Imhof, *Grundriss der offentlichen Vorlesungen über die Experimental-Naturlehre* (1794–5), I, 1, 7, who defines physics in the old sense and lectures on physics in the new; in Beccaria's *Institutiones in physicam experimentalem,* for which see Tega, *Rev. crit. stor. fil.,* 24 (1969), 193 and in C. A. Guadagni, *Specimen experimentorum naturalium* (1779²), who defines physics inclusively and then restricts himself to a very narrow experimental physics, namely mechanics, hydrostatics, pneumatics, optics, for 'ad haec potissimum referri potest' (p. 5, 10–11).

30. Lichtenberg, *Briefe,* II, 220, 306; Herrmann, *NTM,* 6:1 (1969), 70–4, 80; see Brunet, *Physiciens* (1926), 93, for a complaint by Musschenbroek about the rapid outdating of physics texts. A measure of the value of Erxleben's text is the contemporaneous and yet very old-fashioned *Abrégé de physique* by the Berlin academician J. H. S. Formey.

31. By 1780 the British had several good texts to choose among, e.g., Adams or Nicholson, and no reason to consult Lichtenberg/Erxleben. For syllabi of lectures on 'experimental philosophy' at Cambridge toward the end of the century see F. S. Taylor in A. Ferguson, *Nat. Phil.* (1948), 152–3.

32. Kästner to J. E. Scheibel, 1 April 1799, in Kästner, *Briefe* (1912), 218. Cf. Kästner, *Selbstbiographie* [1909], 13, comparing the seriousness of the French officers who attended his courses during the Seven Years' War with the lightness of the German students.

for his extreme sleight of hand,' showed 'physical and magnetic recreations' that always amused the court; the other two, Brisson and Sigaud, were more serious physicists.[33]

Another force besides the demonstration experiment making for specialization of physics was applied mathematics. All our modernizing textbook writers advocated the use of mathematics in physics. 'sGravesande went so far as to place natural philosophy among the branches of mixed mathematics; for physics, he says, comes down to the comparison of motions, and motion is a quantity. 'In Physics then we are to discover the laws of Nature by the Phenomena, then by Induction prove them to be general Laws; all the rest is handled Mathematically.'[34] Musschenbroek and Desaguliers sound the same theme, and even Nollet, although he does without equations.[35] In fact the nature of their primary readership—university students with little mathematics and a general public with none—precluded elaborate proofs or geometrical deductions. Even the best texts do not use calculus; the experiments they serve up are designed not for quantitative analysis, but to assist, convince, and divert students who could not follow mathematical demonstrations.

Nevertheless the expectation that physics should be mathematical helped to redefine the traditional boundary between natural philosophy and mixed mathematics. Dutch and English Newtonians laid claim to optics, mechanics, hydrostatics, hydrodynamics, acoustics, and even planetary astronomy. By 1750 these subjects were recognized as constituting a special borderline, or, as we should say, interdisciplinary, group of 'physico-mathematical' sciences, or even 'mathematical physics.'[36] In each of these sciences, according to d'Alembert, one develops mathematically a single, simple generalization taken from experience as, for example, hydrostatics from the experimental proposition, 'which we would never have guessed,' that pressure within a liquid is independent of direction.[37] To be sure, there were few mathematical physicists—about one for every twenty pure mathematicians, according to an estimate of the early 1760s[38]—and they did not always pay court to experimentalists;

33. Libes, as quoted by Silliman, *Hist. Stud. Phys. Sci.*, 4 (1974), 143; *Almanach dauphin,* as quoted by A. Franklin, *Dict.* (1906), 570; cf. Fourier, *MAS,* 8 (1829), lxxvi; Torlais, *Hist. med.* (Feb. 1955), 13–25.

34. 'sGravesande, *Math. Elem.* (1731^4), I, viii–ix, xii–xiii, xvi–xvii. Cf. Brunet, *Physiciens* (1926), 48–54; Ruestow, *Physics* (1973), 132.

35. *Ibid.*, 133–6; Desaguliers, *Course* (1763^2), I, v; Nollet, 'Discours,' in *Leçons,* I (1764^6), lviii, xci.

36. D'Alembert in *Encyclopédie* (1778^3), XXV, 736, art. 'Physico-mathématiques;' *Prel. Disc.* [1751] (1963), 54–5, 152–5; Karsten, *Phys.-chem. Abh.* (1786), I, 151. Cf. the title of Grimaldi's masterpiece, *Physico-Mathesis de lumine* (1665).

37. D'Alembert in *Encyclopédie* (1778^3), XIII, 613, art. 'Expérimental.' This example became a commonplace among mathematical physicists, e.g., Lagrange, *Mécanique analytique* (1788), in *Oeuvres* (1867), XI, 193; Bossut, *Traité* (an IV^2), I, xix, xxiv, who stresses the difficulty of getting the initial generalization.

38. Beccaria to Boscovich, 31 May 1762 (Bancroft Library, U. California, Berkeley): 'Per venti puri matematici si stenti a trovare un fisico matematico.' Cf. Lambert to Karsten, 15 Sept. 1770

d'Alembert, for example, conveived that a subject once mathematized had no further need of the laboratory.[39] The essential point is not that few complete physicists could be found, but that the ideal had been recognized. The physicist, says Pierre Prevost, should be able to 'calculate, observe and compare.' But, as it is very difficult to excel at everything, and science, like all else, advances by division of labor, the physicist should emphasize either the experimental or the mathematical branch of his discipline.[40]

Since electricity was always considered a physical science, its place in the body of knowledge and its treatment varied with the fortunes of physics as a whole. It first received extended treatment in 1600, as a digression in a book about magnets. It found a place in scholastic compendia either near the magnet, as an example of attraction, or among complex 'minerals,' where the chief electric body, amber, was treated. The Dutch Newtonian texts consider electricity under 'fire,' a reclassification required by their dropping mineralogy and advised by the observation, early in the eighteenth century, of electric discharges in evacuated tubes. In the 1740s, owing to the discovery of spectacular phenomena easily reproduced, electricity became the leading branch of experimental physics, and the most popular source of diverting, and sometimes vapid, demonstrations. ('Electricity can sometimes become weak enough to kill a man.'[41]) As a serious study it commanded many monographs, and won its own extensive, independent section in the textbooks of natural philosophy. Soon it required its own texts, the best of which, Cavallo's *Complete Treatise on Electricity* (1777), spread into three volumes octavo in its fourth edition of 1795.

None of this was mathematical. Electricity, in common with other new experimental sciences of the seventeenth century, proved more difficult to quantify than the traditional subjects of 'physical mathematics.' In the mid-eighteenth century the great quantifier d'Alembert, despairing of yoking electricity to his favorite discipline, left it to the experimenters: 'That is mainly the method that must be followed with phenomena the cause of which reason cannot help us [find], and among which we see connections only very imperfectly, such as the phenomena of magnetism and of electricity.' About twenty years later, in 1776, Lichtenberg, professor of pure and applied mathematics at the University of Göttingen, allowed that the non-mathematical experimenter had done his share: electricity, he said, 'has more to expect from mathematicians than from apothecaries.' Another ten years and another quantifier and

(Lambert, *Deut. gelehrt. Briefw.*, IV:2 [1787], 277): 'Seit vielen Jahren brachte junge Leute von Universitäten kaum etwas mehr als die Mathesin puram mit.'

39. See d'Alembert's encyclopedia articles just cited; *infra,* i.5; Hankins, *D'Alembert* (1970), 94–6. For this attitude d'Alembert was criticized by Lalande (letter to Boscovich, 27 April 1767, in Varićak, Jug. akad. znan. i umjetn., *RAD,* 193 [1912], 239), and Euler by Clairaut (letter to Boscovich, *c.* 1764, *ibid.*, 222): 'Combien un able géomètre qui veut tout tirer de la théorie sans avoir recours aux expériences, peut s'écarter du vrai dans les sciences physico-mathématiques.'

40. Prevost, *Recherches* (1792), vii.

41. Lichtenberg in Erxleben, *Anfangsgründe* (1787⁴), 468, in reference to negative electricity. Cf. Musson and Robinson, *Science* (1969), 85.

organizer, W. C. G. Karsten, professor of mathematics and physics at the University of Halle, considered electricity a part of mathematical physics, although one 'not so entirely mathematical' as mechanics or optics.[42]

Karsten's classification corresponded to contemporary usage. The grouping of electricity (as experimental physics) in the class of mathematics by the Paris Academy in 1785 has been mentioned. A similar but subtler transformation occurred at the Petersburg Academy. From 1726 to 1746 its journal (*Commentarii,* later *Acta)* had two classes, mathematics and physics; the former included analytical and celestial mechanics, the latter everything from optics and hydraulics to botany and astronomy. Electricity was accordingly and, for the time, appropriately classed as physics. In 1747 a new class was added, 'physico-mathematics,' which took optics, hydraulics, heat, electricity, magnetism, and, increasingly, analytic dynamics; 'physics' retained, among the physical sciences, only meteorology, mineralogy, and chemistry.[43] The arrangement persisted until 1790, when 'mathematics' and 'physico-mathematics' united.[44] These moves corresponded to key conceptual innovations in the study of electricity, which it shall be our pleasure later to examine.

2. OCCULT AND OTHER CAUSES

The Aristotelian physicists concerned themselves with the true causes of things. Where the corpuscular or Newtonian philosopher saw few causes or none at all, the peripatetic could distinguish four general categories and several subspecies, one of which, termed 'occult,' became a password among the modernizing philosophers of the seventeenth and eighteenth centuries. To despise occult causes, to insist upon cleansing physics of them, was forward-looking; to accuse the enemy of advocating such bugaboos was always a good thrust in head-to-head philosophical combat. Cartesians and Newtonians flung the charge not only at peripatetics, but also at one another.[1] Much of our story turns upon the notion of occult cause.

ARISTOTELIAN NATURAL PHILOSOPHY

Aristotle's physics, as inherited by the Renaissance, was enriched or, as some said, polluted by the conflicting interpretations of scores of schools of

42. D'Alembert in *Encyclopédie* (1778³), XXV, 736, art. 'Physico-mathématiques'; P. Hahn, *Lichtenberg* (1927), 41; Karsten, *Phys.-chem. Abh.* (1785), 151.

43. Owing to this change, Kästner found himself reviewing 'physicomathematics' from St. Petersburg for the 'physics' section of the *Commentarii de rebus ad physicam et medicinam pertinentibus.* The Leipzig doctors who ran the *Commentarii* could not take physics with mathematics, 'a severiori enim mathesi medici nostri abhorrent.' Kästner to Heller, 1 Jan. 1752, in Kästner, *Briefe* (1912), 21.

44. Cf. the reorganization of the Royal Society of Science, Göttingen, in 1777–8: the omnibus class 'physics and mathematics' was divided into two, 'physics' getting chemistry and the biological sciences, 'mathematics' the usual mixed mathematics and electricity. Ak. Wiss., Gött., *Novi comm.*, 1 (1778), iii–iv, xvi–xx.

1. Cf. Genovese, 'Disputatio' (1745), quoted by Garin, *Physis,* 11 (1969), 220: '[Cartesianism] cessit Newtonianismo, iisque armis victus est quibus ille peripatetismum fugaverat.'

philosophy. One therefore cannot declare unambiguously the principles of sixteenth-century peripatetic philosophy. The school to which we shall subscribe is the Collegium conimbricense, the Coimbra Jesuits, who published commentaries on the Aristotelian texts in the first years of the seventeenth century. These commentaries recommend themselves on several grounds. They are authoritative and erudite; they stay close to the ancient texts; and they were very widely used. Descartes, among many others, learned his physics from them.[2]

The four scholastic categories of cause, among which we seek the occult, are the material, efficient, formal and final. They are more easily illustrated than defined. In Aristotle's own example of the making of a statue, the material cause is the bronze; the efficient, the sculptor's art; the formal, the statue's final figure; and the final, earning the sculptor a living, honoring the party sculpted, edifying the public.[3] The statue is an affair of art. In most cases of interest to the physicist, however, in natural processes, the number of causes reduces to three, the formal and final coinciding, or even to two, when the efficient cause is the nature or form of the body undergoing change.[4]

In Aristotle's philosophy, each individual is what it is in virtue of its 'form,' its defining principle, the sum of its 'actual' properties. 'Actual,' *actu*, signifies properties currently realized or activated as distinguished from potential ones; an animal now has the tendency to grow old, potentially of being old. Although each individual has but one form, Aristotle separates the characteristics it embraces into two groups, the 'substantial' or 'essential,' and the 'accidental.' Essential characteristics are those by which an individual belongs to a species; they explain why the world contains *kinds* of things—dogs, stars, marble, men. Accidental characteristics differentiate individuals of the same species one from another; they make it possible to distinguish between this dog and that, or between Plato and Socrates. Size, shape, color and 'attitude,' for example, are usually accidents, so that an individual six feet tall, thin, black and silent is no less a man than chubby, white, chattering Socrates. The form of an individual is the sum of its actual properties; the form is *not* the individual, however, and indeed has no separate existence except in the mind of the philosopher.[5]

A second principle, 'prime matter,' likewise incapable of independent existence, is necessary to bodies. Prime matter is the principle of materiality and potentiality; it reifies a given form to constitute an 'actual' body; and it readily exchanges one form for another to bring about change.[6] Just how one form

2. Descartes, *Corresp.* (1936), III, 185; Gilson, *Index* (1912), iv; *infra,* ii.1; *supra,* i.1, n. 1.

3. *Phys.*, II.3, 194^{b}16–195^{a}; *Metaphys.*, Bk. Δ, 1013^{a}25–1014^{a}15. Although these texts do not specify final or formal causes in the case of the statue, they imply those given here. Cf. Coll. conim., *Comm. in Phys.*, I (1602), cols. 327–9, 396–8.

4. *Phys.*, II. 7, 198^{a}25–28.

5. The Intelligences that regulate the motions of the heavens, the pure Form or Unmoved Mover of the world (*Metaphys.*, Bk. Λ, 1073^{a}–1074^{b}35), and the angels of the schoolmen all excepted. Cf. Coll. conim., *Comm. de Caelo* (1603^{2}), 267–8.

6. *Phys.*, I.7, 190^{b} 15–30, I.9, 192^{a}25–30, II.1, 193^{b}20; *Metaphys.*, Bk. Z, 1029^{a}20–30; *Gen. and Corr.*, I.3, 318^{b}18; *Cat.*, Chapt. 4, 1^{b}25–4^{b}20. Cf. Coll. conim., *Comm. in univ. dial.*, I (1607), 336–42. The degree of potentiality of matter has been differently interpreted according as one

succeeds another became a tough knot for the peripatetics. Some sixteenth-century philosophers, holding tight to the Aristotelian definition, continued to ascribe a single form to each individual, and referred the introduction of new forms to the stars or to God. Others, departing far from their original, in effect resolved an individual into a collection of independently-existing forms contained in an independently-existing piece of matter, like so many marbles in a box. The replacement of one of these 'substantial forms' by another amounted to no more than a change of place.[7] In this debased condition, with reified individual qualities, the theory could give an easy, empty, explanation of everything.

In certain sorts of change, called 'natural' by peripatetics, form can play the part of efficient as well as of formal and final cause. A standard example is the growth of plant or animal. An acorn—or better, this acorn—has, at this instant, in consequence of its form, the power to develop into an oak; a power that will become the efficient cause of development whenever artificial impediments to its action—being out of ground, being deprived of nutrients—disappear. The formal cause of growth is likewise form, understood not as the form of this acorn, but as the form of the oak to which it tends. This final form, when interpreted as the goal of growth, is also the final cause. Note that the acorn, or any plant or animal, has its power to change, or, to use the school term, its 'mover,' within it. Note also that this power, which is different for each natural species, *is not further analyzable*.

Precisely the same account can be given of the fall of a rock or the ascent of fire. Among their essential properties earth and fire possess, respectively, the qualities gravity and levity; when unconstrained, earth moves to the center of the world and fire towards its circumference. The form of a rock separated from the main body of the earth has among its accidental characteristics actually 'being on this shelf' and potentially being in any number of other places. The rock's form does not regard these possibilities indifferently: when the accidental constraint disappears, when the rock falls from the shelf, the element of gravity in its form moves it directly towards the center of the world. This is not a case of action at a distance, which Aristotle would not allow in the material world.[8] The center of the universe does not draw the stone: the stone 'knows' from the relevant element in its form, 'being on the shelf,' that it is separated, and it moves itself towards full actuality, it propels itself towards the ground, whenever possible.[9]

---

takes the texts of the *Physics* or the *Metaphysics* as fundamental. The difference is perhaps consequential for Renaissance scholastic physics; cf. Coll. conim., *Comm. in Phys.*, I (1602), cols. 205–6, 228–9, and de Vries, *Schol.*, 32 (1957), 161–85.

7. Dijksterhuis, *Mechanization* (1961), 281–4; Reif, *JHI*, 30 (1969), 26–7.

8. *Phys.*, VII.2, $243^{a}1$–5, $244^{b}1$–$245^{a}20$. Cf. Thomas Aquinas, *Commentary* (1963), 436–9; van Laer, *Phil.-Sci. Prob.* (1953), 80–94; Coll. conim., *Comm. in Phys.*, II (1609), col. 311.

9. The rock's 'knowledge' ultimately comes from the place it occupies; Aristotle's sublunary space is, as it were, full of sign posts directing rocks downward and fire upward. Cf. *On the Heavens*, IV.3, $310^{a}14$–$311^{a}15$; *Phys.*, II.1, $192^{b}12$–15.

Opposed to the natural motions of organic growth and free fall are 'violent motions' that carry an object against the tendency of its form: flinging a javelin, compressing air, killing an animal, brainwashing a philosopher. In such cases the efficient cause necessarily lies outside the object moved. The same is true of another class of motions, which may be called 'indifferent,' motions to which the essential form offers neither encouragement nor resistance: displacing a rock horizontally, heating or cooling water, moistening or drying mud.

The last two qualities, the heating power (hotness) and the moistening (wetness) are, with their contraries coldness and dryness, the chief agents of change in the sublunary world. Aristotle considered them unique in combining the two attributes he deemed necessary for such agents: they are *tangible* and hence notify the philosopher of their working, and they come in contrary pairs each member of which can *act* upon the other. (Aristotle arbitrarily makes hotness and wetness active, and coldness and dryness passive.) The last criterion is of capital importance. Gravity and levity, for example, do not constitute an agent-patient pair. If one places a hot body in contact with a cold one, or a wet in touch with a dry, the first pair become lukewarm and the second damp. A rock, however, does not share its gravity with the shelf supporting it; however long they remain in contact the rock will sink and the shelf float.

The bodies constructed by the union of the fundamental qualities with prime matter are the 'elements' of the inorganic world. Aristotle accepted the view, already ancient in his time, that precisely four such elements existed, air, earth, fire and water; and he associated them with the fundamental qualities in such a way that, as observation showed, any two elemental bodies could interact. This condition required that each element be associated with a pair of fundamental qualities. The affiliations chosen by Aristotle, fire (hot, dry), air (hot, moist), water (moist, cold), earth (cold, dry), remained standard. This account does not, however, exhaust the essences of elemental bodies, for fire has levity as well as hotness and dryness, and earth has gravity as well as coldness and dryness. Gravity and levity, although invariably associated with earth and fire, cannot be derived from the four active qualities: all six are irreducible, singular powers.[10]

There is another capital distinction to be drawn between gravity / levity on the one hand and the active qualities on the other. The qualities—and 'secondary qualities' like hard / soft, rough / smooth, and brittle / malleable, which Aristotle supposes compounded of them—immediately identify themselves to the sense of touch. Gravity / levity do not. To be sure, a rock held in the hand gives one a sense of its gravity. But this sense records the force exerted to prevent the rock's natural motion; it in no way differs from felt resistance offered to any other push, and it vanishes when the rock rests on the ground. Our sense of touch alone cannot inform us of the gravity of the largest boulder.

10. *Gen. and Corr.*, II.1, 329^{a}25–331^{a}5; *Meteor.*, I.3, 341^{a}25; *Cat.*, Chapt. 8, 9^{a}30–9^{b}10. Cf. Coll. Conim., *Comm. in Phys.*, I (1602), cols. 392–4; *Comm. de Caelo* (1603^{2}), 424–9.

Similarly we have no sense of our own gravity, or of electricity. The scholastic philosophers distinguished clearly between these last qualities and the active ones. In their terminology, the active qualities and their compounds are manifest, and the gravitational *hidden* or, to say the worst, *occult*.

Philosophers admitted an occult gravitational quality to account for the apparent directed self-movement of heavy bodies. Magnetism presented a similar problem. Iron flies to the lodestone in roughly the same manner as a rock moves to the center of the world, driven by its peculiar self-actualizing form. Sublunary place, the 'sphere of influence,' as it were, of the world's center, directs the self-motion of the rock, while the magnet's characteristic quality, diffused through space, confers on the iron, or actuates, a power of self-motion, and guides it to union with the lodestone. Or so magnets operate according to Aristotelian commentators from Averroes to the Coimbra Jesuits. St. Thomas in particular took pains to explain the induction of self motion in iron, and to distinguish it in detail from free fall: gravity acts towards a point, from any distance; magnetism moves towards a body, and can be induced only over short distances.[11] Other scholastics, attending to the Aristotelian principles that the mover must be conjoined to the mobile and, except for souls and heavenly intelligences, can move only by being moved,[12] tried to explain how the lodestone could 'diffuse' its power to the iron without appearing to affect ('move') the intervening medium. 'They say [it is the testimony of Jean Buridan] that the magnet alters the air or water that it touches and propagates to the iron a quality which, because of a natural affinity between the iron and the magnet, attracts the iron but nothing else.' The magnet works just like that peculiar fish of St. Albert's, which numbs the hands of fishermen by doing something to the water.[13]

The account transmitted by Buridan is an adaptation of the medieval theory of 'multiplication of species,' according to which all bodies in the universe impress their peculiar qualities and powers (species) upon, and diffuse (multiply) them through, the surrounding medium. The multiplied species affect a body according to its nature. Consider the exemplar of multiplication of species, the propagation of light. An incandescent source (lux) imprints its species (lumen) on any transparent medium which, however, does not itself therefore become incandescent or colored; the lumen becomes manifest only at the surfaces of opaque bodies, or within translucent ones. Celestial influences operate in the same manner as light, but with greater discretion: they act preferentially upon certain special materials, which thereby may be made into

11. Thomas Aquinas, *Commentary* (1963), 433; Daujat, *Origines* (1945), I, 49–78; Urbanitsky, *Elektrizität* (1887), 10, 103–4; T. H. Martin, Acc. Pont. nuovi Lincei, *Atti*, 18 (1865), 99, 105.

12. *Phys.*, 241^{b}25–242^{a}25.

13. Buridan, *Quaestiones super VIII libros Physicorum*, VII, 4, quoted in Daujat, *Origines* (1945), I, 72. That St. Albert's fish, the torpedo, stuns by electricity was a discovery of the 18th century (*infra*, iii.4).

medicines or talismans. The magnet is still more exclusive, for its species work visibly only upon iron, steel, and other lodestones.[14]

The medieval account of magnetism remained standard until the middle of the seventeenth century. One finds it, for example, in Nathanael Carpenter's influential *Geography Delineated*; in the widely used monograph of Vincent Léotaud, who followed St. Thomas' model despite his claim to expound a 'new magnetic philosophy;' and—an excellent testimonial to its persistence—it has left a trace in the anti-peripatetic atomistic compendium of Gassendi.[15] One also finds it constantly put forward as the paradigm of attractions, as the cleanest case of local motions effected by occult qualities: 'in the magnet God has offered to the eyes of mortals for observation qualities which in other objects he has left for discovery to the subtler research of the mind.'[16]

A convenient and representative baroque exposition of the results of this 'subtler research of the mind' may be found in a *New philosophy and medicine concerning occult qualities* published in Lisbon in 1650. Its author, Duarte Madeira Arrais, was a distinguished physician trained at the University of Coimbra.[17] His book therefore has a double authority: first, because it is informed by the commentaries of the Coimbra Jesuits; second, because its subject, occult qualities, figured prominently in medical theory in connection with growth, nutrition and the efficacy of poisons and purgatives.[18]

Madeira assumes that all philosophers admit the existence of the four elementary active qualities, of the secondary tactile qualities compounded from them, of 'manifest super-elemental qualities' like light, sound and impetus, and of the vital powers of animals. In addition, he says, there is a class of 'occult super-elemental qualities' like the virtues whereby the magnet draws iron, the remora stays ships and purgatives expel foul humors. These virtues are called 'occult' because, 'though manifest to the intellect, they are not apparent to the senses,' *sub humanos sensus non cadunt*. They must be super-elemental because 'remarkable effects' like the attractions of remoras and magnets, or the shock inflicted by electric eels, cannot arise from elemental qualities. Not only are hotness, dryness, coldness and moistness in any degree incompetent to produce magnetic qualities: they also act less rapidly and efficiently than, say, the virtues of magnets or scorpions.[19] As for the details of occult action, Madeira

14. R. Bacon, *Opus maius* (1900), II, 407ff.; Crombie, *Grosseteste* (1962²), 211–12; Coll. conim., *Comm. in Phys.*, I (1602), col. 394, II (1609), cols. 309–12; *Comm. de Caelo* (1603²), 196–9.

15. Carpenter, *Geographie* (1635²), 54–5; Léotaud, *Magnetologia* (1668), 31–3; Gassendi, *Opera* (1658), I, 347.

16. Dee, *Prop. aph.* (1568²), §xxiv.

17. Barbosa Machado, *Biblioteca* (1930²), I, 715–16.

18. Morhof, *Polyhistor* (1747⁴), II, 305: 'Nulla autem fecundior his qualitatibus occultis magis est, quam ars medica.'

19. Madeira Arrais, *Novae phil.* (1650), 1–19. The Coimbra Jesuits also say that the facts compel the philosopher to introduce occult qualities: 'Nec enim semper effecta ad quattuor primas qualitates, ut falso quidam opinantur, referri queunt.' Coll. conim., *Comm. in Phys.*, II (1609), col. 318.

generally follows the lead of St. Thomas. The occult quality, diffused about the substance that bears it, awakens a self-acting potency in appropriate neighboring bodies, which move themselves as their forms require.[20] Activated iron flies to the seat of its occult trigger, the magnet; a purgative stimulates a 'directive quality' in the affected humor, which guides it into the intestines.

Often the carriers of the stimulating power bear an external analogy or similitude to the substances they excite; but this similarity is not necessary, and where it does exist it is a formal, not an efficient cause.[21] Madeira does not believe in the doctrine of signatures, the notion that the outward appearance of herbs and stones, rightly read, reveals their powers and purposes. He sets few constraints against multiplication of occult qualities. He himself is circumspect. Others, however, had long since undermined the explanatory value of occult qualities by invoking them to resolve all the difficult phenomena, real or imaginary, treated in natural philosophy.

SYMPATHIES AND ANTIPATHIES

In the occult as elsewhere familiarity breeds contempt. It may be useful to identify a magnetic virtue the possession of which distinguishes a closed group of interacting substances; but when one ascribes several irreducible special qualities to every stone or plant or drug, one has a science of words, not of things.[22] 'The learned doctor asks me the cause and reason why opium puts people to sleep. A quoi respondeo / quia est in eo / virtus dormitiva / cuius est natura / sensus assoupire.' (Opium is a soporific because it contains a dormative virtue.) Thus Molière's candidate in medicine, answering his examiner in empty fractured Latin, to the great applause of the faculty. The same point was made by several sober physicists for whom Francesco Lana, a Jesuit obliged to teach the philosophy of Aristotle, may be allowed to speak. Ask most natural philosophers, says Lana, the cause of any natural phenomenon; 'they can only reply that it happens by an occult cause, that such is the nature of that substance.' And if you persist, and ask, say, how the occult cause whereby the magnet draws iron and not straw differs from that whereby amber draws straw and not iron? 'They reply, this is the nature of amber, and that the nature of the magnet.' Such people, according to Lana, bring disgrace and ruin to natural philosophy.[23]

The most extravagant occult qualities were the sympathies and antipathies

20. Cf. Cabeo, *Meteor.* (1646), I, 31–2; and, on the definition of occult qualities ('spectant ad facultates incognitas, causasque habeant incompertas'), Gassendi, *Opera* (1658), I, 449.

21. Madeira Arrais, *Novae phil.* (1650), 335, 352, 405–19.

22. To use the conceit in Fontenelle, 'Éloge' of du Hamel, *Oeuvres* (1764), V, 80: 'des idées anciennes et des nouvelles, de la philosophie des mots & de celle des choses, de l'École & de l'Académie.' Cf. Sprat, *Hist.* (1967), 113, 336; Flourens, *Fontenelle* (1847), 53–4. For an example of peripatetic evasion at its worst see Middleton, *Br. J. Hist. Sci.*, 8 (1975), 148.

23. Molière, *Le malade imaginaire* (1673), 3e intermède; Lana Terzi, *Prodromo* (1670), Proem., 3–4. Cf. Boyle, 'Occult Qualities,' unpublished Ms. quoted by M. B. Hall, *Boyle on Nat. Phil.* (1965), 59.

invoked especially by the hermetic philosophers and physicians of the sixteenth and seventeenth centuries. Gaspar Schott, S. J., an authority on sympathetic action, explains it in these words: 'Sympathetic effects arise from a friendly affection, or coordination and innate relation, of one thing to another . . ., so that if one is acting, or reacting, or only just present, the other also acts or is acted upon.' The operative quality or affection is not further analyzable. 'It originates directly from the particular temperament of each thing, being nothing but a certain natural inclination of one thing towards another.' An antipathy works in the same way, with aversion in place of inclination, just like—here Schott offers the inevitable analogy—magnetic attraction and repulsion.[24]

Sympathies and their opposites explained the fabulous as easily as they elucidated the natural. To Madeira the ship-staying power of the remora and the iron-pulling virtue of the magnet are equally acceptable, and he no more doubts the 'virtues of the tree of life of the Garden of Eden' than he does the efficacy of scammony or cinchona. Less critical writers speak readily and knowledgeably of the glance of the basilisk, the attraction of the weasel for the toad, the generation of minerals by celestial influence, fetal imprints produced by maternal appetites, and the curative power of the powder of sympathy.[25] This last extravagance represents the nadir of the doctrine of active qualities.

The seventeenth century credited Paracelsus with the invention of a powder or salve which could heal at a distance. No doubt this salve, made among other things of skull moss, mummy, and the fat and blood of a dead man, would be most beneficial when used as directed, viz., smeared upon a stick or napkin previously dipped in the wound and kept far away from the patient. According to the Paracelsians, the virtue of the salve, drawn out and fortified by celestial influences, flies to the injury by 'magnetic sympathy,' say between the separated particles of blood or, according to the great magus Robert Fludd, between the necrotic ingredients of the salve and the living flesh (opposites attract).[26] Many physicians and alchemists endorsed or improved the salve and its theory. One influential promoter was Sir Kenelm Digby, whom some contemporaries considered a respectable philosopher.[27]

Digby was an English Catholic who learned his philosophy from a Jesuit named Thomas White, alias Blacklow, and spent his early manhood serving the Stuarts in diplomatic missions on the Continent.[28] In 1623, at the age of

24. Schott, *Thaumaturgus* (1659), 368–70; *Phys. cur.* (1662), 1285ff. Cf. Kircher, *Magnes* (1641), 644.

25. Mousnerius (Fabri) lists these curious items, which he says are commonly alleged in favor of sympathetic action, in *Metaphysica* (1648), 283–5. Fabri rejects them all, including the remora, which he calls a 'mere fable.'

26. Debus, *J. Hist. Med.*, 19 (1964), 390–2, 404–11, quoting Fludd's *Philosophia moysaica* (1638). Cf. Thorndike, *History*, VII, 503–6.

27. Digby knew and was esteemed by Descartes, Mersenne, Fermat, John Wallis and Athanasius Kircher. See Gabrieli, *Digby* (1957), 197, 230–2; Petersson, *Digby* (1956), 120–8; Dobbs, *Ambix*, 18 (1971), 1–25.

28. Digby, *Two Treat.* (1645^2), 180: 'To him [Thomas White] I owe that little which I know; and what I have, and shall set down in this discourse, is but a few sparks kindled by me at his great fire.'

twenty, he met a monk in Italy who gave him a secret recipe for the powder, which he was to dissolve in water and blood from the wound and set aside in a basin exposed to sunlight. Digby made the secret public in 1657, in a strange address to a 'solemn assembly' in Montpellier, where he had gone for the waters.[29] The disclosure included a theory of the cure, and, by way of illustration, several bizarre examples of natural sympathetic phenomena.

All bodies, Digby says, perspire when bombarded by light. The perspiration, or effluvium, which is characteristic of the emitter, tends to diffuse towards kindred substances, which preferentially absorb it. 'The reason hereof is the resemblance, and sympathy, they have one with the other.'[30] Everyone knows, for example, that a greater stench will attract a lesser. 'Tis an ordinary remedy, though a nasty one, that they who have ill breath, hold their mouths open at the mouth of a privy, as long as they can, and by the reiteration of this remedy, they find themselves cured at last, the greater stink of the privy drawing unto it and carrying away the lesser, which is that of the mouth.'[31] Similarly, in Digby's cure, the blood particles fly sympathetically to the wound whence they came, bringing with them the subtlest atoms of the healing powder of sympathy.[32]

Digby's tract was often reprinted, translated, and glossed.[33] The powder of sympathy and the similar technique of 'transplantation'—the transfer of poison to a sympathetic imbiber, for example, to a hair of the dog that bit you —also found favorable or agnostic treatment in eighteenth-century encyclopedias.[34] Such survivals occur almost exclusively in a medical context. By 1700 natural philosophy had largely freed itself from the animistic sympathies, from the innumerable occult qualities that had posed, according to the literary executor of the seventeenth century, 'the most vexed question of the age.'[35] Peripatetic physics, found guilty, among other reasons, by association, also fell victim to the purge.

The implication of guilt by association was commonly employed by corpuscular philosophers. Robert Boyle, for example, often found it advantageous to conflate hylomorphism and hermetic animism.[36] Up-to-date peripatetics met this gambit by denouncing hermeticism as loudly as the corpuscularians did,

29. Petersson, *Digby* (1956), 265–6.

30. Digby, *Late Disc.* (1658), 5, 11–12, 68. Digby says (*ibid.*, 68–75) that these sympathies derive from similarities in density and particle shape; but in practice they are non-mechanical occult qualities, *pace* Dobbs, *Ambix*, 18 (1971), 11, 25.

31. Digby, *Late Disc.* (1658), 76–110.

32. *Ibid.*, 133–41. Union of the atoms with sunbeams improves their efficacy.

33. At least 25 editions of *Late Disc.* were published by 1700. Physicians disputed the efficacy of the cure throughout the century. Petersson, *Digby* (1956), 272–4, 326; Schreiber, *Gesch.*, II:2 (1860), 416.

34. For Chamber's *Cyclopedia* see Shorr, *Science* (1932), 29–31, and A. Hughes, *Ann. Sci.*, 7 (1951), 354–6; for Zedler's *Universal Lexicon* (1732–50), Shorr, *op. cit.*, 60–8, 71–2; for Diderot's *Encyclopédie* (1751ff.), Thorndike, *Isis*, 6 (1924), 379–82.

35. Morhof, *Polyhistor* (1747^{4}), II, 303.

36. Boyle, iii.1.

and by disavowing the miracle mongers, the hacks and novelty hunters, the dupes of Arab commentators, all who had muddied the pure Aristotelian water. Some reforming peripatetics adopted the tactics of the enemy. Niccolò Cabeo, S. J., cracked down on occult qualities and endorsed an eclectic experimentalism; his colleague, Honoré Fabri, put forth as 'purified' Aristotle a physics stained with the thought of Descartes.[37] But many physicists brought up on, and sympathetic to, Aristotelian principles, judged themselves unequal to purging the contaminated peripateticism of their day, and reluctantly embraced the radical alternative of corpuscularism.

An outstanding example of the frustrated Aristotelian reformer is the Minim monk, Marin Mersenne, the confidant of Descartes, the correspondent, guide and goad of much of learned Europe in the 1630s and 1640s. Mersenne, educated like Descartes in the Jesuit college at La Flèche, began his career hoping to rid traditional natural philosophy of hermetic mumbo-jumbo, of astrological influences, of bits and pieces borrowed from cabalists and magicians. A 'new Aristotle' would answer the 'grunting of the German beast' (Paracelsus), subdue the *cacomagus* (Fludd), secure religion against the animists and the pantheists, and, above all, distinguish the natural from the supernatural, blunt superstition and confound scepticism.[38] Eventually, to meet the threat, Mersenne gave up altogether the search for physical causes in the Aristotelian sense: and from the mid-1630s, under the inspiration of Galileo and others, he taught that 'true physics' could only be a descriptive science of motions. The rejection of essences, or rather of the claim to know them, sunk Mersenne's old program for reform along with the abuses he aimed to correct. He avoided shipwreck by jumping to the good ship 'Mécanisme,' which, although unknown and perhaps unknowable in its inner workings, at least sailed according to discoverable laws.[39]

Mersenne was not alone in identifying the Fludds and not the school philosophers as the greatest threat to true physics in the seventeenth century.[40] The same contempt for hermeticism spices the reply of the Oxford mathematicians, Seth Ward and John Wilkins, to the attack on the universities made by the Fluddist John Webster in 1654. Webster blasted the schools for despising the 'noble and almost divine Science of natural Magick,' which he understood to rest upon the doctrine of signatures, and for sticking to Aristotle. Wilkins and Ward replied that Aristotle was 'one of the greatest wits, and most useful that ever the world enjoyed,' and recommended his books, 'the best of any

37. Cabeo, *Meteor. comm.* (1646), I, 254; Fabri, *infra,* ii.1.

38. Lenoble, *Mersenne* (1943), 9–10, 29, 95, 147.

39. *Ibid.*, 361. "Les hommes ont introduit la sympathie et l'antipathie, et les qualitez occultes dans les arts et dans les sciences pour en couvrir les deffauts, et pour excuser leur ignorance . . . : car lors que l'on connoist les raisons de ces effets [magnetism, electricity] la sympathie s'évanoüit avec l'ignorance, comme ie demonstre dans le tremblement des chordes qui sont à l'unisson.' Mersenne, *Harmonie universelle,* quoted by Lenoble, *ibid.*, 371–2.

40. E.g., Gassendi, *Opera* (1658), III, 236–7, 251. Cf. the sudden conversion of Walter Charleton from hermeticist to corpuscularian in the early 1650s; Gelbart, *Ambix,* 18 (1971), 149–68.

Philosophick writings,' as correctives to the nonsense of the disciples of Hermes and Pythagoras.[41] Not that Wilkins and Ward were peripatetics; indeed, they were enthusiastic moderns, 'Copernicans of the Elliptical family.' That did not, however, blind them to their kinship to Aristotle and Ptolemy, or to their distance from true revolutionaries like John Webster.[42]

The role here assigned to hermeticism is much more modest than that claimed for it in some recent writing on Renaissance science. The hermetic magus, it is said, must necessarily have learned the sympathies of things by experience; moreover, the need to control and manipulate astrological influences, special qualities of bodies, and the harmonies of the world, directed him to the laboratory. 'In this way Paracelsian mysticism acted as a powerful stimulus towards the new observational approach to science.'[43] 'It is the Renaissance magus, I believe, who exemplifies the changed attitude of man to the cosmos which was the necessary preliminary to the rise of science. . . . The Renaissance magus was the immediate ancestor of the seventeenth century scientist.'[44] Although the claim may hold for certain traditions of alchemy and medicine that became something like chemistry and physiology in the seventeenth century, it fails for most of physics and mixed mathematics.

Perhaps the most powerful support for controlled and careful experiment in physics towards the end of the sixteenth century was the example of mixed mathematics, the requirements and achievements of architecture, fortification, navigation, Tychonian astronomy, optics. Galileo taught these subjects, which he had studied at a Florentine trade school set up specially to teach applied mathematics.[45] William Gilbert, the first important electrician, also taught mathematics; and the artisans whose methods he may have followed in his fundamental work on electricity and magnetism likewise had an interest in, and urgent need for, practical mathematics.[46] The number magic of cabalists and hermeticists did not contribute significantly to surveying, cartography, architecture, or exact astronomy; the *De occulta philosophia* (1533) of the celebrated magus H. Cornelius Agrippa did not encourage 'within its purview the growth of those mathematical and mechanical sciences which were to triumph in the 17th century.'[47] Numerology, like hermetic animism, was antithetical to the application of mathematics to practical problems.[48] Agrippa himself spurned the grubby calculations of astronomers: 'I omit [he says] their vain disputes about Eccentricks, Concentricks, Epicycles, Retrogradations, Trepidations, accessus, recessus, swift motions and circles of motion, as being the

41. J. Webster, *Acad. exam.* (1654), 68, 76–7; Wilkins and Ward, *Vind. acad.* (1654), 5, 46, and 22–3, where they spoof hermetic jargon.
42. *Ibid.*, 29; Debus, *Science* (1970), 37, 42, 48, 57–60.
43. Debus, *J. Hist. Med.*, 19 (1964), 391.
44. Yates in Singleton, *Art* (1968), 255, 258.
45. Geymonat, *Galileo* (1965), 7–10.
46. Cf. Zilsel, *J. Hist. Ideas,* 2 (1941), 1–32.
47. Yates in Singleton, *Art* (1968), 259.
48. Cf. Strong, *Procedures* (1936).

works neither of God nor Nature, but the Fiddle-Faddles and Trifles of Mathematicians.'[49]

An apparent exception to this antithesis is John Dee, an Elizabethan magus, alchemist, and navigational authority on whom recent hermetic champions heavily rely.[50] Dee wrote a mysterious alchemical book, the *Monas hieroglyphica,* and spent his declining years interviewing angels; but it was not his occultism that inspired his few—and, for their content, unimportant—contributions to applied mathematics, but his study early in life with the great geometers, cartographers, and instrument makers of Louvain, Gemma Frisius and Gerard Mercator.[51]

3. CORPUSCULAR PHYSICS

From a logical point of view corpuscularism was an extreme form of Aristotelian natural philosophy that recognized very few real qualities and—in its purest form—but one means of action, namely pushing, in the corporeal world. This radical parsimony proved immensely fruitful. The qualities supposed primary, like extension, motion, figure, impenetrability and inertia, which everyone understands intuitively, provided the basis for a 'comprehensible' and quantifiable physics, while most of the Aristotelian real characteristics were declared 'secondary,' creations of the perceiving mind, the business of the psychologist. Making do only with inert, sub-microscopic corpuscles and their motions, the revolutionary philosopher of the seventeenth century proposed to describe all the workings of lifeless matter and much of the operation of plants and animals.

The cynosure of the corpuscularians was Descartes, who tried to anchor his physics on unshakeable foundations. Everyone knows how he strove to doubt everything, but could not bring himself to doubt the existence of himself doubting; how the 'clarity and distinction' of his apprehension of the existence of his doubting mind became the touchstone for the truth of other propositions; and how, as the first of the propositions so secured, he gave the existence of an omnipotent God incapable of deceiving him about the truth of propositions he perceived clearly and distinctly. From this heady metaphysical journey Descartes brought back the clear and distinct principle of the equivalence of matter and extension (wherefore a void space is a contradiction in terms) and several

49. Agrippa, *Vanity* [1530] (1676), 86. The hermeticist distaste for exact computation and applied mathematics is emphasized by Debus, *Ambix,* 15 (1968), 15–25, who quotes van Helmont (p. 24): 'The Rules of Mathematicks, or Learning by Demonstration, do ill square to Nature. For man doth not measure Nature; but she him.'

50. Yates in Singleton, *Art* (1968), 262, 264; French, *Dee* (1972), 160–87.

51. Heilbron in Shumaker and Heilbron, *John Dee* (1978), 34–49. It is gratifying that Paolo Rossi, who has stressed magical elements in the thought of modernizing seventeenth-century philosophers like Bacon, now considers Yates' teachings to be wrong-headed and even mischievous. Rossi in *Reason* (1975), 259–64.

rules of motion, true perhaps of the world in which he found them, but mainly false in ours.[1]

In 1644 Descartes published his substitute for Aristotle, the *Principia philosophiae*, which begin with the metaphysical underpinnings just sketched, exhibit the rules of motion, and then apply—or appear to apply—them to the problems of physics in the widest sense. The freshness and ingenuity of the book made a great impression. To be sure it had a few passages not altogether intelligible, but the misunderstanding was more likely to be the fault of the reader than of the writer. Had Descartes not gone far beyond the ancients in mathematics, and said so? Was it not presumptuous to expect to be able to understand him in everything, even when, as in the *Principia*, he used no mathematics? Descartes anticipated his readers' difficulties. He advised them to read his book straight through, again and again, like a novel: 'on taking it up for the third time I dare say that you will find the solution of most of the difficulties previously noticed; and if a few remain, you can clear them up by rereading yet again.'[2] The quaintness of Descartes' formulations must not be allowed to obscure the influence of his physics nor the fact that its form—applications of firmly grounded rules of motion—is precisely that of Newton's.

Among the most puzzling and difficult problems faced by reductionists like Descartes was the explanation of attraction, particularly that of the magnet.[3] His ingenious solution, which formed the basis of magnetic theory for over a century,[4] exploits most of the characteristic features of his system. Cartesian lodestones owe their efficacy to a system of channels or pores which provide one-way passage for particles of appropriate shape. The pores may be likened to threaded gun barrels fitted with diaphragms to insure unidirectional flow. A magnet possesses two sets of pores aligned with its polar axis, each set admitting an opposite flow of minuscule, twisting, screw-like particles of precisely the size and pitch to wriggle through them. These helical or 'channelled' particles are necessary by-products of the creation of the universe, as appears from the following considerations.

At the beginning, we may imagine, the universe was an undifferentiated continuous bloc of matter: it was extended, and nothing more. God divided this matter into equal microscopic cubes, which he gave a powerful tendency to rotate about their centers. Since no void can exist—extension *is* matter—realization of the God-given rotational urge can occur only if the edges of contiguous cubes are ground to rubble, 'freeing' the potential spherulae they

1. *Principia* (1644), in *Oeuvres,* IX:2, 27–38.

2. *Ibid.*, 11–12.

3. Cf. Mersenne's continued struggle with it; Lenoble, *Mersenne* (1943), 366–7.

4. The most striking evidence being the winning of the Paris Academy's three prize competitions of the 1740s for the best 'explanation of the attraction of the magnet' by three embroiderers of Descartes' theory; L. Euler, AS, *Pièces,* 5 (1748), 3–4; Dutour, *ibid.*, 51–114; D. and J. Bernoulli, *ibid.*, 117–44.

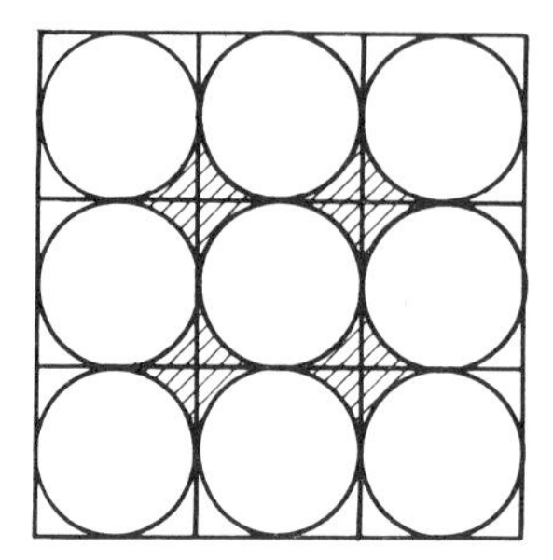

1.1 Sphericles of the second matter within their parent cubes; the shaded portions are cross-sections of future channeled particles.

circumscribe (fig. 1.1). The rubble and the little balls Descartes calls 'first' and 'second' matter, respectively; they consist of the same material, differing only in size, shape, and—an added quality—mobility.[5]

We are coming to the channelled particles. We have assumed that the spherulae remain where freed, so that their centers form a cubic lattice. Another, closer-packed arrangement is possible (fig. 1.2) and, according to Descartes, is favored by the mobility of the first matter. The close arrangement makes channelled particles. The rubble filling the interstices has, as its narrowest cross-section, the shaded lozenge of fig. 1.1. Imagine that a few layers of balls are exactly superposed on those of fig. 1.1: pores with lozenge-shaped cross-sections result. Suppose that because of their shapes and relative rest the

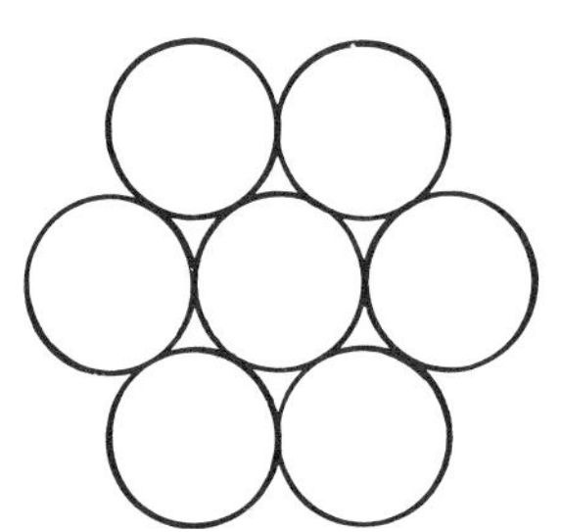

1.2 Close-packed arrangement of the sphericles.

first matter in the pore forms a stable entity, which will resemble a grooved bar (fig. 1.3);[6] and suppose further that the movement of neighboring globules tends to twist the layers which molded the bar. The grooves become threads, right- or left-handed according to the direction of the twist. The headless cylindrical screws are channelled particles.[7]

The most general cosmic processes produce magnetism. The scrapings of mobile first matter, displaced during the packing of the second, combine to form an agitated spherical body, or sun, held together by the surrounding spherulae. The latter, together with the remaining interstitial first matter, pre-

5. *Principia,* in *Oeuvres,* IX:2, 126–9. In practice Descartes also endows his matter with inertia and, via impenetrability, with something like repulsion. Cf. Carteron, *Rev. Phil.*, 47 (1922), 261, 491–3.

6. Characteristically Descartes ignores the transverse pores; his purpose is to show how channelled particles might be formed, not to demonstrate that, in every case, they are so formed.

7. *Principia,* in *Oeuvres,* IX:2, 153–6.

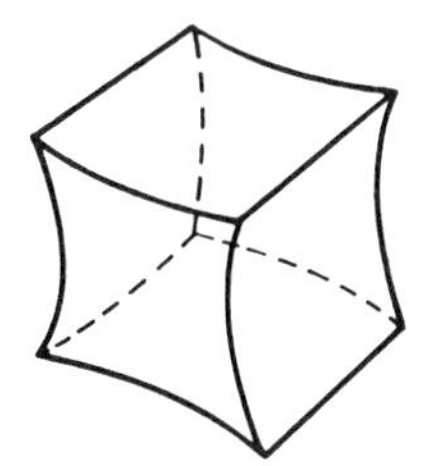

1.3 A channelled particle.

serve their motion by circulation in a vortex about the sun, a vortical or re-entrant curvilinear motion being the only type possible in Descartes' plenary universe. The channelled particles, adopted to potential pores in the vortical medium, or sky, move easily, tending towards the sun along the axis of the vortex and towards the circumference in the plane of its equator, where the centrifugal tendency is a maximum. The constant axial bombardment eventually opens two sets of threaded solar pores.

Meanwhile, channelled particles unable to penetrate the sun tend to settle on its surface. A completely covered sun, no longer able to maintain the surrounding whirlpool, might be pulled into a neighboring sun's swirl, where, if captured, it would play the part of a planet. Such a system might itself be captured, the old sun becoming a planet, and the old planet a moon. This was the origin of our earth and her satellite, which is retained by a little eddy in the vast vortex carrying the solar system. Our local eddy also causes the tides, the fall of heavy bodies, and, through the circulation of the channelled particles, magnetism.[8]

The threaded axial pores, which date from the earth's sunny past, are identical with the magnet's. The channelled particles traverse the earth as pictured (fig. 1.4a), emerge to be deflected by the air, whose pores are unfit to admit them, and, by bombardment, tend to orient magnets to receive them. Their course determines dip and declination, and their duality, or double-handedness, explains polarity.[9] To understand magnetic attraction and repulsion, note that the channelled particles tend to congregate where they find ready passage, namely about a lodestone, which they enclose in a mini-vortex. When two magnets, with contrary poles opposing, come so close together that the channelled particles issuing from the one just manage to reach the other before the air deflects them, they continue along the path of least resistance, enter the pores before them, and unify the magnetic mini-vortices (fig. 1.4b). The resulting vortical flow drives the air from the gap between the magnets; and the displaced air, circling to their back sides, pushes them to union. Magnetic 'attraction' is nothing but mechanical impulse.

A magnet draws iron nails in much the same way: its vortex threads them,

8. *Ibid.*, 157–74, 194–200, 209–10, 225–31. Cf. Scott, *Sci. Work* [1952], 167–94; Aiton, *Vortex Th.* (1972), 30–64.

9. Only one set of pores (right- or left-handed) in fact is needed, but it might be difficult to explain why it alone was produced.

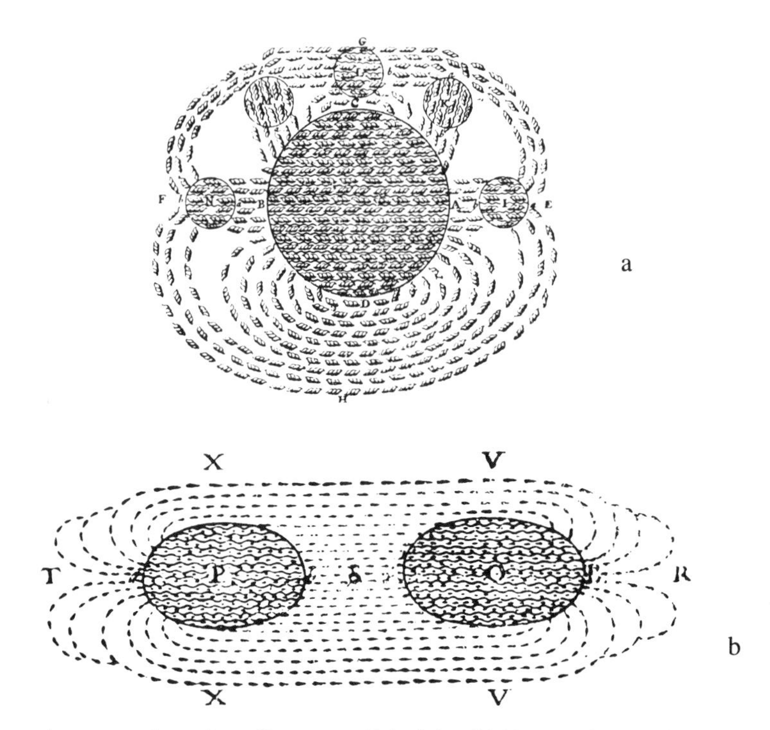

1.4 Cartesian magnetism, from Descartes, Principia *(1644): (a) the earth's magnetic vortices ABFGE and BAEHF, with oriented lodestones at I, K, L, M, and N (b) the joint vortex TXVRVX of lodestones P and Q with opposite poles facing.*

orients the diaphragms in their pores, which have the same shape as a lodestone's, and renders them magnetic in a sense contrary to its own. As for the repulsion between like poles, it occurs because the channelled particles issuing from them, being unable to enter the oppositely threaded pores of the opposing magnet, require space to execute their gyrations. And thus, by mobilizing the history and present economy of the universe, does Descartes rescue the archetype of actions at a distance from the grip of occult qualities.[10]

SPREAD AND REFINEMENT OF CARTESIAN PHYSICS

Interpreted on its own terms, as a secure metaphysical system, Cartesian natural philosophy menaced not only the received physics but also established religion. The eradication of most substantial forms and the elimination of all actions save pushes were not the only, nor the chief, irritants. In his account of the origin of our planet Descartes implied that there are many earths and many suns, and that the universe has no bounds. Moreover, despite a clever equivocation, his story of the

10. *Principia,* in *Oeuvres,* IX:2, 271–305.

capture of the moon by the earth, and of both by the sun, strongly endorsed, if it did not prove, the theories of Copernicus.[11] All this opposed both Aristotle and the Bible. Then there were delicate questions like the miracle of the eucharist, about which peripatetics could mumble apparently meaningful words ('Transsubstantiation saves the accidents while changing the essences of the bread and wine'), while Descartes, who had no such distinctions, struggled to show that, in this particular at least, his philosophy was compatible with mystery.

Above all Cartesian philosophy, with its methodical doubting, was subversive in spirit, the rallying point of novelty hunters, of modernists, of philosophical revolutionaries of all kinds. Alarmed authority responded first in the Netherlands, in the 1640s, when Descartes' teachings began to enter the universities, and especially the medical faculties, at Utrecht and Leyden. Their academic senates and curators, stirred up by their theologians, forbade the teaching of 'absurd, paradoxical, or novel doctrines,' of 'dangerous, new, or anti-Aristotelian theses,' that is, of Descartes.[12] In making, and then ignoring, these condemnations the Dutch were in advance of their time. In the 1660s prohibitions and injunctions against Cartesian philosophy were enacted all over Europe, testifying at once to the rapid spread of 'opiniones novae' and to the slow realization that they could not be stamped out by edict.

In August, 1662, the faculty of arts at the University of Louvain, agitated by the papal legate, called upon its professors to advise its students that though Descartes' writings 'might be well-considered in many points regarding natural philosophy, they contain some opinions not sufficiently in conformity with the sound and ancient doctrine of the Faculty.' Shortly thereafter the Faculty of Theology condemned six Cartesian propositions put forward in a thesis for a doctorate in medicine.[13] These moves perhaps inspired that of the Congregation of the Index, which in 1663 damned and prohibited certain of Descartes' writings 'until corrected.'[14] Lutheran theologians were no less vigilant than the Roman Catholics and the Dutch Calvinists. In Sweden, where Descartes had died of the cold in 1650, they fought grimly against the new opinions, which they almost succeeded in prohibiting at the University of Uppsala in 1664.[15]

The French also began to persecute Cartesians in the 1660s. In 1667 the Crown ordered the Chancellor of the University of Paris not to deliver his scheduled oration in honor of the reburial of the remains of the philosopher, just arrived from Sweden. A straw in the wind. In 1671 the Court, speaking through

11. The equivocation: since a body can be said to move only when it changes position relative to an object of reference, choose the 'object' to be the surrounding subtle matter; the earth is then at rest, because it remains in contact with the same portion of its vortex. *Oeuvres,* IX:2, 109. Prominent Cartesians of the second generation, e.g., Rohault and Malebranche, came out unequivocally for Copernicus; cf. Robinet, *Malebranche* (1970), 235.

12. Ruestow, *Physics* (1973), 36, 43–4; Lindborg, *Descartes* (1965), 30, 38–9.

13. Monchamp, *Hist.* (1886), 347–8, 354–69.

14. *Ibid.*, 338, 389–91, with text of the condemnation.

15. Lindborg, *Descartes* (1965), 93–6, 115–16.

the Archbishop of Paris, commanded the theologians of the Sorbonne to see to the enforcement of a ridiculous edict pronounced by the Paris Parlement in 1624, which prohibited the teaching of any philosophy but Aristotle's. After consideration the Parlement decided not to repeat its former folly. The famous spoof of Boileau may also have helped.[16] 'Whereas for several years an unknown person named Reason has tried to enter the schools of the University by force and to evict one Aristotle, the ancient and peaceful tenant of the said schools, the court . . . maintains and preserves the said Aristotle in the full and peaceful possession of his rights in the said schools, orders that he will always be taught and followed by the regents, doctors, masters of arts and professors of the said University, who, however, are not obliged to read him, or to know his language or his opinions.' The King did in fact succeed in driving the upstart temporarily from the higher schools of his realm.[17]

In this act of preservation he was ably assisted by the Jesuits. Already in 1650, in response to complaints about the teaching of 'new opinions' from several of its provincial administrators, the ninth general congregation of the Society drew up a list of proscribed propositions, 96 in all, of which fifteen touched upon Cartesianism.[18] The twelfth congregation (1682) repeated the prohibition against inculcation of novelties; the fourteenth reaffirmed it (1696); so did the fifteenth (1706), which pointed out thirty obnoxious doctrines drawn primarily from Cartesian physics, and the sixteenth (1732), which boiled the thirty down to ten.[19] Meanwhile the provinces clamored for the vigorous enforcement of the proscriptions.[20]

The repetition of these complaints and injunctions testifies to their ineffectiveness. A Jesuit professor who persevered in flagrantly disobeying them might be transferred to a third-rate school, or relieved of teaching duties;[21] but many advocated Cartesian doctrines without suffering more than an occasional reprimand. Even the Jesuit father who may have written the notice of Descartes for the Index, Honoré Fabri, was suspected, and quite rightly, of harboring Cartesian sympathies.[22] And why not, if the new philosophy contains anything useful, and one takes precautions? 'Just as formerly God allowed the Hebrews to marry their captives after many purifications, to cleanse them of the traces of infidelity, so after having washed and purified the philosophy of Monsieur Descartes, I [it is Gaston Pardies, S. J., professor of mathematics at Louis-le-

16. Cousin, *Frag. phil.*, III (1866^5), 300–17.

17. *Ibid.*, 318–22; Mouy, *Développement* (1934), 170–1 (Boileau's *arrêt*); Lamprecht in Columbia U., *Studies*, III (1935), 196n.

18. Monchamp, *Hist.* (1886), 204–9. Cf. L. C. Rosenfeld, *Rev. Rel.*, 22 (1957), 14–40.

19. Pachtler, *Rat. stud.* (1887–94), III, 122–7, gives all forty propositions.

20. *Ibid.*, 121–3; Sortais, *Cartésianisme* (1929), 20.

21. *Ibid.*, 42–4. Perhaps the most interesting case is Y. André, transferred from Louis-le-Grand to Hesdin (Artois), then, since he persisted, removed from teaching (1713), finally locked up in the Bastille (1721); he thereupon apologized and returned to teaching mathematics. *Ibid.*, 23, 28–32.

22. *Ibid.*, 49–50; *infra*, ii.1.

Grand, an ornament of his Society] could very well embrace his opinions.'[23]

This liberty, and perhaps more, was virtually conceded by the general congregation of 1706; for the last of its 'prohibitions' allowed 'the defense of the Cartesian system as an hypothesis whose postulates and principles are well integrated [recte cohaerent] with one another and with their consequences.' Descartes became more and more acceptable to the Society in the eighteenth century; he may have erred, but he was a veritable Loyola compared with Voltaire.[24] The Jesuits came to teach Descartes' physics, and to esteem his mind. 'Even when Descartes lets us see the weaknesses and limits of human nature, he shows a penetrating, luminous, broad, methodical and systematic mind . . ., almost compelling us to follow him or, at least, to admire him.'[25]

While the Jesuits were painfully assimilating Cartesian physics, much of the rest of learned Europe adopted it. At the Dutch universities, after the troubles of the 1640s and 1650s, Cartesian fellow travellers captured the teaching of physics and several chairs in the medical schools. They retained their hold until the advent of the Newtonians.[26] The Swedes fought to a standstill, and agreed that Cartesian physics could be taught, but Cartesian hermeneutics proscribed, an accommodation that perhaps prepared the way for the flowering of Swedish physics in the eighteenth century.[27] In Italy, after ferocious battles in the 1660s in Naples, where Cartesian physics had been introduced by lawyers and radical physicians,[28] Descartes made his way into the northern universities. In 1700 his followers Guido Grandi, an excellent mathematician,[29] and Michelangelo Fardella, a great friend of the French Cartesians,[30] captured chairs of philosophy at

23. Pardies, *Lettre d'un philosophe à un Cartęsien* (1672), in Ziggelaar, *Pardies* (1971), 119. Like Fabri, Pardies defended himself against charges of Cartesianism by the unpersuasive argument that he rejected Descartes' laws of motion (as did everyone else). *Ibid.*, 80, 115.

24. Sortais, *Cartésianisme* (1929), 89, 92; Werner, Ak. Wiss., Vienna, Phil.-Hist. Cl., *Sitzb.*, 102 (1883), 681–2; Bouillier, *Hist.* (1854), I, 557–80.

25. Regnault, *Origine* (1734), II, 358–9. Cf. P. Baudory's speech at Louis-le-Grand in 1744: Descartes 'erravit aliquando, quod humanum est, at non erravit errantium auctoritate, quod turpe ac imbecillum est.' Schimberg, *Education* (1913), 520.

26. Ruestow, *Physics* (1973), 73–88; Monchamp, *Hist.* (1886), 345, 476–83.

27. Lindborg, *Descartes* (1965), 325–38, 348–9. Conversely, the intellectual backwardness of Spain, according to the eighteenth-century reformer B. J. de Feijóo, owed not a little to an irrational fear of Descartes; as late as 1772 the Royal and Supreme Council of Castile ordered the University of Salamanca to have nothing to do with him. Browning in Hughes and Williams, eds., *Var. Patt.* (1971), 358, 365.

28. Fisch in *Science* (1953), I, 529, 544–5, 553–4; Berthé de Besaucèle, *Cartésiens* (1920), 3–17.

29. *Ibid.*, 55–7. Tenca, Ist. lomb., Cl. sci. mat. nat., *Rend.*, 83 (1950), 494–7, and Pavia, Coll. Ghisl., *Studi,* 1 (1952), 21–4; Grandi estimated that his Jesuit teacher of mathematics, Saccheri, was 'seven-eighths Cartesian,' the remaining part being reserved for good points in Aristotle and Gassendi.

30. Cromaziano, *Restaurazione* (1785–89), II, 84–5; Berthé de Besaucèle, *Cartésiens* (1920), 65–85; Werner, Ak. Wiss., Vienna, Phil.-Hist. Cl., *Sitzb.*, 102 (1883), 77, 132–3.

Pisa and Padua, respectively. Others brought the cause to Bologna, Rome and Turin.

Cartesian physics was taken up in England in the 1640s by the Cambridge philosopher Henry More, a Platonist who, for a time, thought that nothing could be finer than the system of Descartes; 'all that have attempted anything in naturall Philosophy hitherto [he said] are mere shrimps and fumblers in comparison to him.'[31] More urged that Cartesian philosophy be taught in all public schools and universities, particularly Cambridge, where he had already made it influential, and where Newton encountered it—and profited from it—in the 1660s.[32] Joseph Glanville, the self-ordained extirpator of dogmatists, the exploder of the 'steril, unsatisfying verbosities' of the schools, had nothing but praise for our prince of systematizers, 'the grand secretary of nature, the marvelous Descartes.'[33] Other fierce anti-peripatetics, like Robert Boyle, endorsed Cartesian corpuscularism as a good antidote to Aristotelian physics. But Descartes' day in England was brief. Even before the big guns of Newton and Locke came into play, the British had shivered before the materialist consequences of Cartesianism and drawn back.[34] More himself in 1668 had rejected Descartes' metaphysics as 'erroneously and ridiculously false,' but retained the physics; in 1671 he had come to regard all parts of the system as 'most impious, most inept, and entirely false.'[35]

Descartes then replaced Aristotle as the foil against which British physics tested its metal. Hooke could praise the 'most incomparable Descartes,' that 'most acute and excellent Philosopher,' for providing the law of refraction, and yet prefer his own ideas about the nature of light. Newton started with Descartes, and soon went far beyond him in optics; it took longer to elude the spell of the great solar vortex, and to replace Cartesian mechanism with the theory and apologetics of universal gravitation.[36] An apt symbol of the relationship between Cartesian and English physics is the text of Rohault in the editions

31. More to S. Hartlib, 11 Dec. [1648], in C. Webster, *Br. J. Hist. Sci.*, 4 (1969), 365. Cf. *ibid.*, 371: 'If Mr. [William] Petty [who did not share More's enthusiasm] should have twice the Age of an ordinarie man, and spend all his dayes in experiencing, he will never bring an instance against Des Cartes his principles of Light, of the Lodestone, of the Rainbow, of the Flux and Reflux of the Sea, etc.'

32. Lamprecht in Columbia U., *Studies,* III (1935), 208–9; C. Webster, *Ambix,* 14 (1967), 153, 156, 167–8; Power, *Exper. Phil.* (1664), sig. c.1^{r}, on the 'ever-to-be-admired Des-Cartes.'

33. Glanville, *Vanity of Dogmatizing* (1661), quoted by Lamprecht in Columbia U., *Studies,* III (1935), 201.

34. Pacchi, *Cartesio* (1973), viii–ix, 231–4, 248–50; cf. Lamprecht in Columbia U., *Studies*, III (1935), 185–7, 199, 229, and Bouillier, *Hist.* (1854), II, 491–8.

35. Lamprecht, in Columbia U., *Studies* III (1935), 219–25. Yet More appears to have taught Cartesian physics as late as 1674 (*ibid.*, 195).

36. Hooke, *Micrographia* (1665), Pref., 46, 54, 57; Whiston, *Memoirs* (1749), 8ff; cf. Koyré, *Newt. Stud.* (1965), 53–114, and the preface to the *PT* for 1693 (*PT,* 17, 581–2): 'Real Knowledge is a nice thing; and as no Man can be said to be Master of that which he cannot teach to another, so neither can the Mind itself, at least as to Physical matters, be allowed to apprehend that, whereof it has not in some sense a Mechanical Conception.'

of Clarke: the theories of the disciple of Descartes provided the stimulus for the criticisms and rectifications of the disciple of Newton.

While Descartes' star sank in England it rose in France, at least outside the universities.[37] The Oratorians, a teaching order second in importance only to the Jesuits, boasted several prominent Cartesians, above all Malebranche. The *noblesse de robe,* from whom Descartes sprang, welcomed his teachings, and played the part in France that physicians did in the universities of the low countries. The abbreviated obsequies of 1667, welcoming home Descartes' mortal remains, and the banquet that followed were attended by members of the Conseil d'Etat and a crowd of prominent lawyers.[38] In the 1650s the group maintained by the Parisian lawyer Henri-Louis de Montmor—a forerunner of the Paris Academy of Sciences—was sufficiently Cartesian to harass members who were not, and secure enough to encourage public lectures on the new physics.

Some of the lectures, particularly Jacques Rohault's, had an immense success. No one needed the old physics any more. Supply followed demand. Publication of texts on scholastic physics almost ceased in France. 'Such texts as do appear [lamented Gabriel Daniel, S. J., in 1690] are treatises on physics that assume the principles of the new philosophy.'[39] Rohault did not gain admission to the Paris Academy, but he led the Cartesians to the gate. After some official persecution, they entered in the person of Rohault's Joshua, Pierre-Sylvain Régis. That was in 1699, the year the Academy reorganized with the Cartesian reformer Malebranche as an honorary member and the moderate Cartesian Fontenelle as its secretary.[40] It remained Cartesian for two generations.

Several important new features entered Cartesian physics with Malebranche, whose life was a model of progressive enlightenment. In 1664, after completing his theology at the Sorbonne without enthusiasm or distinction, he ran across Descartes' physiological fragment, the *Traité de l'homme.* It was the first work of the master that he had seen. He read it straight through, 'with such excitement,' says Fontenelle, 'that it gave him palpitations of the heart, which obliged him sometimes to interrupt his reading.' It pushed him from scholastic darkness into the light of mathematics and physics. Ten years later the fruit of his learning and meditation appeared as the *Recherche de la verité* (1674–5).[41]

37. Cartesian physics officially entered the colleges of the University of Paris in 1721. Lantoine, *Hist.* (1879), 138–9, 150–1.

38. Cousin, *Frag. phil.*, III (1866[5]), 302; Bouillier, *Hist.* (1854), I, 254–5. 'Cette philosophie a été incroyablement applaudie partout . . . surtout parmi les nobles'; Pardies, c. 1670, quoted by Ziggelaar, *Pardies* (1971), 79.

39. Mouy, *Développement* (1934), 98–113; R. Hahn, *Anatomy* (1971), 6–8; *infra,* ii.4; Daniel, *Voyage au monde de Descartes* (1690), quoted by Sortais, *Cartésianisme* (1929), 59.

40. Mouy, *Développement* (1934), 145–7, 166–79.

41. Fontenelle, 'Eloge' of Malebranche, in Malebranche, *Oeuvres,* XIX, 1000. Grandi too came to mathematics from Cartesian philosophy; Tenca, Ist. lomb., Cl. sci. mat. nat., *Rend.*, 83 (1950), 495.

During the 1690s Malebranche climbed still further by mastering the Leibnizian calculus, an instrument more effective in hunting out the truths of physics than even the geometry of Descartes. He was joined in these studies by a group of strong mathematicians, later his colleagues in the Academy, all also Cartesian in physics, of whom the best known are Pierre Varignon and the Marquis de l'Hospital. Fontenelle shows us each of these men traversing the same road to Enlightenment: from scholastic night through Cartesian day to the blazing light of the differential calculus.[42] Their example and their work were to influence the mathematical physicists of the mid-century, d'Alembert, Maupertuis, and even Euler.[43]

Among Malebranche's modifications of Cartesian physics were a continuing repair of the laws of motion[44] and the discovery of several orders of whirlpools that Descartes had overlooked. Malebranche announced his discovery at the first session of the renovated Academy, in an important paper on optics. Since, he said, light is a pressure transmitted through the globules of the second matter, those globules cannot be hard seeds, as Descartes taught, but elastic balls; otherwise rays of different colors—pressure waves of different frequencies, in Malebranche's model—could not cross without destroying one another.[45] Nothing could be easier than to make the globules elastic: merely imagine them to be so many minute vortices in exceedingly rapid motion, striving to expand owing to 'centrifugal force,' but restrained by the similar tendencies of their neighbors. Much can be explained by these mini-vortices. Malebranche accounts for all elasticity and for solidity or hardness, which had given Descartes much trouble;[46] later he worked the mini-vortices into the theory of planetary motions, and the explanation of refraction;[47] and some of his followers found in them the explanation of electricity.

To understand the hold of Cartesian natural philosophy on the European mind one must understand it not as an ontology, but as an epistemology. Descartes pointed the way not to apodictic truth, but to intelligible physics: 'In the true philosophy,' says Huygens, 'one considers the cause of all natural effects in terms of mechanical motions. This, in my opinion, we must certainly do, or else renounce all hopes of ever comprehending anything in physics.'[48] And that

42. Robinet, *RHS*, 14 (1961), 238–43, and *RHS*, 12 (1959), 5–8, 15–16.

43. Hankins, *J. Hist. Ideas*, 28 (1967), 194, 201–5, and *D'Alembert* (1970), 17, 21, 118–20.

44. Mouy, *Développement* (1934), 292–304.

45. Malebranche, *MAS* (1699), 22–32; Robinet, *Malebranche* (1970), 277–8, 285–94; Mouy, *Développement* (1934), 289–91, 305–10; Duhem, *Rev. met. mor.*, 23 (1916), 77–9, 89–91. The objection seems to be that a hard globule must move as a unit, with one motion at a time, while a compressible ball can transmit many different pushes simultaneously. Cf. Fontenelle, *HAS* (1799), 19.

46. Mouy, *Développement* (1934), 282–8; Malebranche, *Oeuvres*, II, 326, III, 272–3 (text of 1712). Cf. Carteron, *Rev. phil.*, 47 (1922), 491–2.

47. Malebranche, *Oeuvres*, III, 283–4, 296–8; Mouy, *Développement* (1934), 310–14.

48. Huygens, *Treatise* [1690] (1945), 3. Cf. Koyré, *Newt. Stud.* (1965), 118, and Huygens' objection to Newton's optical theory of 1672: why suppose seven or eight colors in the spectrum

is all one must do to be a Cartesian physicist: the details of Descartes' mechanical pictures, his fanciful laws of motion, his metaphysical underpinnings, none of this need be—or often can be—accepted. As Fontenelle put it: 'Il faut admirer toujours Descartes, et le suivre quelquefois.'[49]

The Cartesian physicists of the late seventeenth Century did not recommend push-pull physics for its truth. Most placed physical truth, the essence of body, the ultimate cause of change, beyond the reach of the human mind. Fontenelle, who liked to poke fun at metaphysics ('which is to most people like an alcohol flame, too subtle to burn wood'[50]), declared that the process of collision, the only allowable interaction between extended bodies, is at bottom unintelligible to us.[51] The weakness of our minds, the false witness of our senses, and our share in Adam's fall showed Boyle that the hypotheses even of the corpuscular philosophy could only be doubtful conjectures.[52] 'Physical demonstrations can beget but a physical certainty.'[53] *A fortiori,* one could never choose between corpuscular explanations equally comprehensive, between Descartes' and Boyle's explanations of the spring of the air, for example; the same clock, after all, can be driven by a spring or by a weight.[54] Glanville hunted out paradoxes to prove the impossibility of ever comprehending matter.[55] Régis, despite his love of system, conceded physics to be 'problematic' and 'uncertain.' So did Huygens.[56]

---

rather than two, of which the remainder can be composed? For then 'it will be much more easy to find an Hypothesis by motion.' Newton, *Papers* (1958), 136; Guerlac in Wasserman, *Aspects* (1965), 327; McDonald, *Ann. Sci.*, 28 (1972), 219–21.

49. Fontenelle, *HAS* (1735), 139. Cf. Fontenelle, 'Digression sur les anciens et les modernes' (1687), *Oeuvres,* IV, 121: '[Descartes'] new way of philosophizing [is] much more valuable than the philosophy itself, a good part of which is false, or very uncertain.'

50. 'Eloge' of Malebranche, in Malebranche, *Oeuvres,* XIX, 1002. Failure to distinguish between agreement with Descartes' system and acceptance of the general principles of his physics has resulted in a large and inconclusive literature (for which see Lissa, *Cartesianismo* [1971], 33–4) on the 'question' of Fontenelle's Cartesianism; Lissa, *ibid.*, 37, concludes judiciously that Fontenelle 'libera dall'architettonica construzione metafisica di Cartesio un dottile strumento di indagine [fisica], il cui spirito è fondamentale antimetafisico.'

51. Fontenelle, 'Doutes sur le système physique des causes occasionnelles' (1686), in *Oeuvres* (1764), IX, 45–6. Cf., Carré, *Philosophie* (1932), 20–1; Marsak, Am. Phil. Soc., *Trans.,* 49:7 (1959), 20–1, 31–2; Robinet, *Rev. synt.*, 82 (1961), 82–3.

52. Boyle, *Reason and Religion* (1675), in *Works,* IV, 164–5; van Leeuwen, *Problem* (1970), 93; Guerlac in Wasserman, *Aspects* (1965), 321–3. Cf. McGuire, *J. Hist. Ideas,* 33 (1972), 523–42, who emphasizes the connection between Boyle's nominalism in science and voluntarism in theology; Burtt, *Met. Found.* (1955), 166–302, who notes (p. 185) Boyle's anticipation of Newton's positivism; and Westfall, *Ann. Sci.*, 12 (1956), 107–10, who points to some passages from Boyle admitting a realist interpretation.

53. Boyle, *Excellency Theol.* (1674), in *Works,* IV, 42.

54. Boyle, *Works,* I, 12, II, 45–6, V, 74–5, quoted by Mandelbaum, *Philosophy* (1964), 90–1. Boyle's attitude towards physical hypotheses may have influenced Locke's; Gibson, *Locke's Th.* (1917), 205–31, 260–5.

55. *Scepsis scientifica* (1665), quoted by Gibson, *Locke's Th.* (1917), 257.

56. Mouy, *Développement* (1934), 147, 153; *DSB,* VI, 608.

Malebranche arrived at a similar view by a straighter path. Having a mind, he said, as good as anyone's, and yet being unable clearly to conceive how one body can act upon another, he declared that they do not, that the very notion was a vulgar error. Created beings cannot act; what we, in our ignorance, believe to be the push of one object upon another, or the stimulation of our senses by a third, are not acts but 'occasions' for acts.[57] God directly causes rebounds appropriate to the pushes and sensations agreeable to the stimuli. Being free from caprice, He acts always—or almost always, to leave room for miracles—in the same manner. The consequences of collisions may be confidently predicted by the laws of mechanics. Such laws, mathematically formulated, are the only true knowledge: 'I do not believe that there is anything useful which men can know with exactitude that they cannot know by arithmetic and algebra.'[58] Since these laws are God's choices, we can not deduce them *a priori,* as Descartes tried to do. But we must always strive to reduce our physics to one of impulse; for although we can have no conception of the 'mechanics' at work, we understand immediately that something must happen in collisions, something 'certain and incontestable,' and we can calculate the result.[59]

This last sentiment suggests why the 'true' physics must be mechanical. Mechanical theories have the advantage over all others of clarity, precision, completeness, and naturalness.[60] They are also relatively intelligible. To be sure the inner nature of a collision is not comprehensible; but there are degrees of unintelligibility, and the less the better. 'It is certain [says Fontenelle] that if one wishes to understand what one says, there is nothing but impulse, and if one does not care to understand, there are attractions and whatever one pleases; but then nature is so incomprehensible to us that it is probably wiser to leave her alone.' Our idea of impenetrability tells us immediately that something must happen when moving balls collide; but nothing suggests that, when mutually at rest and widely separated, they must attract one another.[61]

Boyle extols the mechanical philosophy first for its 'intelligibleness or clearness' (as against 'intricate' disputes of the peripatetics, the 'darkness and ambiguity' of the spagyrists); second, for its economy (requiring only matter and motion); third and fourth, for its radical simplicity (matter and motion being the simplest 'primary' concepts); and last for its comprehensiveness. He dismisses forms and qualities not because they are wrong or 'self-repugnant,' but because

57. Malebranche, *Oeuvres,* III, 203–12, 217–18. For the theological connections of the doctrine see Rodis-Lewis, *Malebranche* (1963), 135–8, 296–300.

58. Malebranche, *Oeuvres,* II, 292g.

59. *Ibid.*, II, 403: 'Il n'y a aucune raison, ni aucune expérience, qui démonstre clairement le mouvement d'attraction.' Cf. Robinet, *Malebranche* (1970), 77; Mouy, *Développement* (1934), 316–18; *infra,* i.4–5.

60. De Volder, as quoted by Ruestow, *Physics* (1973), 94–5.

61. Fontenelle, 'Eloge' of R. de Montmort, in *Oeuvres,* VI, 26; cf. Marsak, Am. Phil. Soc., *Trans.,* 49:7 (1959), 13–14.

'we conceive not, how they operate to bring effects to pass.'[62] 'It is more certain to reason from mechanical and intelligible principles than to depend upon novelties which are not expressed in ideas familiar to the mind.'[63] We shall take this insistence on mechanical explanation, on invoking 'mechanism as immediate cause of the phenomena of nature,' as the distinctive mark of Cartesian physics, and shall regard it as an epistemological condition, a requirement for securing the simplest, the most intelligible, the most fruitful, and the most satisfying of natural philosophies.[64] Or so it was for those who held with Fontenelle that 'nature is never so admired as when she is understood.'[65]

PRAECEPTOR GERMANIAE

Although Dutch Cartesianism bubbled over from Holland into a few neighboring Calvinist universities like Duisburg, the philosophy of Descartes in the strict sense never prospered in Germany.[66] The Jesuit schools fought it bitterly. The Lutheran universities of the north opposed it as subversive of sound theology. At Marburg, in 1688, they smothered a Cartesian work by a colleague, a professor of medicine. In the 1690s the theologians at the avant-garde University of Halle harassed the professor of 'new philosophy and mathematics.'[67] Although resistance to Descartes weakened in the German Protestant schools after 1700, that did not establish his philosophy, at least under his name. It was squeezed out between the school philosophy and the teachings of Leibniz, as modified, expanded and systematized by the 'preceptor of Germany,' Christian von Wolff.

The system owed much to Descartes, whose promise of a mathematical philosophy, and its extension to theology, Wolff hoped to realize.[68] He went to the University of Jena expressly to study mathematics under G. A. Hamberger, and began his own teaching career at Halle in 1706 as professor of that 'unknown and unusual' subject.[69] Soon he added physics, then metaphysics, philosophy, theology, law, to the increasing annoyance of his colleagues. He wrote text books on all these subjects, and in his own language, which obliged

62. *Excellency Corp. Phil.* (1674), *Works,* IV, 72. Cf. van Leeuwen, *Problem* (1970²), 75, 106.

63. Nollet, *Leçons* (1747–8), II, 477; cf. Nollet to Bergman, 20 Sept. 1766, in Bergman, *For. Corresp.* (1965), 285: 'Des causes mécaniques, qui sont les seules capables d'étendre les progrès de la physique expérimentale.'

64. Cf. Mairan, 'Eloge' of Molières, *MAS* (1742), 200; Bouillier, *Hist.* (1854), II, 569, 573–5.

65. 'Préface sur l'utilité des mathématiques et de la physique' (1733), in *Oeuvres,* V, 11.

66. Bouillier, *Hist.* (1854), II, 404–5; Paulsen, *Gesch.*, I (1896), 519; Bartholmess, *Hist.* (1850), I, 101.

67. Hermelink, *U. Marburg* (1927), 306–16, 330–1; Förster, *Übersicht* (1799), 49–50, 54.

68. Vleeschauwer, *Rev. belg. phil. hist.,* II (1932), 659–63, 676–7; Wolff, 'Lebensbeschreibung,' in Wuttke, *Wolff* (1841), 114, 121.

69. *Ibid.*, 146; Gottsched, *Wolff* (1755), 9–13; Vleeschauwer, *op. cit.*, 666–7.

him to invent a German philosophical vocabulary.[70] The mathematics book, first published in Latin in 1713, covered the full range of pure and applied subjects; it remained standard for half a century, and is still useful for its annotated bibliography.[71] The physics books stressed experimental confirmation and described in detail the apparatus necessary to achieve it.[72] They also had a wide circulation; they played the same role in Protestant Germany in the eighteenth century that Melanchthon's had in the sixteenth and seventeenth.[73] By 1715 Wolff was recognized as an authority in his own right and as the intellectual heir of Leibniz. The universities of Wittenberg and Jena begged him to add his light to their faculties; the city of Bologna sought his advice on water control; and the Czar of the Russias consulted him about the educational backwardness of his people.

Wolff's rational theology did not please everybody. When he went so far as to say a kind word about Confucius his enemies cried atheism and brought the matter before the King of Prussia, Frederick William I. The King ordered Wolff to leave his dominions forthwith (1723); the philosopher immediately found refuge and a higher salary at the University of Marburg, whence he continued to issue texts, now in Latin, on all respectable subjects.[74] His reputation, enhanced by martyrdom in the cause of academic freedom, brought him offers from the Petersburg Academy of Sciences, the universities of Göttingen and Utrecht ('under such circumstances as no professor in Holland had'), and even Halle, to which Wolff returned in 1740.[75] During the Marburg exile Wolff's philosophy spread to the Petersburg Academy and the universities of Leipzig, Jena, Tübingen and Würzburg, to mention only those most seriously infected. Count von Manteuffel set up an influential group called the Alethophiles to promote Wolff in Berlin.[76] And, doubtless most gratifying of all, the Marquise du Châtelet, sometime mistress of the great puffer of English philosophy, Voltaire, announced that Leibniz' was the only metaphysics that satisfied her and hired a Wolffian to initiate her into its mysteries.[77] Wolff himself sought a correspondence with the lady, the instrument, he hoped, for the conversion of the French. She gave him more than he could have hoped, an *Institutions de physique* prefaced by an abridgment of his philosophy, a clear and, what was unusual, a concise account that won the approval of both Wolff and

70. Gottsched, *Wolff* (1755), 30, 35–6, 42, 48, 51, and Beylage, 9–14, 17.

71. One finds Wolff's mathematics recommended into the 1770s.

72. Wolff, *Ausf. Nachr.* (1733²), 474, 476, 479; *supra,* i.1.

73. Paulsen, *Gesch.*, I (1896), 527; cf. Bartholmess, *Hist.*, I (1850), 99–100, 103.

74. Gottsched, *Wolff* (1755), 57–72; Förster, *Übersicht* (1799), 95–6, 140.

75. Gottsched, *Wolff* (1755), 90–1, 100, 102, and Beylage, 50, 67; Wolff in Wuttke, *Wolff* (1841), 154–5, 165. Negotiations for the return were begun under Frederick William I and completed under his successor, Frederick the Great.

76. Gottsched, *Wolff* (1755), 104, 120–1; *infra,* ii.2. Cf. Paulsen, *Gesch.*, I, 546, for the Wolffian foundations of the University of Erlangen (1743).

77. Letter to Frederick the Great, 25 April 1740, in Du Châtelet-Lomont, *Lettres* (1958), II, 13; Du Châtelet-Lomont, *Institutions* (1740), 13. Cf. Barber, *Leibniz* (1955), 127–40.

the Alethophiles. 'What is certain, what leaps to the eye,' wrote Manteuffel, 'is that she has given up all the chimeras of her friend Voltaire, whom she far surpasses in precision and clarity of ideas.'[78]

Wolff's severe rationalism rests on two principles, one logical, the law of contradiction, the other psychological, the law of sufficient reason. The latter plays the same part in his system that the principle of clarity and distinction of ideas plays in Descartes': a phenomenon is satisfactorily explained, its existence fully understood, when its 'sufficient reason' can be given. All contingent truths depend on this proposition. The principle of sufficient reason, says the Marquise du Châtelet, is the only 'compass able to guide us through the shifting sands of [metaphysics],' the only 'thread that can lead us in these labyrinths of error.'[79] What suffices for one philosopher, however, seldom contents another, and, in practice, the touchstone of Wolffian truth was the satisfaction of Wolff's reason.

That powerful instrument disclosed that the world consists of distinct, unextended 'units' or 'elements,' unselfconscious Leibnizian monads. Unfortunately, one cannot fully reduce physical phenomena to these elements; in particular, as unphilosophical mathematicians like Euler liked to point out, one cannot understand how they might make up an extended body. Wolff dismissed such objectors as metaphysically illiterate ('Euler is a baby in everything but the integral calculus'[80]) and built his physics not upon his elements, but upon what he called the basic 'phenomena' of matter: extension, inertia, and moving force.[81] The fundamental objective of physics is to reduce other phenomena to the basic set: 'per extensionem, vim inertiae et vim activam omnes corporum mutationes explicari possunt.' Everything works mechanically; the physical world is nothing but a clock; physics is 'mechanical philosophy.'[82]

The physicist should have nothing to do with occult qualities, for which, by Wolff's definition, he can give no sufficient, that is, mechanical reason. He must reject action at a distance, attractions understood as a primitive force. One can assign no reason that A and B, separated in space, should act upon one another; matter can act only by impact, by immediate contact and mutual 'obstruction;' where there appears to be attraction, as in the cases of electricity and magnetism, we must assume the existence of an unseen, mediating, material emanation.[83] As Mme du Châtelet put it: attraction is 'inadmissible, since it

78. Wolff in Wuttke, *Wolff* (1841); 178; Wolff to Manteuffel, 7 June 1739, in Ostertag, *Phil. Geh.* (1910), 38, 40. Cf. Droysen, *Zs. franz. Spr. Lit.*, 35 (1910), 226–38.

79. Du Châtelet-Lomont, *Institutions* (1740), 13, 22, 25.

80. Wolff to Manteuffel, 4 Aug. 1748, in Ostertag, *Phil. Geh.* (1910), 147–8. Cf. *ibid.*, 75: Euler 'understands not the least little things in philosophy.'

81. Wolff, *Cosmologia* (1732), §§226, 296, 298; the problematic standing of these 'basic phenomena' is discussed by Campo, *Wolff* (1939), I, 223–5, 230, 250–1.

82. Wolff, *Cosmologia* (1737^2), §§74, 75, 79, 117, 127, 138.

83. *Ibid.*, §§133, 320–5, 149: 'Qualitas occulta dicitur ea, quae sufficiente ratione destituitur, cur subjecto insit, vel saltem inesse possit.' Cf. Leibniz to Wolff, 23 Dec. 1709, in Leibniz,

offers nothing from which an intelligent being can understand why the velocity and direction—the determinations of the being under discussion [motion acquired under the suppositious distance force]—are such rather than otherwise. Not even God could say how a body acted upon at a distance would move.' Sufficient reasons are mechanical causes. Hence, said the marquise, sounding the call to the colors, all true physicists should rally to a search for a mechanical explanation of gravity.[84] Wolff himself would have done so, successfully we are told, had he been able to perfect his own general physics.[85] He did hint that one should follow up the Cartesian accounts of magnetism, and he repeatedly pointed to electrical experiments as evidence that apparent attractions are mediated by a 'subtle matter.'[86]

The fact that Wolff could not stop to carry out his reductionist program was perhaps more an encouragement than a disappointment to others. On the one hand one could do physics as he did, 'acquiescing,' as he often said, 'in the phenomena;' here he limited himself to a description in terms of physical concepts such as cohesion and elasticity, inadmissible as fundamental powers but very useful as 'proximate causes,' 'physical' as opposed to 'mechanical' principles.[87] On the other hand one could seek deeper, mechanical explanations, assured by Wolff himself that the goal was possible and worthy of attainment. The Wolffian physicist, the last heir of the mechanical philosophy of Descartes, long protected Germany from the irrational and slipshod methods of Sir Isaac Newton.

4. ATTRACTION IN NEWTON

The 'Preface' to Newton's masterpiece, *Principia mathematica philosophiae naturalis* (1687), formulates its mission as follows: 'The whole business of philosophy seems to consist in this—from the phenomena of motions to investigate the forces of nature, and then from these forces to demonstrate the other phenomena.' What this means is plain from the body of the book. 'To investigate the forces of nature' means to infer mathematical propositions about forces, somehow known to exist; 'to demonstrate the other phenomena' means to compare quantitative data with logical consequences of the propositions. If the procedure succeeds, the propositions, according to Newton, must be regarded as true; for '[they] are deduced from Phenomena & made general by

---

Briefw. (1860), 113: '[Actio in distans] pugnat tamen cum magno illo Principio Metaphysico . . . , quod nihil sine ratione sive causa fiat.'

84. Du Châtelet-Lomont, *Institutions* (1740), 47–9, 328–34. Precisely the same line of reasoning appears in D. and J. Bernoulli, AS, *Pièces,* 5 (1748), 119.

85. Gottsched, *Wolff* (1755), 151.

86. Wolff, *Cosmologia* (1737²), §320; Wolff to Manteuffel, 8 March and 4 Oct. 1744 ('zur Zeit noch keine warscheinlichere Erklärung [of electricity] als durch die vortices cartesianas gefunden waren') and 8 Nov. 1747, in Ostertag, *Phil. Geh.* (1910), 65–7, 137–8.

87. Wolff, *Cosmologia* (1737²), §§235–8, 241, 292; cf. Campo, *Wolff* (1939), I, 240, 244–5.

Induction: w[hi]ch is the highest evidence that a Proposition can have in this philosophy.'[1]

In the special case of gravity Newton infers the accelerations of planets towards the sun and of satellites towards their primaries by examining Kepler's empirical laws in the light of certain mathematical principles about centripetal forces. That is investigating the forces of nature. He then applies the gravitational law to deduce the motions of the planets, the comets, the moon and the sea, which he shows agree most beautifully with the best data available.[2] In keeping with his conception of natural philosophy, Newton believed that he had thereby demonstrated the existence of gravity, of a centripetal force acting between every pair of material particles in the universe and causing their mutual accessions, or acceleration, in accordance with a simple mathematical expression. As to the ultimate nature of this force, its seat and mechanism, Newton—or, rather, the author of the *Principia,* for there were several Newtons—declined to speculate: since the phenomena did not settle whether gravity was innate to material particles, or a property of the space between them, or the result of a pressure from an interplanetary medium, he refused overtly to frame—or feign—an hypothesis. 'To us it is enough that gravity does really exist, and act according to the laws which we have explained, and abundantly serves to account for the motions of the celestial bodies, and of our sea.'[3]

Newton's view of the theory of universal gravitation failed to convince his contemporaries. Indeed, his apologetics are not very plausible. It requires more than a mathematical midwife to deliver from the phenomena the proposition that all pairs of particles in the universe—or even in the solar system—mutually gravitate. One needs besides some laws of motion and, in particular, the principle of rectilinear inertia, assumed to hold true of every bit of heavenly and earthly matter. But neither this principle, nor the universal applicability of any such principle, is evident, or perhaps even probable. Galileo inclined toward circular inertia and the Aristotelians made qualitative distinctions between sublunary and celestial behavior the basis of their physics.[4] One might object to Newton that, perhaps, the planets do not move like separated terrestrial objects; perhaps they go naturally in circles, which unknown, non-mutual, non-central agencies distort into the postulated Kepler ellipses? Newton saw this loophole. To close it he laid down three rules, which obliged the philosopher to ascribe similar effects to similar causes and to regard as *universal* those qualities of matter found to belong to, and to be unalterable in, bodies accessible to experiment. These rules have proved very, though not invariably fruitful, but they do

1. Newton to Cotes, 1712, in Koyré, *Newt. Stud.* (1965), 275. Cf. Newton, *Math. Princ.* (1934), 400: 'In experimental philosophy we are to look upon propositions inferred by general induction from phenomena as accurately or very nearly true.' This is the fourth *regula philosophandi,* which first appeared in *Princ.* (1726³); Koyré, *Newt. Stud.* (1965), 268–71.
2. *Math. Princ.* (1934), xviii.
3. *Ibid.*, 547. This is from the General Scholium, which first appeared in *Princ.* (1713²).
4. Cf. M. B. Hesse, *Forces* (1961), 146–7.

not secure the *existence* of universal gravity.[5] All one can say is that bodies act *as if* they are endowed with Newton's gravitational force; and even that is not strictly correct, as the precession of Mercury's perihelion seems to show.

Newton claimed that his powerful and fruitful, but nonetheless hypothetical theory of gravitation was a direct induction from the facts, and free from speculation. He had made precisely the same claim about his first scientific work, the paper on optics of 1672. He said that he had proved that white light is physically a composition of 'rays' differently refrangible, whereas his experiments had demonstrated only that a different index of refraction characterized each spectral color obtained from white light passed through a prism. In the ensuing many-sided controversy, Hooke hit upon an unexceptionable alternative hypothesis, that white light consists of a complex aether pulse, unanalyzed until its passage through the prism. Newton never recognized that his representation of light, though unexceptionable as a mathematical description, became as hypothetical as Hooke's when interpreted physically. He still claimed in the last edition of the *Opticks* (1717–8), which maintained that white light is a physical mixture of colored rays, that his purpose was 'not to explain the Properties of Light by Hypotheses, but to propose and prove them by Reason and Experiments.'[6]

It is not strange that Newton's first readers, who understood correctly that he believed in his universal centripetal accelerations, did not interpret his procedures in the later standard instrumentalist sense. Moreover, the intensity of his belief predisposed them to disregard his occasional disclaimers and to conclude —wrongly this time—that the *Principia* advanced a particular view of the cause of gravity. Its frequent references to mutual and equal attractions, to bodies drawing one another across resistanceless spaces, to powers exercised in proportion to mass, to accelerative forces diminishing as the square of the distance, made natural the inference that Newton held gravity to be an innate property of bodies, and to act immediately at a distance.[7]

One example is Proposition VII of Book III. It argues that 'all the parts of any planet A gravitate towards any other planet B,' a formulation which, as Newton's editor Roger Cotes told him, seems to imply the hypothesis that the power of gravitating resides *in* the several parts of matter.[8] Newton's early

5. *Math. Princ.* (1934), 398–400. The first two rules, re the multiplication of causes, appeared as 'hypotheses' in *Princ.* (1687¹); the third first occurs in *Princ.* (1713²), where all three are called rules. Cf. Koyré, *Newt. Stud.* (1965), 261–8; Cohen, *FN*, 575–9; McGuire, *Cent.*, 12 (1968), 233–60; *St. Hist. Phil. Sci.*, 1 (1970), 3–58; and *Ambix*, 14 (1967), 69–95.

6. *Opticks* (1730⁴), 1; Sabra, *Theories* (1967), 233, 273–97. A concept of the constitution of light akin to Hooke's figures in the discovery of the diffraction of X rays. Heilbron, *Moseley* (1974), 66–7, 71.

7. Cf. Koyré, *Newt. Stud.* (1965), 137, 149, 153–4; McGuire, *Arch. Hist. Exact Sci.*, 3 (1966), 206–48; Guerlac, *Newton* (1963), 10, 25. Buchdahl, *Minn. Stud.*, 5 (1970), 216, points to apparently realist references to gravity in *Principia*.

8. *Math. Princ.* (1934), 414; Edleston, *Corresp.* (1850), 153, 155. Cf. Hollman, Ak. Wiss., Gött., *Comm.*, 4 (1754), 224–6, who points to the same proposition.

English followers, his 'rash disciples' as Aepinus called them,[9] either misunderstood him to have considered gravity an essential quality of matter or, having understood his hedges, nonetheless believed gravity to have 'as fair a claim to the Title' of essentiality as any other property. The rashest, Cotes, professor of astronomy at Cambridge, provoked Continental natural philosophers already sufficiently aroused by insisting, in the preface to his edition of the *Principia,* that Newton's gravity had the same ontological status as the irreducible properties of Cartesian matter: 'either gravity will have a place among the primary qualities of bodies, or Extension, Mobility and Impenetrability will not.'[10] The same sort of irritant also appeared in the less official writings by John Keill, John Freind and George Cheyne, who thought that 'all the Particles of Matter endeavor to *embrace* one another.'[11]

Huygens and Leibniz had already criticized the *Principia* interpreted as Newton wished, as a book about effects, not causes: excellent mathematics, they said, but no physics; an *asylum ignorantiae,* a refuge for those too lazy or too ignorant to work out a clear, proper, mechanical account of gravitation.[12] The queries of the Latin *Opticks,* and the still stronger assertions of the disciples, who raised gravity to a quality inherent in matter, called forth wider opposition. The Leibnizians protested with particular vigor; for at the same time that the rash disciples insisted upon the primitivity of attractions, they were trying to appropriate to Newton all credit for inventing the calculus.[13]

In the *Acta eruditorum* of 1710 an anonymous Leibnizian, none other than Wolff, blasted Freind for taking attraction as a primitive force; Freind replied; and Wolff blasted him again, from a carefully fortified position. An attractive force may be admitted, he said, as a *pis aller,* as a phenomenon needing explanation; but if, with Freind, one supposed it innate and non-mechanical, which is to say inexplicable and unintelligible, we return to 'occult qualities,' to the

9. *Tent.* (1759), 5–6; Hutton, *Dict.* (1815), and *Encycl. Brit.*[3], allow the charge in their articles 'Attraction.'

10. *Math. Princ.* (1934), xxvi; Koyré, *Newt. Stud.* (1965), 159, 273–82; Cotes to Clarke, 25 June 1713, in Edleston, *Corresp.* (1850), 158–9. A primary quality is inherent and universal, but not necessarily essential, such that matter could not exist without it; God might have been able to make matter extended and mobile but not gravitating, or gravitating according to a law other than Newton's. The rash disciples explicitly made gravity primary but not essential; Kant was to go all the way. Cf. Tonelli, *RHS,* 12 (1959), 225–41.

11. Cheyne, *Philosophical Principles* (1715–6), as quoted in Koyré, *Newt. Stud.* (1965), 156, 282 and Bowles, *Ambix,* 22 (1975), 21; Keill, *PT,* 26 (1708), 97–110; Schofield, *Mechanism* (1970), 41–5; Freind, *Prael.* (1710), 4: 'Datur vis attractrix, seu omnes materiae partes à se invicem trahuntur.'

12. Koyré, *Newt. Stud.* (1965), 117–23, 140, 264–5, 273–4; Alexander, *Leibniz-Clark Corresp.* (1958); Guerlac, *Newton* (1963), 7, and Guerlac in Wasserman, ed., *Aspects,* 329. Cf. the review of *Principia* in *J. des sçavans,* 16 (1688), 328: 'Mr. Newton n'a qu'à nous donner une Physique aussi exacte qu'est [sa] Mechanique. Il l'aura donné quand il aura substitué de vrais mouvemens en la place de ceux qu'il a supposez.'

13. Wolff's shock at the Latin *Opticks* and *Principia* appears from his letters to Leibniz of 17 Aug. 1710 and 11 Dec. 1712, in Leibniz, *Briefw.* (1860), 124–5, 154.

'cant of the schools,' to 'sounds without content.'[14] A literal attraction, an action at a distance, exceeds the power of creatures, which, according to Leibniz and Wolff, can produce local motion only by pushing. Hence it implies either a perpetual miracle or a spiritual, non-mechanical agency. But neither alternative is admissible. Good philosophers do not call upon God to save the phenomena, nor invoke invisible and intangible, or 'inexplicable, unintelligible, precarious, groundless and unexampled' means of communication. 'Nobody'—it is still Leibniz—'nobody would have ventured to publish such chimerical notions . . . in the time of Boyle.' Alas! sighed Wolff, Newton was never more than a beginner in philosophy, and too weak to oppose the shameful quirks and crotchets of his disciples.[15]

In France, the learned press, the *Journal des sçavans* and the *Journal de Trévoux,* rallied support for the true, or Cartesian, physics. In the Paris Academy the mathematician Joseph Saurin, who in 1703 had parried a thrust at the doctrine of vortices, led the cause. Have nothing to do, he warned, with English gravity, one of those scholastic occult qualities. 'We need not flatter ourselves that, in physics, we can ever surmount all difficulties; but let us always philosophize from the clear principles of mechanics; if we abandon them, we extinguish all the light available to us, and we sink back into the old peripatetic darkness, from which Heaven preserve us.'[16] Fontenelle, who could never distinguish between innate gravity and sympathies, horrors, and 'everything that made the old philosophy revolting,' fired away at Newtonian occultism even as he commemorated the death of its founder: 'The continual use of the word Attraction, supported by great authority, and perhaps too by the inclination which Sir Isaac is thought to have had for the thing itself, at least makes the Reader familiar with a notion exploded by the Cartesians, and whose condemnation has been ratified by all the rest of the Philosophers; and we must now be upon our guard, lest we imagine that there is any reality in it, and so expose ourselves to the danger of believing that we comprehend it.'[17]

Newton had not found an adequate defense: the rules of philosophizing of the second edition of the *Principia* (1713) and its general scholium, hinting at a possible aether mechanism for gravity and claiming to advance no hypotheses, scarcely altered its character, while the explicit admission into the *Opticks* of

14. [Wolff], *Acta erud.* (1713), 307–14, a response to Freind's answer (*PT,* 27 [1710], 330–42) to the *Acta*'s review (*Acta erud.* [1710], 412–16) of Freind's *Praelectiones.* Wolff's authorship of the reviews appears from his letters to Leibniz of 6 June and 16 July 1710 (Leibniz, *Briefw.* [1860], 119–22).

15. Leibniz (1716) in Alexander, *Leibniz-Clarke Corresp.* (1958), 92, 94; Wolff to Manteuffel, 19 April 1739 and 30 Oct. 1747, in Ostertag, *Phil. Geh.* (1910), 61, 133–4. Cf. Malebranche to Berrand, 1707, in Malebranche, *Oeuvres*, xix, 771–2: 'Quoique M. Newton ne soit point physicien, son livre [the *Opticks*] est très curieux et très utile à ceux qui ont de bons principes de physique.'

16. Saurin, *MAS* (1709), 148; cf. *ibid.*, 133–4, and Fontenelle, *HAS* (1737), 115–17; Brunet, *Maupertuis* (1929), I, 21, and *Intro.* (1931), 9, 28–9.

17. Fontenelle, *Elogium* (1728), 12 (Newton, *Papers* [1958], 454); Flourens, *Fontenelle* (1847), 130–5.

1704—and, in larger measure, into its Latin translation of 1706—of microscopic forces acting directly at a distance only worsened the situation.[18] And, when he came to approach the accusation directly, Newton was unable to distinguish sharply between his qualities and those of the peripatetics. The latter, he says, are 'supposed to lie hid in Bodies, and to be the unknown Causes of manifest Effects;' his are 'manifest Qualities, and their Causes only are occult.'[19] These distinctions do not, and did not, persuade. The rash disciples saw no point in them: 'If the true Causes are hid from us, why may we not call them occult Qualities?'[20] As for the enemy, Fontenelle, who had his schooling from the Jesuits, rightly observed that Newton's specification was precisely that of a scholastic occult quality. The same point was made by the Jesuit Regnault, in a bit of dialogue he arranged between the principals. It runs like this. Newton: 'Attraction is a cause that I do not know; but after all it is the cause of sensible effects, of phenomena.' Descartes: 'There you are, back to occult qualities; for occult qualities were just the unknown causes of manifest effects.'[21]

As for Newton's attempted evasion—only the cause of gravity is occult—it scarcely reassured continental philosophers that the English would, or that anyone successfully could, seek a mechanical cause of Newton's mutual gravitation. Newton may have written that gravity might be effected by impulse; but 'could Sir Isaac think that others could find out these *Occult causes* which he could not discover? With what hopes for success can any other man search after them?'[22] By beginning wrong, by starting from a convenient mathematical fiction rather than from intelligible first principles, Newton has ended wrong: since, according to the Cartesians, no mechanical cause of gravitation was possible on his theory,[23] he perforce had introduced an occult one.

NEWTON OF THE *OPTICKS*

The tight, self-justifying, towering mathematician of the *Principia* seldom appears in the more open, accessible and even romantic author of the *Opticks*. The book, whose main business is to elaborate Newton's unhypothetical doctrine of light, ends with a set of imaginative conjectures, of 'bold and eccentric thoughts,' of hypotheses not permitted others, but allowable to Newton because he calls them 'queries.'[24] The first edition (1704) poses sixteen, all dealing with light; the Latin translation (1706) and the second English edition (1717–8)

18. Koyré, *Newt. Stud.* (1965), 156–60.

19. Newton, *Opticks* (1952), 401 (Query 31, first published in *Optice* [1706], as Query 23).

20. Keill, *Intro.* (1720), 3–4; cf. d'Alembert, *Prel. Disc.* (1963), 82.

21. Regnault, *Origine*, III (1734), 66–7. Cf. the 'Disputatio physicohistorica' in Musschenbroek, *Elementa physicae* (Naples, 1745), I, 59, 73.

22. Fontenelle, *Elogium* (1728), 21; Newton, *Opticks* (1952), 376. Cf. Koyré, *Newt. Stud.* (1965), 147–8.

23. E.g., Huygens. *Oeuvres*, XXI, 471: 'Je crois voir clairement, que la cause d'une telle attraction n'est point explicable par aucun principe de Méchanique.'

24. Priestley, *Experiments and Observations on Different Kinds of Air* (1781^3), I, 258, quoted by Cohen, *FN*, 191; Koyré, *Arch. int. hist. sci.*, 13 (1960), 15–29.

contain 23 and 31, respectively, and touch upon all branches of chemistry and physics. In their final form the queries are notable for their advocacy—as far as the interrogative form would permit—of a sometimes contradictory world redundantly filled with several aethers and with particles that act upon one another at a distance.

Newton's last guesses at the structure of the world, guesses that profoundly influenced eighteenth-century physics, owed their complexity to a double difficulty he never quite resolved. The first related to God's providence: should one assume that He continually and *directly* preserves His creation, or that He assigned its general maintenance to appropriate secondary, physical agents? Secondly, regarding the latter possibility, what sorts of agents ought one to consider?[25] Newton's earliest public answer to this question, which he sent the Royal Society in 1675, but which was not published until 1744,[26] supposed much the same range of mechanisms as appeared in the last edition of the queries. We are to imagine several distinct aethers, each very subtle and elastic, and 'some secret principle of unsociableness' and the reverse, whereby particles, both of aether and of grosser bodies, selectively flee and approach one another.

A good model of aether action, Newton says, is the behavior of the electrical vapor condensed in his telescope lens; just as the vapor spreads from and returns to the glass, driving light bodies before it, may not a thin, tenacious and springy local aether, constantly imbibed by our earth, 'bear down' upon objects in its path and so cause terrestrial gravity? And, just as friction elicits the electrical effluvia, may not the imbibed gravitational aether, transformed in the 'bowels of the earth,' appear again as air, gradually ascending and rarefying until it 'vanishes into the aetherial spaces?' The sun, too, may fancy this aether, which, in its rush to serve as 'solar fewel,' might push against the planets and retain them in their courses. Still more spirits are required to move the planets and their secondaries: Newton, not yet (1675) free from Descartes, assigns the job to 'aethers in the vortices of the sun and planets,' mutually unsociable aethers, moreover, to prevent their mingling and reciprocal destruction. There is also the special optical medium, which stands rarer in optically denser bodies, and which reflects, refracts and diffracts light corpuscles according to the gradient of its density and the direction of its vibrations.[27]

Although Newton took the trouble to sharpen these ideas,[28] they did not long dominate his physics. His capital discovery—which surprised and perplexed him—that the inverse-square law held precisely for planetary motions,

25. Cf. McGuire, *Ambix,* 15 (1968), 154–208.

26. A table of dates of composition and publication of Newton's work on physics is given in Thackray, *Atoms* (1970), 12.

27. These speculations were first published by Birch, *Hist.* (1756–7), III, 249–60 (Newton, *Papers* [1958], 179–90); for Newton's electrical experiment, *infra*, iii.2, and for his earlier, quasi-Cartesian astronomy, Whiteside, *Br. J. Hist. Sci.*, 2 (1964), 117–37.

28. E.g., Newton to Boyle, 28 Feb. 1678–9, in Newton, *Papers* (1958), 70–3.

showed him that the celestial spaces must be resistanceless, aetherless voids. Since he thought it philosophically absurd and, what is worse, conducive to atheism, to ascribe to bodies the capacity to attract one another directly over sensible distances, he inclined to scrap secondary gravitational agents altogether, and to make God the immediate and omnipresent cause of the mutual accessions of bodies.[29] Now the chief reason for rejecting the gravitational aether—to clear the interplanetary spaces of material obstacles—did not apply to the optical aether, which did not need to extend far beyond the bodies whose interaction with light it mediated. Nonetheless Newton temporarily dropped the optical medium of the 1670s and, in the queries to the *Opticks* of 1704, assigned its functions to short-range forces by which the particles of bodies acted at a distance upon light 'rays' or corpuscles. Moreover, in the *Optice* of 1706, he expanded his earlier hints about the social intercourse of particles, explaining chemical phenomena in terms of specific, elective, microscopic attractions and repulsions between ultimate corpuscles of bodies.[30] Whether he conceived these forces to be innate, or direct manifestations of God's continuing activity does not appear; in either case the *Optice* nicely complements the *Principia,* which together constitute the most consistent of the world pictures that Newton decided to make public.

Neither Newton's opponents, nor his own restless intellect, nor, indeed, the advancement of knowledge, allowed him to stop here. When, after 1706, experiments showed that electricity,[31] which Newton had earlier associated with a special aether, might figure in the production of light, he conceived that the electrical effluvia might be the backbone of the frame of the world, the hidden bond between the attraction of the *Principia* and the light of the *Opticks*. In the General Scholium of 1713, partly in response to the criticisms of Leibniz and Wolff, he cautiously returned to secondary causes, and hinted at a new insight into the operations of a 'certain most subtle spirit which pervades and lies hid in all gross bodies.' Except for gravity, which Newton did not mention explicitly, this spirit shouldered all the tasks performed by the multiple aethers of 1675: By its 'force and action . . . the particles of bodies . . . attract one another at near distances and cohere, if contiguous; and electric bodies operate to greater distances, as well repelling as attracting the neighboring corpuscles; and light is emitted, reflected, refracted, inflected and heats bodies; and all sensation is excited . . .' This advertisement of the 'electric and elastic spirit,' which sits incongruously on the last page of the revised *Principia,* was amplified in the

29. Newton seems to have held this position for about 20 years after the publication of the *Principia*. McGuire, *Ambix,* 15 (1968), 154–208.

30. Cf. Koyré, *Newt. Stud.* (1965), 137; Lohne, *Arch. Hist. Exact Sci.*, 1 (1961), 400–2; Rosenfeld, *ibid.*, 2 (1965), 365–86; McGuire, *ibid.*, 3 (1966), 231–2, and *Ambix,* 15 (1968), 155–7, 161–4; Guerlac, *Newton* (1963), 5–13, 27–35. The change from the agnosticism of the *Principia* to the advocacy of distance forces in the *Optice* was echoed in J. Harris, *Lex. techn.* (1704–10); Bowles, *Ambix,* 22 (1971), 23–9.

31. *Infra*, iii.2; Guerlac in Hughes and Williams, *Var. Patt.* (1971), 156–7.

new queries to the *Opticks* of 1717–8, where a multipurpose aether uneasily shares the universe with interparticulate forces acting directly at a distance.[32]

The 31 queries of the last redaction divide into four almost equal groups. The first seven, identical with those in the earlier editions, refer the reflection, refraction, diffraction and emission of light, and the production of heat, to direct short-range attractions and repulsions between the particles of bodies and the rays of light. Queries 8–16, again largely unchanged, mention neither forces nor aethers explicitly.

The next set, queries 17–24, those newly composed in 1717, reinstate a subtle, elastic, active medium, rarer in optically denser media, to whose density gradients Newton refers refraction, diffraction and—though with less conviction—gravity. The optical mechanisms plainly conflict with those of the first set of queries. Moreover, they cover less ground, as they make no provision for reflection or emission, which the earlier scheme plausibly attributed to particle-ray forces directed oppositely to those responsible for refraction and inflection. And, of course, they do not avoid action at a distance, for, as Newton hints, one 'may'—indeed, must—suppose the aether to 'contain Particles which endeavor to recede from one another.'[33] As for the gravitational mechanism, the argument from resistance still applies, and all the more so as the optical medium must become denser the rarer the spaces it occupies.

The final set of queries, 25–31, being slight reworkings of those first published in 1706, return us to the aetherless world of short-range interparticulate forces. They argue against luminiferous and gravitational media, reattribute optical phenomena to particle-ray forces, and admit a host of interparticulate attractions and repulsions to explain cohesion, capillarity, elasticity, selective chemical combination and the power whereby 'Flies walk upon the Water without wetting their Feet.'[34]

Newton's stated purpose in posing his occasionally contradictory queries was to encourage the 'inquisitive' to search farther, to find and ultimately to quantify the short-range forces—acting directly and reciprocally among particles of light, aether and matter—through which God vicariously operates the universe. His own positive achievements would serve as guide and goal in this search, the *Opticks* showing how to design apt experiments and to reason semi-quantitatively about them, and the unique *Principia* illustrating the final steps toward a mathematical physics. Perhaps no one before the mid-eighteenth

32. The qualifiers 'electric and elastic' first appeared in the English *Princ.* (1729); their authority is an autograph interlineation in Newton's copy of *Princ.* (1713²). See Hall and Hall, *Isis*, 50 (1959), 473–6, and Koyré and Cohen, *Isis*, 51 (1960), 337.

33. Heimann and McGuire, *Hist. St. Phys. Sci.*, 3 (1971), 242–3, observe that the aether presented no difficulty of action at a distance to Newton, for the mode of existence of the repulsive force was precisely to fill the space between aether particles; in several manuscripts (*ibid.*, 244), Newton expressly says that the aether is nonmaterial and nonmechanical. Cf. Laudan's comments in *Minn. St.*, 5 (1970), 234–8.

34. Cf. Koyré, *Arch. Hist. Exact Sci.*, 13 (1960), 15–29; Newton, *Opticks*, 339–406.

century tried to implement Newton's arduous program.[35] The queries served instead as a quarry of qualitative images in the style of Descartes' *Principia philosophiae*. The distinctions Newton had tried to draw between fact, theory, hypothesis and query meant little to worshipful, uncritical sectaries who—being ignorant of or indifferent to the difficulties that had bothered him—all too often thought that 'Sir *Isaac* was *infallible* in everything that he *proved* and *demonstrated,* that is to say, in all his Philosophy.'[36]

5. FORCE AMONG THE EARLY NEWTONIANS

'Sir Isaac Newton has advanced something new in the latest edition of his *Opticks* which has surprised his physical and theological disciples.' Thus the London *Newsletter* of December 19, 1717[1], pointed to the new aether queries, which made patently inconsistent the corpus the disciples had undertaken to defend. They escaped from their predicament by the simple expedient of ignoring the revived aether, which does not figure significantly, if at all, in the authoritative texts of Keill (1720), Pemberton (1728), Desaguliers (1734), and 'sGravesande (1720), or in the influential popularizations of Algarotti (1737) and Voltaire (1738).[2] Not until the 1740s did Newton's aether become important for physical theory.[3]

Even without the aether Newton's expositors had some tidying up to do: freeing the word 'attraction' from the rash interpretation of Keill and Freind and the ambiguous usage of Newton; answering the Cartesian insistence upon the epistemological superiority of 'mechanical' explanations; and clarifying the number and interrelations of the many short-distance attractions and repulsions mentioned by Newton from time to time. The tone for much of this work was set by 'sGravesande.

'sGravesande's definition of gravitation returns to the most positivistic of Newton's: gravity is not an occult cause but a manifest effect.[4] 'When we use the Words Gravity, Gravitation, or Attraction,' says Desaguliers,[5] 'we have a Regard not to the Cause, but to the Effect; namely to the Force, which Bodies have when they are carried towards one another.' Attraction signifies no more than that, if left to themselves, bodies would move toward one another, 'force'

35. An exception must be made for optical theory, as developed by Robert Smith; cf. Steffens, *Development* (1977), 28–48. Smith was Cotes' cousin and his successor at Cambridge.

36. B. Martin, *Suppl.* (1746), 26. Cf. Casini, *L'univ.-macch.* (1969), 10.

1. Quoted in Kargon, *Atomism* (1966), 138.

2. Cf. Thackray, *Atoms* (1970), 104–6; Guerlac in Hughes and Williams, *Var. Patt.* (1971), 158; Schofield, *Mechanism* (1970), 19, 24, 29–30, 36.

3. *Infra,* i.6.

4. 'sGravesande, *Math. El.* (1731[4]), II, 207; Keill, *Intro.* (1745[4]), 4, adopts the same formulation ('So likewise we may call the Endeavor of Bodies to approach one another Attraction, by which word we do not mean to determine the Cause of that Action'), as does, e.g., Boerhaave's translator, Shaw, in Boerhaave, *New Method* (1741), I, 156n ('[Attraction signifies] not the cause determining the bodies to approach . . . , but the effect, i.e., the approach').

5. *Course* (1763[3]), I, 6–7.

meaning to Desaguliers and the Dutch Newtonians either momentum (or kinetic energy) or the inertia that maintains motion *(vis insita),* not a physical cause acting between bodies.[6]

The distinction was important tactically, as a help in routing the 'Army of Goths and Vandals in the Philosophical World,'[7] the Cartesians who persisted in imagining that (to speak with 'sGravesande) 'because we do not give the cause of . . . Attraction and Repulsion, . . . they must be looked upon as Occult Qualities.'[8] A good example is James Jurin's parry of the charge of occultism hurled at Newtonians by Wolff's disciple G. B. Bilfinger. Jurin, a frequent invoker of attractions, appealed to Book I of the *Principia:* 'I [Newton] use the words attraction, impulse, or propensity of any sort towards a centre, promiscuously and indifferently, one for another, considering those forces not physically but mathematically: wherefore the reader is not to imagine that by these words I anywhere take upon me to define the kind, or manner of any action, the causes or the physical reason thereof . . .'[9]

This defense was not made without cost, for it turned out that the quantities of fundamental interest to physicists were not forces as represented by Desaguliers and 'sGravesande, i.e., as macroscopic effects, but forces as microscopic causes, such as the suppositious mutual pull between all pairs of particles of matter. To study these quantities one had to do as Newton did in the *Principia*: one had to compute the force-effects to be expected from assumed force-causes, and to confirm the latter by the former; inhibitions against supposing force-causes delayed the fruitful application of the scheme. Perhaps for this reason the first attempt of the Newtonians to obtain a 'law of force' for the interaction of two magnets failed.[10]

There were two steps in the answer to the Cartesian insistence on the priority of mechanical explanations. First, a Lockean element: we know nothing but our ideas, derived ultimately from our unreliable sense experience; these ideas at best correspond to, but do not reach, the ultimate nature of things; in particular, we have no notion of the essence of matter.[11] 'Nothing can be present to our Minds besides our Ideas, upon which our Reasonings immediately turn,' says

6. *Ibid.*, I, 45: 'Motion is that Force with which Bodies change their Place. . . . Force and Motion mean the same thing': cf. 'sGravesande, *Math. El.* (1731[4]), I, 20–1. See also the examples in Ruestow, *Physics* (1973), 126–7, who, however, misinterprets them as statements about causes. For Musschenbroek's views, which were less instrumentalist than 'sGravesande's, see Hollman, Ak. Wiss. Gött., *Comm.*, 4 (1754), 228–30; Crommelin, *Sudh. Arch.*, 28 (1935), 138–9.

7. Desaguliers, *Course* (1763[3]), I, 21.

8. 'sGravesande, *Math. El.* (1731[4]), I, 17–18. Cf. Desaguliers. *Course* (1763[3]), I, 21.

9. Jurin, *CAS,* 3 (1728), 282–3; Newton, *Math. Princ.* (1934), 5–6. Cf. Jurin, *PT,* 30 (1717–19), 743; Voltaire, *Lett. Phil.* (1964), II, 27, 39–41; Boss, *Newton* (1972), 112–15.

10. *Infra,* i.7.

11. E.g., Locke, *Essay* (1894), I, 391–2, 410–15; II, 191–225. This was of course a standard theme in the Enlightenment; cf. Hume, *Enquiry* (1894), 32–3, and *Treatise* (1888), 16; Cassirer, *Philosophy* (1951), 53–6, 74; Hankins, *D'Alembert* (1970), 79–80, 107–110; Heimann and McGuire, *Hist. St. Phys. Sci.*, 3 (1971), 267.

'sGravesande; and these ideas do not take us to the bottom of things. 'What substances are, is one of the Things hidden from us. We know . . .some of the Properties of Matter; but we are absolutely ignorant, what Subject they are inherent in.'[12] We must content ourselves with such general regularities, or 'laws of nature,' as we can uncover: for 'we are at a loss to know, whether they flow from the Essence of Matter, or whether they are deducible from Properties, given by God to the Bodies, the world consists of, but no way essential to Body; or whether finally those Effects, which pass for Laws of Nature, depend upon external Causes, which even our Ideas cannot attain to.'[13] This general line, called 'modest' by its friends and 'pyrrhonistic' by its enemies,[14] became standard in early Newtonian apologetics: it is found not only among the Dutch,[15] but also in Pemberton, Maclaurin and even Keill,[16] and in the first Newtonian texts of Italy and Germany.[17]

The second step in responding to the Cartesians was to make a virtue of necessity. We know nothing of the essence of matter? Then we no more know how motion is transferred in collisions than how it is caused in attraction. One may recall that Fontenelle had formerly met this objection with the aristocratic notion of degrees of unintelligibility. His opponents now responded with the democratic doctrine of equality of incomprehension. 'The Cartesians reproach the Newtonians for having no idea of attraction. They are right, but there is no basis for their judgement that impulse is any more intelligible.'[18] 'Impulsion is a principle at least as obscure as that of attraction.'[19] 'The action of one body on another by contact is as inexplicable as *actio in distans*.'[20] We are therefore

12. 'sGravesande, *Math. El.* (1731⁴), I, xiv, x–xi.

13. *Ibid.*, I, xii; cf. Strong, *JHI*, 18 (1957), 68–9.

14. Cf. Tega, *Riv. crit. stor. fil.*, 4 (1969), 195; Nollet, 'Discours,' in *Leçons*, I (1743), lxiv.

15. Cf. Brunet, *Physiciens* (1926), 42–3, 63–75, 89–92; Ruestow, *Physics* (1973), 122–31.

16. Pemberton, *View* (1728), 'Intro.'; Maclaurin, *Account* (1750²), 115–17; Keill, *Intro.* (1745⁴), 11: 'We shall not here give a Definition of Body, taken from its intimate Notion or Essence, wherewith we are not perfectly acquainted, and perhaps never shall be.' Cf. Strong, *JHI*, 18 (1957), 59, 66, 76; and, for an elegant later version of the argument, Nicholson, *Intro.* (1790³), I, 3–5.

17. E.g., the strongly positivistic notes ('Nullatenus rerum operationes intelligimus,' etc.) added by Giuseppe Orlandi to the first Neapolitan edition of Musschenbroek's *El. phys.* (1745), quoted by Garin, *Physis*, 11 (1969), 214n; the methodological portions of Beccaria's unpublished 'Institutiones in physicam experimentalem' (c. 1750), quoted by Tega, *Riv. crit. stor. fil.*, 24 (1969), 194–5; and Lichtenberg's conventionalist annotations to Erxleben's *Anfangsgründe*, for which see Herrmann, *NTM*, 6:1 (1969), 81.

18. Condillac, *Traité des systèmes* (1749), quoted by I. Knight, *Condillac* (1968), 73. Cf. Voltaire, *Lett. phil.* (1964), II, 27; 'Vous n'entendez pas plus le mot d'impulsion que celui d'Attraction, et si vous ne concevez pas pourquoi un corps tend vers le centre d'un autre corps, vous n'imaginez pas plus par quelle vertu un corps en peut pousser un autre.'

19. Paulian, *Dictionnaire*, I (1773²), 188–9, art. 'Attraction.' Cf. *Encyclopédie méthodique*, art. 'Attraction.'

20. Playfair to Robison, 28 June 1773, citing Hume, in Olson, *Isis*, 60 (1969), 95, a favorite argument among the Scots (Heimann and McGuire, *Hist. St. Phys. Sci.*, 3 [1971], 295). Cf. Bailly

well-advised to be agnostic about the cause of gravity, and to proceed to the true business of the philosopher, an enquiry into its laws.[21]

THE FRENCH MATHEMATICIANS

The enquiry was rapidly advanced, the Cartesians confounded, and the cause of gravity triumphant, though still unknown, when the mathematicians of the Paris Academy—particularly Maupertuis, Bouguer, Clairaut and, somewhat later, d'Alembert—allowed themselves to be seduced by unsettled problems in the mathematical philosophy of Newton.[22] Two such problems deserve notice. The earlier concerned the shape of the earth, said to be oblate (shorter in polar than equatorial diameter) by Newton, and oblong (longer in the poles than the equator) by those who believed in measurements made by Jacques Cassini, head of the Paris Observatory. In 1728 Maupertuis, like the Dutch Newtonians a decade earlier, made a tour of the natural philosophers of London, and returned enthusiastic for the principles of Newton.[23] Four years later, in perhaps the first open exposition of Newton's astronomy in France,[24] Maupertuis argued for flattened poles. To resolve the controversy the Academy sent an expedition to Peru in 1735 and to Lapland in 1736.[25] The northern expedition, led by Maupertuis, unambiguously confirmed the shortening of the earth's axis, 'simultaneously flattening the poles and the Cassini.'[26] The mathematical young Turks thereupon took the offensive, and openly urged the adoption of attraction, understood as a mathematical hypothesis, on the sole ground that it saved the phenomena.[27]

Maupertuis hoped to slip his pioneering Newtonian expositions into the Cartesian Academy by oiling them with the arguments of the Dutch Newtonians, who accordingly thought his work the best physics France had yet produced.[28] We know only a few properties of things, and nothing of the essence

---

to Le Sage, 1 April 1778 (Prévost, *Notice* [1805], 299–300): 'Le phénomène de la communication du mouvement, quoiqu'aussi incompréhensible que celui de la pésanteur, est de notre connoissance plus intime.'

21. 'sGravesande, *Math. El.* (1731⁴), II, 215–16; Maclaurin, *Account* (1750²), 156.

22. Cf. Fontenelle, *HAS* (1732), 112: 'invité sans doute par une occasion d'employer la plus subtile géométrie.'

23. Maupertuis met Clarke, Pemberton, and Desaguliers, and was elected FRS; Brunet, *Maupertuis* (1929), I, 15.

24. D'Alembert, *Prel. Disc.* (1963), 89; Brunet, *Maupertuis* (1929), I, 22. In fact Bouguer, *Entretiens* (1731), has priority, as he claimed (*Entretiens* [1748²], 4), but his work was not published until after Maupertuis', and, as a dialogue, lacks system.

25. Todhunter, *Hist.* (1873), I, 93–102, 231–48; Brunet, *Maupertuis* (1929), I, 33–58, and *Clairaut* (1952), 30–53.

26. Maupertuis had become the tutor and technical advisor of the first continental expositors of Newton, Voltaire, Algarotti, and Mme du Châtelet (Brunet, *Maupertuis* [1929], I, 23–7); Voltaire, *Corresp.*, II, 377–89, 392, 400, 405–6 (letters of 1732).

27. According to Brunet, *Maupertuis* (1929), I, 188, Maupertuis used to give a dinner for the young Newtonians on the days of meetings of the Academy of Sciences, to which they would repair full of 'good spirits, presumption, and strong arguments.'

28. Voltaire to Maupertuis, 1738, quoted in Brunet, *Maupertuis* (1929), I, 58, 61.

of matter; attraction is just such a property, a fact not a cause; it would be 'ridiculous' dogmatically to exclude it from consideration. As for being intelligible, it is not less so than impulsion, for we have no conception of the true cause of either.[29] Assuming attraction (here Maupertuis in practice means force-cause), we may calculate many things of interest; and the law of squares is really a very nice law, mathematically speaking.[30]

In all this there is an air of apology. Maupertuis later called attention to the 'circumspection with which I presented the principle [of gravity], the timidity with which I dared hardly compare it with impulse, the fear I had when giving the reasons that had led the English to reject Cartesianism.' Despite his caution his performance made him many enemies, or so he said.[31] Bouguer supplied a similar argument with similar care, and ended with a proposal for peaceful coexistence: since Cartesians and Newtonians agree in admitting gravity as a fact and ignorance of its cause, they might easily be brought to cooperate with one another. 'For they need only say (the Newtonians perhaps without much believing it and the Cartesians without much hope) that the words attraction and weight shall signify only a fact while they wait for the discovery of its cause.'[32]

Maupertuis' younger colleague Clairaut dispensed with epistemological balm. Having swallowed gravitational attraction in Lapland ('revolting,' he said, when taken literally, but worth exploring as an hypothesis), he came before the Academy in 1739 with an essay on refraction based upon Newton's short-range attractive force. 'I've no wish at all to establish attraction as an essential property [he said]; I've no opinion on a question that is beyond my powers. My sole purpose is to show you how Newton uses attraction when he tries to explain refraction . . . I demand that you do me the kindness of listening.'[33] And they listened, much to the surprise of the President of the Royal Society of London: 'I remember the time when whoever would have spoken of attraction, as is now [1740] done before the philosophes, would have been as little noticed as someone wishing to resolve all difficulties by occult causes.'[34] Fontenelle had also to admit and lament the success of the attractionists. In 1737, in his Eloge of Saurin, he recalled the late mathematician's prayer ('may

29. Maupertuis, *Discours sur les différentes figures des astres* (1732), in *Oeuvres* (1756), I, 90–104; Brunet, *Maupertuis* (1929), II, 44–5, 346–7, 350, 357; Brunet, *Intro.*, 210–14; Koyré, *Newt. St.*, 162–3.

30. The shift to force-cause is evident in Maupertuis, *MAS* (1732), 343–62, which treats attraction 'geometrically, i.e., as a quality, whatever it may be, the phenomena of which may be calculated when one assumes it spread uniformly throughout all parts of matter, and acting in proportion to its quantity.' The law of squares is nice because a gravitating sphere, the most regular of bodies, then acts, as a whole, according to the same law as its parts; *ibid.*, 347; Brunet, *Maupertuis* (1929), II, 365–7.

31. Quoted by Bouillier, *Hist.* (1854), II, 561–2.

32. Bouguer, *Entretiens* (1748²), 24–5, 30–1, 47–9. Cf. Nollet, *Leçons* (1743–8), II, 476, and Abat, *Amusemens* (1763), 415–24: Newtonians like 'sGravesande and Maclaurin, who admit the possibility of an impulse theory of gravity, are in much the same case as the Cartesians.

33. Clairaut, *MAS* (1739), 263; Brunet, *RHS*, 4 (1951), 138–9, and *Clairaut* (1952), 58–9.

34. Hans Sloane to the abbé Bignon, 16 Oct. 1740, in Jacquot, RS, *Not. Rec.*, 10 (1953), 95.

God preserve us from . . . peripatetic darkness,' i.e., Newtonian attractions), and added: 'Would anyone have believed that it ever would be necessary to pray Heaven to preserve the French from prejudice in favor of an incomprehensible system, and moreover a foreign one; the French, who so love clarity, and who are so often accused of liking only what originates with themselves?'[35] But by then the battle was lost.[36]

The second intriguing Newtonian problem was the computation of the motion of the moon. In a curious case of simultaneous non-discovery, Clairaut, d'Alembert and Euler, each following his own method of approximation, calculated a value for the precession of the moon's apogee just half that observed. At a dramatic meeting of the Academy in November, 1747, Clairaut announced that the law of gravity failed for the moon, and proposed to save the phenomena by adding a little universal r^{-3} force to the canonical inverse square.[37] Bouger thereupon suggested that different portions of planets might attract according to different laws of distance, some by the square, others the cube, etc., Newton's law being a fortuitous average. The Cartesians had long regarded the possiblity of such fiddling as a good reason to reject Newton's procedures.[38]

Newton found a defender in Buffon, who, without bothering to calculate, held that nothing so simple as gravitational attraction could obey a compound law: should an r^{-3} term be needed, it betrayed the existence of another kind of force, perhaps magnetic, a possibility that momentarily intrigued d'Alembert. As for Clairaut, he dismissed Buffon's arguments as ignorant and metaphysical. In the end neither an alteration in Newton's law nor the introduction of an ad-hoc force proved necessary. Clairaut, d'Alembert and Euler had each made a mistake, as Clairaut was the first to discover, in 1749; whatever gravity might be, it operated according to the law of squares.[39]

The episode had instructive consequences. For the non-mathematical Buffon it confirmed a realist, and even a romantic interpretation of the gravitational force. He was to teach that there was but a single primitive force in nature, 'a power emanating from the divine power,' the cause of organization from the original chaos, the agent of all physical and chemical phenomena, viz., New-

35. Fontenelle, *HAS* (1737), 117; *supra,* i.4.

36. Brunet, *Intro.*, 338–41; Wolff to Manteuffel, 19 April 1739 ('in Paris they think there are no other philosophies than the Cartesian and Newtonian') and 11 July 1748 ('the so-called Newtonian philosophy has been well-received in France, Italy and Holland') in Ostertag, *Phil. Geh.* (1910), 61, 147. According to d'Alembert, *Traité de dynamique* (1743), the Cartesians were then 'much weakened'; in its second edition (1758), they 'hardly exist.' Guerlac in Wasserman, *Aspects* (1965), 318.

37. Clairaut hoped that the new law would also clear up small discrepancies between his calculations of the earth's shape and geodesic measurements. Brunet, *Clairaut* (1952), 82–3.

38. Fontenelle, *HAS* (1732), 113, 116; Bouguer, *Entretiens* (1748^2), 51–5, where Clairaut's lunar force is linked to Newton's short-range (r^{-3}) cohesion.

39. Brunet, *Clairaut* (1952), 82–7; Hankins, *D'Alembert* (1970), 32–5; Clairaut to Cramer, 26 July 1749, in Speziali, *RHS,* 8 (1955), 227; Maheu, *RHS,* 19 (1966), 221–3.

ton's inverse-square attraction. Apparent deviations from r^{-2} must be referred to shapes of the particles of matter and to posterity, which has the job of inferring the true shapes from the apparent forces.[40] Among mathematicians, however, the interaction with the moon advanced an instrumentalist interpretation of Newton. It showed that, as a calculating tool, the hypothesis of universal gravity withstood the finest tests available; and it suggested that, in case of failure, mathematicians would not scruple to treat the law of squares as an approximate and amendable description. 'It is extraordinarily difficult, or perhaps entirely impossible, to demonstrate the truth or falsity of Newton's law of attraction from the motion of the moon.'[41]

Despite their success with Newton's gravitational theory, the Parisian mathematicians did not hurry to analyze the phenomena of experimental physics in terms of attractions. Their attitude at the mid-century may be found in repetitious prefaces and philosophical essays by d'Alembert, and especially in his 'Preliminary Discourse' to the *Encyclopédie* of Diderot. D'Alembert has no trouble with gravity interpreted as an effect the cause of which we not only do not know, but are under no obligation to seek; so interpreted, he says, Newton's 'Theory of the World (for I do not mean his System) is today so generally accepted that men are beginning to dispute [his] claim to the honor of inventing it.'[42] This acceptance, he admits, came reluctantly, forced by youthful (French!) geometers concerned to establish precise relationships and willing, for the purpose, to suppose forces not evidently reducible to impulse.[43]

Although d'Alembert taught that the goal of physics was just such relationships,[44] he expressly appealed to philosophers to resist transposing attractions from celestial bodies to those about us. His reasons: firstly, these imitations of the gravitational theory will be less precise than their model and consequently the need for admitting them less evident; secondly, their laws would differ from gravity's, 'and it is not natural to think that attraction, if a fundamental [*primitif*] principle, is not uniform and absolutely the same for all the parts of matter;' and, thirdly, the most obvious candidates for Newtonian treatment, the phenomena of electricity and magnetism, 'seem to arise from an invisible fluid, and must make us question whether a similar fluid is not also the cause of other attractions observed between terrestrial bodies.'[45]

40. Buffon, 'Sec. vue' (1765), in *Oeuvres* (1954), 37, 39; 'Tr. aim.' (1788), in *Oeuvres* (1836), III, 76. Cf. Thackray, *Atoms* (1970), 159–60, 205–20.

41. Mayer to Euler, 4 July 1751, in Kopelevich, *Ist.-astr. issl.*, 5 (1959), 281–2. Cf. Bouguer to Euler, 30 May 1751, in Lamontagne, *RHS*, 19 (1966), 227.

42. D'Alembert, *Prel. Disc.* (1963), 81; *Elémens de philosophie* (1759), in *Oeuvres* (1805), II, 379, 420–1.

43. *Prel. Disc.* (1963), 90, 21; cf. Hankins. *D'Alembert* (1970), 24–5, 165–6.

44. *Prel. Disc.* (1963), 22; Grimsley, *D'Alembert* (1963), 240–1, 257.

45. *Elémens* (1759), in *Oeuvres* (1805), II, 421, 426. For the same reason Buffon also specifically exempted electricity from his reductionist program (*Oeuvres* [1835–6], III, 82–3) and even Boscovich hesitated (Costabel in Conv. Bosc., *Atti* [1963], 215).

There are two points of great interest here. For one, the argument implied—what turned out to be the case—that mathematicians would not introduce actions at a distance into electrical theory until electricians had exploded the effluvial system. For another, it showed that, despite their positivist talk, the French mathematicians had not yet grasped the full power, or admitted the legitimacy, of instrumentalism. D'Alembert repeats a thousand times that the first principles of things are unknowable, that we have no distinct ideas of matter or of anything else, that we are 'condemned to be ignorant of the essence and interior constitutions of bodies.'[46] Moreover, he asserts just as often that exact relations among the manifest properties of bodies constitute 'nearly always' the highest knowledge we can attain, the limits prescribed to our understanding, and consequently the only goal we should set ourselves in physics.[47] These propositions, together with the success of the gravitational theory and d'Alembert's insistence on the need to quantify all the physical sciences,[48] would endorse the ascription of promiscuous attractions to matter. But d'Alembert would not—or could not—bring these ingredients together.

He began to warn against multiplying forces just after 1750. Simultaneously his momentary friends Diderot and Buffon, vexed by the unnatural abstractions apparently necessary to quantify physics, started up their program for the peaceful extermination of mathematicians ('in less than a hundred years there will not be three geometers in Europe'[49]). It is tempting to construe d'Alembert's warnings as attempts at accommodation, as efforts to bridle the unwholesome analytics of the mathematical physicist: 'having driven out the spirit of system [he wrote], the spirit of calculation may dominate a bit too much in its turn.'[50] In fact d'Alembert agreed neither with Buffon that mathematical physics would work only for a very small range of phenomena, nor with Di-

46. E.g., *ibid.*, 440; *Prel. Disc.* (1963), 9, 25, 87; *Encyclopédie,* art. 'Cause'; d'Alembert to Voltaire, 29 Aug. 1769, in Voltaire, *Corresp.*, LXXII, 284; d'Alembert to Formey, 19 Sept. 1749, in Formey, *Souvenirs* (1797²), II, 363: 'L'impénétrabilité, l'essence de la matière, la force d'inertie, etc., sont pour tous les hommes des énigmes inéxplicables,' a proposition approved by Formey, *Abrégé,* I (1770), 286: 'La manière dont les propriétés résident dans un sujet, est toujours inconcevable pour nous.' Cf. Grimsley, *D'Alembert* (1963), 224–6, 240; Hankins, *D'Alembert* (1970), 99, 129–30, 152–3, 158; Guerlac in Wasserman, *Aspects* (1965), 330–1.

47. E.g., *Prel. Disc.* (1963), 22; *Traité* (1744), iii.

48. Among other reasons, to isolate 'fugitive and hidden effects' perhaps undiscoverable by observation alone. *Ibid.*, iv–vi; *Recherches,* I (1754), iii–iv; *Elémens* (1759), in *Oeuvres* (1805), II, 407–8; cf. Pemberton, *View* (1728), 17. D'Alembert anticipated that the process of quantification would take several centuries at least (*Réflexions sur la cause générale des vents* [1746], in *Oeuvres* [1805], XIV, 13); often one can only collect facts and hope, as in the case of electricity (*Prel. Disc.* [1963], 23–4, 29). Cf. Hankins, *D'Alembert* (1970), 88, 95.

49. A. M. Wilson, *Diderot* (1957), 187–98. Cf. Lagrange to d'Alembert, 21 Sept. 1781, Lagrange, *Oeuvres,* XIII, 368: 'Physics and chemistry now offer riches more brilliant and easier to exploit than the deep and depleted mine of mathematics (*ibid.*, 99) . . . ; it is not impossible that the places for Geometry in the Academy will eventually go the same way as the chairs in Arabic at the Universities today.'

50. D'Alembert, *Elémens* (1759), in *Oeuvres* (1805), II, 466–7; cf. Hankins, *D'Alembert* (1970), 75–6, 89–92; Lagrange to d'Alembert, 15 July 1769, Lagrange, *Oeuvres,* XIII, 140–1.

derot that the mathematician could succeed only by crude oversimplification.[51]

D'Alembert's warnings concerned the nature of allowable mathematical reductions; like most of his colleagues he did not incline to violate sound philosophy by multiplying abstractions unnecessarily.[52] The spirit of Descartes continued strong in them, especially in d'Alembert, who conceived the laws of motion and collision to be necessary consequences of an essential impenetrability of matter.[53] He could not abandon hope for an impulse theory of gravitation.[54] The manifold special attractions and repulsions of the *Opticks*, apparently unrelated to one another or to universal gravity, ran counter to Cartesian austerity. 'Is this language [the force talk of the *Opticks*] really that of good physics? Should we not worry that by accustoming ourselves to it, and by using attractions and repulsions in all sorts of ways, we will neglect investigations necessary to the advancement of knowledge, and so lose the opportunity of making discoveries?'[55] Attractionists do not examine phenomena whose causes are unknown with due attention: 'Sibi adeo, aliisque, ad ulteriorem veritatis cognitionem viam hoc ipso occludant.'[56]

6. FORCES AND FLUIDS

The early expositors, Keill, Freind, 'sGravesande, and Pemberton, blunted the problem of the multiplicity of forces by working primarily with attractions. Keill and Freind considered only attractions in their accounts of Newtonian short-range forces;[1] Keill very sparingly used repulsions in his text on true physics; 'sGravesande mentioned a few, such as the forces between oil and water and between mercury and iron, but he did not dwell upon them;[2] and Pemberton ignored them altogether. Their resistance to repulsion may be ascribed in part to the want of a universal repellent case comparable to gravity: 'If the laws of attraction were not better demonstrated than those of repulsion, I would never have become a Newtonian.'[3] Again, the multiple 'evident' attrac-

51. Roger, *Diderot St.*, 4 (1963), 230–1; Guerlac in Rockwood, *Becker's Heav. City* (1958), 23–4, wrongly associates d'Alembert's warnings with a concern for increased experimentation.

52. Cf. Paulian, *Dictionnaire* (1773²), arts. 'Attraction' and 'Répulsion': before admitting forces one must show that no extrinsic mechanical cause will do; multiplying forces is bad philosophy; 'nous sommes fâchés que le grand Newton ait insinué cette manière de procéder en physique dans plusieurs endroits de son optique.'

53. *Prel. Disc.* (1963), 17–18, 21; *Elémens* (1759), in *Oeuvres* (1805), II, 462; cf. Hankins, *D'Alembert* (1970), 87, 165–6. No doubt this belief and the Cartesian positions ('Mechanics is the base of all natural philosophy,' etc.) that he takes in several articles in the *Encyclopédie* (Briggs, *Col. U. Stud.*, no. 3 [1964], 42), conflict with d'Alembert's occasional skepticism (e.g., Hankins, *D'Alembert* [1970], 129).

54. *Prel. Disc.* (1963), 83; *Elémens* (1759) in *Oeuvres* (1805), II, 420–1.

55. Nollet, *Leçons* (1743–8), II, 479–80. Nollet accepts Newtonian gravity ('rien n'est si beau'), but no more.

56. Hollmann, Ak. Wiss., Gött., *Comm.*, 4 (1754), 244.

1. Keill, *PT*, 26 (1708–9), 97–110; Freind, *Praelectiones* (1710), 'Praef.,' 2–5.

2. 'sGravesande, *Math. El.* (1731⁴), I, 15. Cf. Thackray, *Atoms* (1970), 104, 118; Schofield, *Mechanism* (1970), 29–30, 43–5.

3. Paulian, *Dictionnaire* (1773²), art. 'Répulsion.'

tions made trouble enough. How are we to understand, or even to represent, the fact that matter particles attract one another according to two or more distinct laws of force, such as those of cohesion and gravitation?[4]

To appreciate the niceness of this problem, consider Newton's theory of matter, to which his expositors tried to hold firm: 'God in the Beginning form'd Matter in solid, massy, hard, impenetrable, movable Particles,' differing in size and shape, but otherwise—as Newton consistently and arbitrarily supposed—homogeneous.[5] From these primitive particles larger ones are compounded, apparently by attraction,[6] so that the smallest particle of, say, gold contains many primitives arranged in an exceptionally stable configuration. The particles move about in obedience to certain 'active principles,' 'such as is that of Gravity, and that which causes Fermentation, and the Cohesion of Bodies.'[7] How does it happen that particular configurations of homogeneous particles are always associated with certain active powers? How does it happen that between the collocations of particles constituting gold cohesion dominates, while between them and the collocations characteristic of aqua regia there is always fermentation; while, again, the principle of gravity acts among them all?

One could always respond to these questions, as Newton did occasionally, by retreating to phenomenology, by making of each 'active principle' a distinct 'law of nature' or 'manifest quality;' all one would need to know about, say, the law of cohesion is that 'all the Parts [of a body] have an attractive Force' that acts strongly at contact and vanishes at the least sensible distance.[8] But most physicists wishing to unify or systematize their concepts of matter preferred to develop one of two other Newtonian representations of the interrelations of forces.

(1) One might say—always reserving the question of the ultimate nature and seat of force—that each primitive particle of Newtonian matter acts according to the same law of force, which changes from attractive to repulsive and back again as the distance increases. 'As in Algebra, where affirmative Quantities vanish and cease, there negative ones begin; so in Mechanicks, where Attraction ceases, there a repulsive Virtue ought to succeed.'[9] Similarly, 'sGravesande, after defining the law of cohesion phenomenologically, states that, beyond its assigned range, the cohesive changes into a 'repellent Force, by which the Particles fly from each other.'[10]

(2) Or one might choose the crude but intelligible alternative of associating the several forces with as many distinct kinds of matter: ponderable matter

4. 'As it is easier to raise most Bodies from the Ground than to break them in pieces; that Force by which their Parts cohere, is stronger than their Gravity.' Desaguliers, *Course* (1763), I, 10.

5. *Opticks*, 400; *Math. Princ.* (1934), Bk. III, Prop. VI, cor. 4, p. 414.

6. *Opticks*, 394.

7. *Ibid.*, 401.

8. *Ibid.;* 'sGravesande, *Math. El.* (1731[4]), I, 11–12.

9. *Opticks*, 395; cf. *ibid.*, 389.

10. 'sGravesande, *Math. El.* (1731[4]), I, 12; cf. Desaguliers, *Course* (1763), II, 366ff; Thackray, *Atoms* (1970), 103.

cohering and gravitating, the 'matter' of heat self-repellent, those of light and, later, of electricity and magnetism, attractive and repulsive according to circumstance. Despite its evident conflict with Newton's matter theory, the last alternative nonetheless could claim Newtonian precedent in the neglected aether of the last edition of the *Opticks*.

A decisive step toward the refinement of these alternatives and the re-revival of the aether was the publication, in 1727, of Hales' *Vegetable Statics*. The volume presents Hales' unprecedented measurements of the amount of 'air' combined in organic and inorganic substances. How is one to understand his result that an apple can yield a quantity of air 48 times its bulk? According to Newton, air consists of particles which repel one another ('how . . .I do not here consider') with a force inversely as the distance;[11] to retain elastic air in a space 1/48th of what it occupies in its normal state would require a pressure of 48 atmospheres. Such a pressure, as Hales observed, would certainly burst the apple. His solution is to distinguish two states of air, one 'elastic,' the other 'fixed,' brought about by the destruction of the original elasticity, which may be restored by heat or fermentation.[12] As for the agent of the fixing, Hales, following Newton, picks sulphurous particles, which have a strong attraction. This attraction does not fix air by *force majeure*: air particles, when bound, lose their repellency, and become attractive of one another.[13] It appears that elastic air repels not only its own particles, but also those of common matter: while sulphur particles try to fix them, they—in violation of Newton's third law of motion—flee the sulphur. In a word, air may be in either an attractive or a repulsive *state*, but not in both simultaneously. As Hales put it, air is 'amphibious.'[14]

THE FORCE OR HOMOGENEOUS SOLUTION

The fundamental importance of Hales' work was immediately recognized by Desaguliers, who abstracted it in the *Philosophical Transactions* and soon was employing the repulsive force it legitimized in an attempt to account for evaporation. Desaguliers distinguished between solids, elastic fluids (airs), and inelastic fluids (waters), characterized, respectively, by particles in a state of attraction, repulsion, and — this is the novelty — attraction and repulsion.[15] Repulsion thereby ascended to the rank of attraction, both now 'first principles

11. *Opticks*, 376, 395–6; *Math. Princ.* (1934), Bk. II, prop. XXIII, pp. 300–2. Cf. 'sGravesande, *Math. El.* (1731⁴), I, 17: '[Air's] Elasticity arises from the force whereby its Parts repel one another.'

12. Hales, *Veg. Stat.* (1727), 93, 119–20.

13. *Ibid.*, 103, 110, 167–71, 179; cf. *Opticks*, 384–5; *FN*, 270–6.

14. *Veg. Stat.* (1727), xxvii, 178–9; cf. Quinn, *Evaporation* (1970), 47–59, and Hamberger, *El. phys.* (1735²), 146, on the difficulty of conceiving simultaneous attractions and repulsions.

15. Desaguliers, *PT*, 34–5 (1727), 264–91, 323–31; *PT*, 36 (1727), 6–22. The theory of evaporation standard to the mid-century assimilated it to chemical solution (of water in air). See Polvani, *Volta* (1942), 213, and Beckman, *Lychnos* (1967–8), 208–10.

of nature;'[16] the chief justification for the promotion, according to Desaguliers, being the behavior of airs and vapors as reported by Hales, and of chemical dissociation as discussed by Newton.[17] Still, Desaguliers did not face up to the problem of reconciling the concept of states of force with the simultaneous exercise of attractions and repulsions implied by the phenomena and required by the principle of equality of action and reaction. It was not until 1739 that he came clear on these matters. He then gave a theory of elasticity of solids that supposed particles simultaneously to attract (the cause of cohesion) and repel (the cause of elasticity) one another.[18] He ended by picturing the particles of matter at the centers of alternating spheres of attractive and repulsive force. Matter thereby remained homogeneous; whether its particles approached or receded from one another was an accident of distance.[19]

The line adumbrated by Desaguliers—the first of the solutions to the problem of the relationship among forces mentioned earlier—received its classical formulation in the hands of Roger Boscovich, S. J. In 1745, at the end of a dissertation on *vis viva,* Boscovich observed that attractions and repulsions, and even collisions, all became intelligible or at least simpler[20] if one conceived each of Newton's primitive particles to be a point, and required any pair of points to interact according to the same spherically symmetric, multivalued law of force, $f(r)$. At vanishingly small r, f is infinitely repulsive, rather than (as with the early Newtonians) strongly attractive, and so plays the part of impenetrability in the usual theories of matter. At a certain distance r_1 this repulsion vanishes, to be followed by an attraction, which extends to r_2, where a second repulsion sets in; after several such oscillations, f settles down to the usual gravitational attraction. Stable configurations of the primitives constitute the smallest particles of chemical substances, as in Newton's scheme. The net force exerted by two such configurations upon one another determines whether, at various distances, they will cohere or flee, combine chemically or ferment, constitute an aeriform vapor or interact magnetically.[21]

The same net forces account for collisions, the Cartesian theory of which Boscovich held to be untenable, as it violated the high principle of continuity and was unintelligible into the bargain.[22] His theory provided continuity, and

16. Desaguliers, *Course* (1763), II, 36; *FN,* 255–6.

17. Desaguliers, *Course* (1763), I, 7, 16–17, and II, 311–12. Cf. Schofield, *Mechanism* (1970), 81–7; *FN,* 236–7.

18. Desaguliers, *PT,* 41 (1739), 175–85.

19. Desaguliers, *Course* (1763), II, 336–50. This account of Desaguliers relies on Quinn, *Evaporation* (1970), 60–110.

20. Boscovich, *Theory* (1961), §4.

21. Boscovich, *De vir. vivis* (1745), §§47, 49. Cf. Costabel, *Arch. int. hist. sci.,* 14 (1961), 3–12; Marković in Whyte, *Boscovich* (1961), 128–9 and 135, emphasizing originality of the thesis that f becomes repulsive as r goes to zero.

22. It was in puzzling over the violation of continuity in the standard account of collision that, he says (*Theory* [1961], §§16–18) he first hit upon his new approach. Cf. *ibid.,* §§73, 127–8; *De vir. vivis* (1745), §41; and Nedeljković, *Philosophie* (1922), 120, 136, 138, 154.

although ultimately perhaps no more intelligibility than one invoking impulse, it had (he says) the very great merit of explaining everything with the same 'felicity:' bodies exchange motions in collision as they do in any other process, through forces that begin to operate before, and ultimately prohibit, contact.[23]

Boscovich applied his theory to many phenomena, including those of electricity; eventually his explanations filled a large book, printed in 1758 and again in 1763, which its author did not deem successful.[24] In fact it had some influence in England, particularly among the Scottish common-sense philosophers[25] and upon Joseph Priestley, who inferred from it—much to Boscovich's horror—the identity of matter and spirit.[26] The theory also intrigued John Michell and William Nicholson, who adopted Boscovich's account of impenetrability and extension in his influential *Introduction to Natural Philosophy*.[27] On the Continent, outside the Jesuit order, Boscovich seems to have had few followers. Although one can find scattered appreciations of his theory,[28] most physicists did not find it useful, and applied mathematicians had no idea how to attack its central problem, finding the form of f.[29]

THE FLUID OR INHOMOGENEOUS SOLUTION

The solution to the problem of the number and interrelation of forces favored by most physicists of the later eighteenth century was reification: the introduction of weightless substances as carriers of the forces associated with heat,

23. *Theory* (1961), §§102–3. Boscovich discusses theories in terms of convenience, utility, fertility, elegance, etc., and the reverse, not in terms of truth, which—as a disciple of Locke—he held to be unattainable in natural philosophy; Nedeljković, *Philosophie* (1922), 13–18, 189. In particular, the point atom and the universal force were not to be taken as 'real' in themselves, but as means of representing the phenomena distinctly (*Theory* [1961], §137); should it prove impossible—which Boscovich very much doubted—to account for the phenomena with only one law of force, he was prepared to admit others (*ibid.*, §§92, 517). Cf. Nedeljković, *Philosophie* (1922), 167–73, 180–2; Marković in Whyte, *Boscovich* (1961), 137.

24. Boscovich to A. Vallisnieri, Jr., 25 Aug. 1772, complaining that his theory has remained 'quasi sepolto' since it runs counter to the common philosophy (Gliozzi in Conv. Bosc., *Atti* [1963], 115–16). Cf. Marković in Whyte, *Boscovich* (1961), 147, and F. M. Fontana to Boscovich, 30 Aug. 1764 (Bosc. Papers, Berkeley): 'Io fin ad ora sono stato contrario all'universalità d'una tal legge [di continuità], stante che il mio lettore di Filosofia l'aura impugnata.'

25. Olson, *Isis*, 60 (1969), 91–103; Heimann and McGuire, *Hist. St. Phys. Sci.*, 3 (1971), 293–5.

26. Boscovich to Priestley, 17 Oct. 1778, in Varićak, Jug. akad. znan. i umjetn., *RAD*, 193 (1912), 208–10; cf. Thackray, *Atoms*, 189–92, and Heimann and McGuire, *Hist. St. Phys. Sci.*, 3 (1971), 270–3.

27. Hardin, *Ann. Sci.*, 22 (1966), 44–5; Nicholson, *Intro.* (1796[4]), I, 7, 15–17. Cf., *Encycl. Brit.*[3], art. 'Earth'; Heimann and McGuire, *Hist. St. Phys. Sci.*, 3 (1971), 275; Schofield, *Mechanism* (1970), 242–6.

28. E.g., Steiglehner, Ak. Wiss., Munich, *Neue phil. Abh.*, 2 (1780), §31, in connection with the repulsion of negatively charged bodies.

29. 'There are indeed certain things that relate to the law of forces of which we are altogether ignorant, such as the number and distances of the intersections of the curve [$f(r)$] with the axis [of r], the shape of the intervening arcs, and other things of that sort; these indeed far surpass human understanding. . . .' Boscovich, *Theory* (1961), §102; cf. Schofield, *Mechanism* (1970), 239–40.

light, fire, electricity, and magnetism.[30] By the end of the century physicists distinguished two electric and two magnetic fluids, light corpuscles, phlogiston, caloric, and perhaps an aether or two. This multiplication of species had several short-term advantages. It immediately explained the existence of a force by the presence of its carrier. It promised that the force could be studied by isolating or concentrating the carrier. And it had a bias toward quantification; at a minimum, the intensity of the force could be made proportional to the 'quantity' or 'intensity' of its fluid. These advantages might be regarded as dearly bought. Both Cartesians and Newtonians of earlier generations would have considered the incoherence of physics—not to mention the inhomogeneity of matter—implied by the representation of specific force carriers as unscientific and weak-minded. The immediate causes of this fall in standards, of this permissive prodigality, were the re-revival of Newton's aether, the general acceptance of a material theory of heat, and the problems and popularity of electricity.

Among Hales' arguments promoting repulsion was the observation that without elastic particles all the parts of matter, being 'endued' with attractions only, would 'immediately become one inactive cohering clump.' It was therefore necessary that the vast mass of attracting matter be everywhere leavened with a 'due proportion of strongly repelling elastick particles.' Hales' experiments identified this yeast with the particles of air, which he had found to be capable both of joining with, and forcing apart, common attractive matter; 'that thereby this beautiful frame of things might be maintained, in a continual round of the production and dissolution of . . .bodies.'[31] To Hales, air is the principle of separation, elasticity, pressure; it plays much the same part in his natural philosophy as the aether occasionally did in Newton's.[32] Since it alone can assume a repulsive state, it must be qualitatively distinct from common matter.[33] Similarly Newton's aether cannot be ordinary matter: being a cause of gravity, it cannot itself gravitate, lest another aether be required to effect its gravitation, and so on.[34]

The parallels between air and aether appear clearly from a letter from Newton to Boyle, published for the first time in 1744. Although written sixty-five years earlier, it turned out to be of immediate scientific interest. It describes an aether that lies in all bodies in amounts inversely proportional to their

30. The process is recognized explicitly by Wilcke, *AKSA,* 2 (1781), 154–5.

31. *Veg. Stat.* (1727), 178. Cf. 'sGravesande, *Math. El.* (1731⁴), I, 17–18, on the dissolution of salt in water: the cohesion between the particles of salt becomes a 'repulsion' in solution, which, in 'sGravesande's language, probably means nothing more than that salt dissolves in water.

32. *Veg. Stat.* (1727), 162. Note that since Hales did not accept a matter of heat, he could not make 'fire' the repulsive principle. Cf. Schofield, *Mechanism* (1970), 75–9.

33. For other dualistic systems of the time see Greene's 'truely English, Cantabridgian and Clarian' *Principles of Philosophy* (1727), and G. Knight, *Attempt* (1748). Cf. Thackray, *Atoms* (1970), 132, 148–9; Heimann and McGuire, *Hist. St. Phys. Sci.*, 3 (1971), 289, 297–301.

34. Newton makes the point explicitly in a draft quoted by McGuire, *Ambix,* 14 (1967), 72–3; cf. *Opticks,* 404, allowing primitives of different densities and forces.

densities. The action of this aether derives primarily from the gradients set up in it across the interfaces between bodies of different densities; for example, the aether just outside the surface of a piece of glass surrounded by air gradually increases from that appropriate to glass to that characteristic of air. When pushing two smooth plates of glass together, one feels a resistance (or repulsion!) from the aether squeezed aside; but once the plates lie flat, the pressure from the circumambient aether holds them firmly together. The aether therefore is the principle both of cohesion and separation; once dissolved in it the particles of vapors 'endeavor to recede as far from one another, as the pressure of the incumbent atmosphere will let them.'[35]

Although this ancient letter conflicted with much in Newton's public writings, including the *Opticks'* aether queries, and although it ended with the usual disclaimer ('I have so little fancy to things of this nature, that, had not your encouragement moved me to it, I should never, I think, have thus far set pen to paper about them'), British natural philosphers took it as evidence that Newton had always believed in, and had virtually demonstrated, the existence of an active, springy, non-material aether. These inferences were drawn by Bryan Robinson, M.D., professor of physics at Trinity College, Dublin, who had taught that Newton's aether operated the nerves and muscles of the body.[36] In 1743 Robinson published a pseudo-mathematical account of the attractive, repulsive, elastic, cohesive and miscellaneous activities of the aether, most of which violate the laws of motion; and in 1745 he issued an aetherial chrestomathy derived from the *Opticks,* the newly published letter to Boyle, and his own work on muscle action.[37] All this publicity had an effect. Beginning in 1745, all significant British electricians postulated a special electrical matter identical with, or similar to, the springy, subtle, universal Newtonian aether.[38] At least one of these electricians, Benjamin Wilson, drew his inspiration directly from Robinson.

Another important carrier of repulsive force was the suppositious 'matter of heat,' or 'elementary fire,' which, in the influential representation of Herman Boerhaave, combined the properties of Newton's aether and Hales' air: a fluid *sui generis,* weightless, universal, penetrating all bodies, expansive, the principle of dilution, fluidity, and fermentation.[39] In Boerhaave's version fire particles exist always and everywhere in the same quantity, and their 'agitation'

35. Boyle, *Works,* I, 70–3, reprinted in Newton, *Papers* (1958), 250–3.

36. Schofield, *Mechanism* (1970), 109–10.

37. *A Dissertation on the Aether of Sir Isaac Newton* (1743), for which see Schofield, *Mechanism* (1970), 110–14; *Sir Isaac Newton's Account of the Aether* (1745), for which see FN, 418–19.

38. Schofield, *Mechanism* (1970), 110, holds that Robinson did for the aether what Keill and Freind had done for short-range forces. Cf. Thackray, *Atoms* (1970), 137–9, and Guerlac in Hughes and Williams, eds., *Var. Patt.* (1971), 160–1.

39. Boerhaave, *New Method* (1741), I, 246–7, 254–5, 287. Cf. *FN,* 226–31; Metzger, *Newton* (1930), 215–24.

gives rise to temperature.[40] This proposition, neither plausible in itself nor easily reconciled with the assumed expansivity of fire, illustrates that even the leading physicists of the early eighteenth century had difficulty thinking exactly and consistently about forces. Boerhaave's doctrine of the uniform distribution of fire was attacked on many sides, perhaps most fruitfully by Joseph Black, who may have marched directly from his criticism to the discovery of specific heats. For our purpose the most interesting objections were those, like Nollet's, that also betray serious confusion in the application of the concept of force.[41]

Boerhaave's doctrine of the materiality and expansivity of fire prospered even where his theory of its distribution failed. His views, first made public in his lectures on chemistry at Leyden, became generally accessible in a pirated edition in 1724, and in 'sGravesande's text of 1731. The official version of Boerhaave's *Elementa chemiae* (1732) appeared in one French and two English editions before 1750; it brought over to the material theory of heat British physicists raised on the kinetic representations of Bacon and Newton, and latter-day Cartesians reluctant to admit special kinds of matter. No doubt the revival of Newton's aether and the work of Hales assisted this reception. By the mid-century most natural philosophers understood heat in terms of Boerhaave's elastic fire-fluid.[42] By then, too, electricians had constructed several theories postulating an electrical matter similar in many respects to elementary fire. Soon, however, the dependence was reversed, and progress in electrical theory, including the handling of forces, guided improvements in the theory of heat.[43]

By 1750 repulsion had been reified in air, aether, fire, and electricity. In the next few decades physicists accepted a second electrical fluid and other force-carrying imponderables like phlogiston, caloric, and the agents of magnetism. Special carriers of attraction likewise multiplied. The number of fundamental fluids became an embarrassment. But none of the many attempts to reduce it by identifying fluids apparently distinct or by introducing other imponderables succeeded.[44] Physics ended the century richer in essences than it had begun, and

40. *Ibid.*, 224–7; Boerhaave, *New Method* (1741), I, 245–6, 249–55.

41. McKie and Heathcote, *Discovery* (1935), 12–13; Nollet, *Leçons* (1743–8), IV, 177–8 (fire acts on all things, but is not acted upon), 185–6 (against Boerhaave's distribution), 203–9 (suggesting how common matter may capture fire).

42. Gibbs, *Ambix*, 6 (1958), 118–19; Nollet, *Leçons* (1743–8), IV, 161–3, 173, 207; J. C. Fischer, *Gesch.* (1801–8), V, 61–9, VII, 523–9; McKie and Heathcote, *Discovery* (1935), 28–9, 93; A. Hughes, *Ann. Sci.*, 8 (1952), 354–7. An interesting example of mid-century heat theory is Wallerius' (*AKSA*, 9 [1747], 272–81): common-matter particles cohere at short distances, but may be driven beyond the 'sphere of activity' of cohesion by subtle, mutually repellent fire particles; they may then repel one another to form a vapor, as Hales showed. The Secretary of the Swedish Academy of Sciences, P. Elvius, pointed out the connection between Wallerius' theory and that of Query 31 of the *Opticks* (*AKSA*, 10 [1748], 8–9).

43. E.g., Wilcke, *AKSA*, 2 (1781), 160–2, an analogy so strict that it appears to require the repulsion of warm bodies; cf. Oseen, *Wilcke* (1939), 250. Nollet, *Leçons* (1743–8), IV, 183, had earlier argued for the universality of fire from the analogy to electricity; cf. A. Hughes, *Ann. Sci.*, 8 (1952), 360.

44. The effort to reduce the number of fluids left a trace in *Encycl. Brit.*[3], art. 'Motion'; see Hughes, *Ann. Sci.*, 7 (1951), 367, and 8 (1952), 338–40. Cf. Lichtenberg's lectures on fire, electric-

more conscious of their hypothetical character. 'The adequacy of a proposed substance to explain a number of natural phenomena can never prove the existence of such a substance.'[45]

Instrumentalism This agnosticism and its implied instrumentalism—by all means invoke imponderables, feign hypotheses, multiply forces, if it is necessary to save the phenomena conveniently—are characteristic of the Newtonianizing physicists of the second half of the eighteenth century. This was the science of men who grew up familiar with attractions and repulsions and the mathematics needed to treat them; who disposed of more and better data than their predecessors, and lacked their epistemological sensibilities. 'I suppose that there is [in 1772] no physicist whom the terms accessus or attraction, and recessus or repulsion, offend, since in so many parts of physics these forces [cohesion, capillarity, dissociation, Hales' experiments] have become familiar.' Thus J. N. de Herbert, born 1725, professor of physics at the university of Vienna and a one-time Jesuit, who set himself the task of showing that electricity agreed with the rest of Newtonian force physics.[46] We have the same, but stronger, from Volta, born 1745: 'The dominion of the principle of mutual forces in chemistry and physics is today [1778] extensive and, in particular, it is becoming continually more evident in the phenomena of electricity.' Attractions so diverse as electricity, gravity, cohesion and the like may momentarily 'terrify the mind,' but soon 'experience domesticates them.'[47] By admitting repulsion between the particles of electrical matter, electrical theory could be made as simple as that of the planets.[48] Where we see no collisions we may assume an attraction.[49]

An instructive contrast with the older generation, the generation of d'Alembert, is afforded by J. H. van Swinden (born 1746), a great admirer of Newton as a methodologist, who earned his doctorate in philosophy at Leyden in 1766 with a dissertation on attractions. Like d'Alembert, and everyone else, he takes attraction to be effect not cause, and to operate everywhere according to inverse squares; and, again like d'Alembert, he excludes electricity, 'which arises from a fluid that we can see, smell and touch.' But van Swinden does not, as had d'Alembert, argue from the supposed mechanism of electricity that other apparent attractions should be referred to impulse; he enthusiastically admits cohesion and capillarity besides gravity, all following r^{-2} and all 'impossible to conceive.' Still, that is no reason to reject them anymore than our inability to

---

ity and the magnet (Hermann, *NTM*, 6:1 [1969], 76); Berthollet to van Marum, 30 July 1795, on 'proof' that electricity contains caloric (Sandoun-Goupil, *RHS*, 25 [1972], 242); and the many examples from Achard, *MAS*/Ber (1779), 27–35, to Voigt, *Versuch einer neuen Theorie des Feuers* (1793), cited in *Encyclopédie méthodique*, 76 (1819), 68–70.

45. Van Marum, 'Lectiones physico-chemicae,' (1793), as quoted by Levere in Forbes, *Marum*, I (1969), 245.

46. Herbert, *Th. phen.* (1778²), Praef., 19.

47. *VO*, III, 236, and *VE*, II, 510–11, respectively.

48. *EO* (Wilcke), Vorrede, sig. ⁺⁺2.

49. Hutton, *Dict.* (1815²), I, 188, art. 'Attraction.'

understand collisions justifies denying the communication of motion.[50] Philosophy suffers from too great a desire to explain everything, *nimia omnia explicandi cupiditas*. Some years later van Swinden admitted electricity among Newtonian forces.[51]

By the last third of the eighteenth century the better physics texts were teaching an open and unfettered instrumentalism. We shall never get to the bottom of things; we should renounce the search for first causes; 'all these things are beyond the reach of our senses, consequently beyond the sphere of our understanding.'[52] 'We can explain nothing in nature completely, we can only derive one phenomenon from another.' Hence we should drop the old program of seeking an intelligible account of gravity and its cogeners: 'As we have no clear conception, or adequate idea, of any mechanical process by which attraction may be caused, all our reasoning on the subject must be not only hypothetical, but visionary.'[53] Under the circumstances one must be content with a physics of 'as if.' 'In explaining electrical phenomena by attraction and repulsion we claim nothing except that the phenomena are such that they would be the same if God had thought it suitable in fact to give the electric fluid the attractive and repulsive force we attribute to it.'[54] These forces are mathematical abstractions, 'ideas in the mind, not in the real world.'[55] The objective is utility: having supposed these forces, one can calculate, one can predict; geometers can compute though they do not understand.[56]

In these epistemological profundities geometers outdistanced philosophers. In 1777 the class of speculative philosophy of the Berlin Academy of Sciences proposed an essay competition on the question, 'What is the *fundamentum virium?*' The geometers complained of the obscurity and uselessness of the question; the King of Prussia, alerted by d'Alembert, ordered his academicians instead to propose, 'Is it useful to a person to be deceived?[57]

What serious opposition there was to instrumentalism among physicists centered on the Eulers, who insisted upon a mechanical account of magnetism and electricity and hoped for one for gravity, and on G. L. Lesage, who found the

50. *Dissertatio* (1766), 9–10, 21–40, 55. Cf. Moll, *Ed. J. Sci.*, 1 (1824), 198, and van Swinden to Deluc, 17 Mar. 1780, Deluc Papers, Box 4 (Yale).

51. Van Swinden, *Oratio* (1767), 16; Ak. Wiss., Munich, *Neue phil. Abh.*, 2 (1780), 3. Compare the earlier argument of Klingenstierna, *Tal* (1755), 26–7: physicists had always supposed that their task was to 'enter into Nature's essential inner constitution': in fact, we should observe and establish rules, and drop Cartesian scruples; in particular, mechanistic electrical theories are mere hypotheses, and always fail before new facts.

52. Beccaria, *Treatise* (1776), 382; cf. Mayer, *Anfangsgründe* (1812^3), 7–8. For the same point in earlier research reports, *EO* (Wilcke), Vorrede, sig. ++2, and J. A. Euler, *MAS*/Ber. (1757), 130. Cf. Lichtenberg to Wolff, 30 Dec. 1784, *Briefe* (1901), II, 174: 'The worst times for physics have been those in which one believed one could decide things which lie beyond the senses.'

53. Respectively, Kästner (1800), in *Briefe* (1912), 224, and Nicholson, *Intro.* (1782), II, 380.

54. Jacquet, *Précis* (1775), 54–5; cf. Haüy, *Exposition* (1787), xvi–xviii.

55. Karsten, *Phys.-Chem. Abh.* (1786), I, 121, 128.

56. Respectively, Haüy, *Traité* (1803), I, viii; *Encyclopédie méthodique*, 76 (1819), 71.

57. Formey, *Souvenirs* (1797^2), II, 366–71.

cause of gravity in a rain of penetrating 'supra mundane' particles. The Eulers maintained a strong front against distance forces, 'mentis deliria,' hallucinations, according to the younger, 'arbitrary' and 'occult' in the opinion of the elder.[58] Their attempt at electrical theory, which will be described in its place, had a little life in Germany.[59] But Leonhard Euler's antique Cartesianism, as expressed qualitatively in his *Lettres à une princesse d'Allemagne* (1768), merely amused the younger mathematicians; 'a great analyst,' they said, 'but a poor philosopher.'[60] As for Lesage, even Euler detested his theories, preferring, he said, to 'admit ignorance about the cause of gravity than to take up such strange hypotheses.'[61] No one did so, although a quantitative account of gravitation can be developed on Lesage's terms.[62] The reason for this neglect, as given by Lesage's sympathetic colleague, J. A. Deluc: the overwhelming prevalence of the idea that 'the essence of forces, or the true differences of things, are beyond [the reach of] human ability.'[63]

7. QUANTITATIVE PHYSICS

In 1750 only a few parts of physics had fallen under the yoke of mathematics: hydrostatics, geometrical optics, much of mechanics, a fragment of pneumatics and thermodynamics. By 1800 the quantification of electrostatics, magnetism and thermodynamics was far advanced, and physical optics would soon enjoy similar preferment. The timing of this quantification owed nothing to the progress of mathematics; not until the turn of the nineteenth century did the electrician or thermodynamicist begin to require mathematical techniques not fully available a hundred years earlier. The quantification of electricity and the simple phenomena of heat awaited, first, a rise in the standards for work in physics and, second, improvements in the power, exactness and reliability of instruments.

RISING STANDARDS

'The determination of the relative and mutual dependence of the facts in particular cases must be the goal of the physicist; and to that effect he requires an

58. J. A. Euler, 'Disquisitio' (1755), 3–4; L. Euler, AS, *Pièces*, 5 (1748), 7, and letter to Müller, 30 Dec./10 Jan. 1761, in *Berl. Petersb. Akad.*, I (1959), 166.

59. *Infra*, iii.3; cf. Achard, *JP*, 21 (1782), 199. In L. Euler's prize-winning essay on magnetism (*supra*, i.3 n. 4) one reads: 'nunquam dubitari quin omnes naturae effectus à causis mechanicis proficiscantur.' AS, *Pièces*, 5 (1748), 4.

60. Lagrange, *Oeuvres*, XIII, 132, 135, 147. Cf. Sarton, Am. Phil. Soc., *Proc.*, 88 (1944), 477.

61. Euler to Lesage, 8 Sept. 1765, in Prevost, *Notice* (1805), 390; Euler to Lesage, 13 Oct. 1761 and 16 April 1763, *ibid.*, 381–4.

62. Lesage thought that his impulsion theory called for a law of gravity of the form $1/(r^2-r)$; Lesage to Boscovich, 20 Sept. 1763, in Costabel in Conv. Bosc., *Atti* (1963), 209–10.

63. Deluc, *Précis* (1802), I, 320. Cf. Deluc, 'Première esquisse du système de M. Lesage' (1781–2), Deluc Papers, Box 73 (Yale): Lesage aims to substitute 'agens physiques' for 'qualités occultes,' viz., attraction and repulsion.

exact instrument that will perform in an exact and invariable manner in every place in the world . . . The history of physics demonstrates a truth now [1782] sufficiently recognized: the physicist who does not measure only plays, and differs from a child only in the nature of his game and the construction of his toys.' Thus F. K. Achard,[1] permanent member of the Berlin Academy of Sciences, sounded a note often heard from continental natural philosophers during the latter eighteenth century. 'Everyone now [1773] agrees that a physics lacking all connection with mathematics and tied to a simple collection of observations and experiments would only be an historical amusement, fitter for entertaining the idle than for occupying the mind of a philosopher.'[2] 'En négligeant le calcul, on fait les expériences sans choix et sans desseins,' said Lambert, on entering Achard's academy in 1765.[3] Academician Le Roy, and professors Kästner, Karsten, and Volta, insist that physics, and especially electrical theory, cannot be advanced further without exact measurement. Lichtenberg does them one better, and advises that all of physics should be reexamined, 'from the ground up, and with all imaginable accuracy, using today's [1784] more complete instruments.'[4]

These statements, which could be multiplied a hundredfold, in themselves bring nothing new: Fontenelle had written that 'physics has substance only in so far as it is founded on geometry;' Boerhaave, in a famous address given in 1715, had insisted that mathematics supplied the only route to useful generalizations in science; and 'sGravesande, as already mentioned, took physics to be a branch of mathematics.[5] But these earlier writers did not practice what they preached.[6] Their successors did. Achard spent days and nights in his laboratory, thirteen in a row once on optical experiments; he left us, among other more useful things, measures of surface tension to four figures and elaborate investigations of comparative electrical conductivities.[7]

Lambert is an interesting case. A self-taught polymath, he took as his main

1. *JP*, 21 (1782), 196.

2. Paulian, *Dictionnaire* (1773²), art. 'Physique.' Cf. the sentiment of the Munich Academy of Sciences (1784): 'Auch zu der Experimentalphysik ist Mathematik und viel Mathematik erforderlich, und diejenigen welche bey jemanden, der keine Mathematik versteht, Experimentalphysik zu sehen glauben, lernen nichts weiter, als was sie von einem Taschenspieler lernen würden.' Westenrieder, *Gesch.*, I (1784), 276.

3. Quoted in Schur, *Lambert* (1905), 9.

4. Le Roy, art. 'Electromètre,' *Encyclopédie;* Karsten, *Phys.-chem. Abh.* (1786), I, 137–8; Kästner, ed. note to Bergman, *AKSA*, 25 (1763), 344–52, on p. 352 ('Yet the theory of electricity will remain uncertain as long as mathematics, the only way to make our knowledge of nature certain, is not applied more fully to it'); Kästner, 'Verbindung' [1768] in *Verm. Schr.* (1783³), II, 359; Lichtenberg, *Briefe* (1901), II, 149; P. Hahn, *Lichtenberg* (1927), 13–15.

5. Fontenelle, 'Préface' (1733), in *Oeuvres* (1764), V, 1–14, and Flourens, *Fontenelle* (1847), 174, 187; Brunet, *Physiciens* (1926), 44–5, 48–9.

6. Cf. Segner's *De mut. aer.* (1733), perhaps the earliest attempt to apply Newtonian theory to atmospheric tides. 'Without mathematics nothing can be done with a difficult physical problem,' he says, and serves up a result off by a factor of 400.

7. Achard, *Chem-phys. Schr.* (1780), 354–67, and *Sammlung* (1784), 20–45, 141–53; Stieda, Ak. Wiss., Leipzig, Phil.-Hist. Kl., *Abh.*, 39:3 (1928), 11–12, 173–4.

line the application of mathematics to physics and even to metaphysics. As a philosopher he worked out an epistemology similar to Kant's; as a physicist he sought effects linked by simple, general, and above all mathematical laws: as an experimentalist he advanced the quantitative study of photometry, pyrometry, hygrometry, and magnetism.[8] He talked as an equal to Leonhard Euler and to Georg Brander, respectively the leading mathematician and the leading instrument maker in Germany.[9] In a word, he was the perfect mathematical physicist: the mathematicians considered him an experimentalist with a 'rare talent for applying calculation to experiments;' the experimentalists thought him a mathematician with an unusual understanding of the behavior of instruments.[10] All of which (we are told) he accomplished by working from five in the morning to twelve at night, with a two-hour break at noon; the common experiences of life were to him so many occasions for calculations, and conversations opportunities for extemporaneous dissertations.[11]

The reduction of experimental data to law, or the deduction of law from first principles, is usually considered the domain of mathematical or theoretical physics. Here the physicists of the second half of the eighteenth century advanced over their predecessors in only a few isolated cases. But in respect of exactness of measurement, which constitutes the basis of quantitative physics, a great change occurred during the latter eighteenth century. Here Achard's measurements of surface tension, made without reference to a mathematical theory of capillarity, are representative. Similar labors occupied much of the life of M. J. Brisson, who in 1787 gave tables of specific weights to several significant figures, 'never entering any result as exact until the results of repeated measurements either showed no differences, or differences small enough to be neglected.' Brisson's attention to quantitative detail, to precautions to be taken, to reliability of instruments, is itself a good measure of the distance between his generation of experimentalists and the preceding one; for Brisson (born 1723) had learned his physics from Nollet (born 1700), who exercised only so much care as produced results, and seldom measured anything.[12]

Our methodologist van Swinden likewise recommended sedulous attention to exact observation, and supplied a heroic example by measuring the magnetic variation every hour of every day for ten years.[13] Another new man was J. A. Deluc (born 1727), who drove himself and his associates to distraction in his attempt to build meteorological instruments that would give reliable, and comparable, quantitative results. Recognizing a kindred soul in Brisson, Deluc oc-

8. Berger, *Cent.*, 6 (1959), 190, 196, 218–19.

9. See Lambert's correspondence with Euler in Bopp, Ak. Wiss., Berlin, Phys.-math. Kl., *Abh.* (1924:2), and with Brander in Lambert, *Deut. gel. Briefw.* (1781), III.

10. The opinions of, respectively, Lagrange, letter to d'Alembert, 3 Oct. 1777, in Lagrange, *Oeuvres*, XIII, 333–4, and of Saussure, *Essais* (1783), ix.

11. Lichtenberg, in Steck, *Bibl. lamb.* (1970^2), xii–xiii.

12. Brisson, *Pesanteur* (1787), ii–iii, xiv–xv; Merland, *Biog. vend.*, II (1883), 11, 21–3; *DSB*, II, 473–5.

13. Van Swinden, *Oratio* (1767), 38–9; Moll, *Ed. J. Sci.*, 1 (1824), 199.

cupied him for eight months calibrating a Deluc thermometer against the last of Réaumur's surviving instruments.[14] With a fellow Genevan, H. B. de Saussure, Deluc liked to dispute about the corrections to be applied to the readings of hygrometers in the third and fourth places of decimals; both shared their odd passion with the public in big books on the errors of barometers, hygrometers and thermometers.[15] Deluc, 'sagacissimo e accuratissimo,' was a byword for precision and dependability; '[he] handles everything like his barometer, precise (so to speak) to the point of error; the least inaccuracy would ruin everything.'[16] It is therefore noteworthy that Deluc often supported imprecise, qualitative and even retrogressive theories, such as those of his friend Lesage.

A taste for fuzzy theory was precisely what precise experiment was calculated to correct. As Desaguliers observed, we are liable to mistake the causes of things unless we 'measure the Quantity of the Effects' each putative cause may produce.[17] It is just the evasions and obscurities made possible by ignorance of geometry that encourage and shelter the concocters of aethers and vortices, of contorted pores, threaded passages, hook-and-eye atoms, of a thousand impossibilities. Or so d'Alembert thought, adding that explanations resting on such fictions are 'so incomplete, so loose, that if the phenomena were completely different, they could very often be explained just as well in the same way, and sometimes even better.'[18] And it is true, as Haüy observed, that a great distance separates the physical theories of the mid-century, like Nollet's picture of electrical action, 'independent of law and rigorous method,' and the new quantitative theories created by Haüy's generation, based on 'exact measurement' and capable of calculating 'the various effects with such precision that they may be predicted.'[19] Nollet lived long enough to witness the Academy's swing against his brand of physics. In the year before his death he wrote his closest collaborator, E. F. Dutour, who had sent him a new paper in the old style: 'It seems to me that you often call on the configurations of the ultimate parts of bodies, on the arrangement of their pores . . ., on an unknown matter to which

14. Varenne de Beost to Deluc, March, April and Oct. 1765, Deluc Papers (Yale); Middleton, *Hist.* (1966), 117–18. Later van Swinden recommended recalibrating all surviving Réaumur and Nollet barometers; letter to Deluc, 12 April 1782, Deluc Papers (Yale).

15. E.g., Saussure, *Essais* (1783), table facing p. 122; Deluc, *Recherches* (1772), I, table facing p. 184. Landriani pointed out to Deluc (letter of 20 Oct. 1788, Burndy Library) that the public was not ready for so much hygrometry, 'owing to the attention it requires and the length of the discussion.' For the aficionado, however, these fat books, especially Saussure's, were 'incomparable' (Volta to Magellan, 28 Oct. 1783, *VO,* VI, 322), a 'masterpiece' (Senebier, *Mémoire* [an IX], 86, who applauds [p. 188] Saussure's 'désir insatiable d'acquérir des connaissances plus exactes').

16. *VO,* IV, 58; Lichtenberg, letter of 1781, in *Briefe,* I, 384–5. Cf. Lesage to Boscovich, 8 May 1772, in Varićak, *RAD,* 193 (1912), 212: 'on peut parfaitment computer sur l'exactitude de ses observations les plus délicates et ses expériences les plus difficiles.'

17. *Course,* I, v; cf. van Swinden, *Oratio* (1767), 38.

18. *Mélanges,* IV, 231, quoted by Hankins, *D'Alembert* (1970), 81; cf. Keill as quoted by Strong, *J. Hist. Ideas,* 18 (1957), 56: 'All these errors [of Cartesians, of course] seem to spring from hence, that men ignorant of Geometry presume to philosophize, and to guess at the causes of Natural Things.'

19. Haüy, *Traité* (1803), I, 337.

you assign a large role, etc. I ought not hide from you that the Academy is getting more and more difficult about this way of philosophizing.'[20]

The emphasis on precise measurement in physics profited from and contributed to efforts to raise the standard of scientific work in the later eighteenth century. The editor of the *Journal de physique* announced that he would not cater to dilettantes or browsers: 'We will not offer lazy amateurs matter purely for amusement, nor give them the sweet illusion that they know something about sciences of which they are ignorant.' Rather, the new journal would meet a need long felt for faster, cheaper, more useful and less parochial communication than the proceedings of learned societies provided.[21] The success of the *Journal* encouraged the foundation of other professional periodicals for short original articles and abstracts of academic memoirs; like the *Journal*, the *Bulletin des sciences* of the Société philomathique and the *Journal of Natural Philosophy, Chemistry and the Arts* (Nicholson's Journal) advertised themselves as international, fast, useful, cheap, and even 'accurate,' at least in the abstracting of papers published by others.[22] The effect of rising standards may also be seen in the establishment of—and the quality of the memoirs published in—such specialized journals as the *Annales de chimie et de physique* (1789) and the *Journal der Physik* (1790).[23]

Simultaneously scientific societies became more particular about their memberships. In 1776 the Council of the Royal Society of London, alarmed at the number of obscure foreign members, declared a moratorium on further admissions.[24] When it was removed election became increasingly more difficult. 'Never before [as van Marum, who sought admission, learned in 1791] has the honor of becoming a member been the object of such aspiration as it is now . . . They are getting very strict about foreign Members.'[25] The Società italiana delle scienze, established in 1782 to overcome the jealousies, infighting and poor communication responsible, according to its founders, for the decay of Italian science, offered membership to any countryman of proven ability, 'recognized for his published work.' Similarly the Hollandsche Maatschappij der Wetenschappen, which began by admitting almost everybody, restricted itself from about 1795 to 'professionals who are professors, or who have acquired their reputations by works which they have published or presented to the Soci-

20. Letter of 13 March 1769 (Burndy): 'Il m'a semblé que vous appelez souvent à votre secours la configuration des parties primordiales des corps, celle de leur porosité. . . . une matière inconnüe à qui vous faites jouir de grands rôles, etc. Je ne dois vous dissimuler que l'Académie devient de plus en plus difficile sur cette manière de philosopher.'

21. *JP*, 1 (1773), i–vii; cf. K. Baker, *RHS*, 20 (1967), 264–7.

22. Soc. phil., *Bull.*, 1 (1791), iii–iv; *J. Nat. Phil.*, 1 (1797), iii; Lilly, *Ann. Sci.*, 6 (1948), 94; Neave, *ibid.*, 417–19.

23. Cf. Kronick, Bibl. Soc. Am., *Papers*, 59 (1965), 28–44.

24. Planta to Cowper, 11 Dec. 1778 (*EO*, I, 312, re Volta's candidacy).

25. Ingenhousz to van Marum, 11 March 1791, in Levere, RS, *Not. Rec.*, 25 (1970), 117, and R. J. Forbes, *Marum*, III, 37; cf. Ingenhousz to Magellan, 2 April 1787, in Carvalho, *Corresp.* (1952), 147.

ety.'[26] Even in the smaller local societies one notes a new seriousness of purpose. The management of the Gesellschaft der naturforschenden Freunde (Berlin), exasperated by a great increase in their scientific correspondence, decided to drop the usual flowery salutations and compliments, and begged their correspondents to do the same, and to come quickly to the point.[27]

INSTRUMENTS

Improved scientific instruments were the material cause and plainest expression of the rising standards and improving accuracy of physics in the later eighteenth century. After 1780 both the quality and quantity of physical apparatus commercially available increased sharply.[28] To take one measure of quantity: the number of *new* British firms making mathematical, optical, and / or philosophical instruments founded per decade remained between 25 and 30 from 1720 to 1780; in the eighties and nineties it averaged 48.[29] The same phenomenon may be followed on a finer scale in Scotland, where, on the average, ten instrument makers were active from 1730 to 1770, as compared to sixteen in the last two decades of the century.[30] In Holland, too, the number of instrument makers in business in the years 1770 to 1800 (about thirty each decade) greatly exceeded the number active earlier in the century (about five each decade, 1700–30, and fifteen each decade, 1730–50).[31] But these numbers give only a pale impression of the growth of the trade. In the first half of the century instrument firms consisted of the owner and a very few assistants; beginning in the 1750s with the shop of George Adams, London establishments grew prodigiously, sometimes—as in the case of the best makers, Peter Dolland, Edward Nairne, and Jesse Ramsden—to a staff of as many as fifty trained artisans.[32] A similar enterprise, employing thirty workers, was set up in Delft in the 1790s.[33]

London manufacturers supplied not only the British, but also much of the world trade in good scientific instruments. To be sure, the Dutch had excellent

26. Soc. ital., *Mem.*, 1 (1782), v–vi, ix; van Marum to Parmenter, 13 June 1817, in Levere, RS, *Not. Rec.*, 25 (1970), 115, and in R. J. Forbes, *Marum*, III, 34, 37; *infra*. ii.2.

27. Ges. naturf. Freunde, Berlin, *Schr.*, 1 (1780), v–x. Cf. Nicholson, *Intro.* (1790^3), I, xi–xiii, on 'the solidity of argument, and precision of expression' of the best English physicists.

28. Cf. Daumas in Crombie, *Sci. Change* (1963), 418–19, and in Singer, *Hist. Tech.*, IV (1958), 403.

29. Compiled from the biographies in E. G. R. Taylor, *Math. Pract.* (1966), 152–353.

30. Bryden, *Scott. Sci.* (1972), 26. From Bryden's table, p. 28, one computes an average life of a little over 20 years for Scottish instrument firms. Assuming a half-life of 10 years for British firms as a whole, and ignoring survivals of firms founded before 1720, one finds the numbers active in each decade from 1740 to 1800 were 49, 54, 53, 55, 74, 85.

31. Compiled from Rooseboom, *Bijdrage* (1950), omitting watchmakers and men known only for a single, non-physical instrument, such as a compass or a telescope. Rooseboom omits those who made only barometers, thermometers, balances, surgical instruments, clocks (*ibid.*, 134–5).

32. Daumas, *Instruments* (1953), 311–20; E. G. R. Taylor, *Math. Pract.* (1966), 43; Bernoulli, *Lettres* (1771), 126.

33. By J. H. Onderwijngaart Canzius, who brought in workers from outside Holland; Rooseboom, *Bijdrage* (1950), 20.

workmen: Jan van Musschenbroek, for example, and Jan Paauw, who made, and indeed made possible, the demonstration apparatus of 'sGravesande and P. van Musschenbroek; and the transplanted German Daniel Fahrenheit, who set up in Amsterdam in 1717. But the Dutch trade did not extend much outside the Netherlands, nor, as the case of Fahrenheit shows, did it always recruit its best makers domestically. In 1790 the leading manufacturer of scientific apparatus in Holland was the Englishman John Cuthbertson, the builder of, among much else, the Teylerian electrical machine.[34]

In France, Nollet had overseen the making of instruments that were perhaps the equal of those of Jan van Musschenbroek and Desaguliers. But Nollet had trouble procuring competent workmen and his successor, Sigaud de Lafond ('as inexact a maker or director of makers of scientific instruments as mediocre physicist'[35]), could not hope to compete with the Dollands and the Ramsdens. Guild restrictions inhibited the development of the French industry, which did not begin to pick up until the 1780s, when Nicholas Fortin began to make precision instruments for Lavoisier, and the Paris Academy set up a 'corps d'ingénieurs en instruments,' which included Fortin, to evade trade regulations and encourage promising artisans.[36] And even then French scientists visiting England could see 'informed artisans unknown [in Paris], and instruments entirely different from ours.'[37]

Except for a very few men such as Lambert's friend Brander, the Germanies had no instrument makers with more than a local clientele before the end of the century; for large pieces and precision work they patronized the English.[38] The experience of Prof. J. G. Stegmann of Marburg, who worked three or four artisans for fourteen years to produce optical instruments 'rather far' below British quality, may be representative.[39] The Italians had to buy almost everything abroad, and bought English when they could afford to. 'The machines from Paris are very mediocre and moreover have suffered greatly in shipment,' Volta wrote of apparatus he had purchased on a trip to France and England in 1781–2. 'Those from London are bellissima, elegant, and arrived in perfect condition.'[40]

We may distinguish three sorts of scientific instruments in the expanded trade

34. Daumas, *Instruments* (1953), 123, 138, 326–33; Crommelin, *Descr. Cat.* (1951), 11–13; Hackmann, *Cuthbertson* (1973), 39–40; Crommelin, *Sudh. Arch.*, 28 (1935), 136–9; Cohen and Cohen-de Meester, *Chem. Week.*, 33 (1936), 379, 391.

35. Lambertenghi to Volta, 8 Nov. 1779 (*EO*, I, 384); Nollet, *Leçons*, I (1743), lxxxvii.

36. Daumas, *Instruments* (1953), 130–7, 339–85. Cf. Bugge, *Science* (1969), 171, for French physics instruments in 1798–9: 'very nice . . . , although the metal work and polish cannot be compared with English work.'

37. Ch. Messier to Magellan, 2 Feb. 1788, in Carvalho, *Corresp.* (1952), 153.

38. Daumas, *Instruments* (1953), 333–6; Hermann, *Phys. Bl.*, 22 (1966), 388–96; Körber, Cong. int. hist. sci., XIIIe (1971), *Actes*, 6, 274–5.

39. Bernoulli, *Lettres* (1771), 46–7; cf. p. 68, on the British monopoly on good optical glass.

40. *VE*, II, 89, 91–2; Bernoulli, *Lettres* (1771), 171–3; Daumas, *Instruments* (1953), 324–5. Cf. Lichtenberg, *Briefe* (1901), I, 386, 389, complaining about the amount of expensive brass on English instruments.

of the second half of the eighteenth century. First, demonstration apparatus, required in quantity to decorate the 'physical cabinets' of wealthy amateurs and to illustrate the lectures of teachers of natural philosophy. A few substantial collections of demonstration apparatus were assembled before 1750, notably by Mårten Triewald at Newcastle, by the Landgrave of Hesse at Kassel and by Voltaire at Cirey,[41] and by the professional lecturers 'sGravesande, Nollet, P. van Musschenbroek, Desaguliers, and perhaps seven or eight others.[42] The great demand came after 1750. In England George III and Lord Bute, in France Louis XVI, the Duc d'Orléans, and the Duc de Chaulnes, in Italy the Grand Duke of Tuscany, inspired both the makers of instruments and the apers of fashion; a great many small cabinets (25 to 75 items, as against 250 to 350 in large holdings) came into existence; almost seventy private collections have been identified, and there must have been many more.[43] At the same time schools, colleges, academies and universities began to establish or augment collections, or to subsidize their professors' purchase of instruments. In so far as these collections included the best contemporary work, they were much superior to those assembled earlier in the century. Here is the estimate of the physicist J. A. Charles, who owned the best and most extensive demonstration apparatus in France in the 1790s (some 330 items), of the instruments of his predecessor Nollet: 'One finds in them neither the elegance of form, nor the beautiful workmanship, still less that severe precision that characterizes the most modern machines.'[44]

The second type of instrument multiplied or improved in the late eighteenth century was the measurer. Here improvement in quality can itself be measured. Perhaps the best-known advance was the Ramsden ruling engine (1773), which could divide an arc accurately into ten-second intervals, as compared with ten minutes and five minutes, the standard divisions of the sectors of 1700 and 1750, respectively.[45] This increase in precision, which depended upon im-

41. Tandberg, Lund, U., *Årssk.*, Avd. 2, 16:9 (1920), 4–5; Kirchvogel, *Phys. Bl.*, 9 (1953), 259–63; Daumas, *Instruments* (1953), 189; Voltaire to B. Moussinot, 5 and 18 May, 1738 (*Corresp.*, VII, 156, 177), regarding the payment of 9–10,000# to Nollet. Triewald's collection (miscellaneous makers) went to the University of Lund, the Landgrave's (largely by J. van Musschenbroek) ultimately to the Hessisches Landesmuseum, and Voltaire's apparently to oblivion.

42. Desaguliers estimated that there existed only 10 or 11 competent and fully furnished lecturers in natural philosophy (according to Daumas, *Instruments* [1953], who dates the estimate 'about 1750,' although Desaguliers died in 1744). For examples of their instruments see Gerland and Traumüller, *Gesch.* (1899), 294–312.

43. Daumas, *Instruments* (1953), 189–94; Torlais in Taton, *Enseignement* (1964), 640–1; Chaldecott, *Handbook* (1951); G. Turner, *Ann. Sci.*, 23 (1967), 213–42 (Lord Bute); *JP*, 9 (1777), 42 (Florence). Many are mentioned in J. Bernoulli, *Lettres* (1777–9).

44. Daumas, *Instruments* (1953), 186n, quoting a note of 1789 on the occasion of the transfer of Nollet's collection (which Brisson, his heir, had sold for 1200# in 1792) to the Conservatoire des Arts et Métiers. Cf. Torlais in Taton, *Enseignement* (1964), 633–4.

45. Daumas, *Instruments* (1953), 249–50, 264–7; Skempton and Brown, RS, *Not. Rec.*, 27 (1973), 240.

provements in lathes, glass-making and metal-working,[46] extended to the measurement of physical quantities: the second half of the eighteenth century enjoyed significantly improved barometers and magnetic needles, standardized thermometers and hygrometers, and a choice of design of a new and characteristic instrument, the electrometer.

Take the case of the barometer. The mathematician Saurin considered it useless to correct the barometers of his day (1720s) for changes in temperature on the ground that the error fell within the limits of accuracy of even the best instruments, about ⅓ of a line (0.7 mm). Perhaps the prime cause of unreliability, air absorbed in the glass or dissolved in the mercury, was eliminated by boiling the mercury before closing the tube; thereafter corrections for temperature (which Deluc found to be about 0.06 lines per degree centigrade), as well as for capillarity and for several subtler effects, became significant. In about 1770 verniers were added; a Ramsden instrument of that date can be read to 0.1 lines. And about 1775 Ramsden introduced an index, which eliminated the effect of parallax and reduced the error in sighting the meniscus by perhaps an order of magnitude.[47] In 1777 the most advanced instruments could be read to a few thousandths of an inch, and the most advanced readers could scoff at the 'gross approximations' that had satisfied their predecessors.[48]

The thermometer had a similar history. In 1731 the usually meticulous Réaumur rejected the suggestion that he use brass instead of paper for his thermometer scales; that, he said, would be pushing accuracy to ridiculous lengths. Not until the 1740s did calibration between fixed points begin to be common, and even then differences in technique, especially in determining the setting for boiling water, created instruments literally incomparable. As late as 1777 the Royal Society found a variation of as much as 3.25 degrees Fahrenheit in the location of the boiling point on their instruments. At about that time, however, the better thermometers were accurately marked to a fifth or a tenth of a degree, and soon good thermometers with comparable fixed points—such as those made by Fortin and Mossy for Lavoisier, which could be read to one hundredth of a degree—became available.[49] In 1787 Charles le géomètre published elaborate formulae for correcting thermometers for dilation of the glass, in order to make their readings, as well as their fixed points, comparable. The same year Saussure climbed Mont Blanc with a perfected thermometer that he liked to read to 1/1000 of a degree and with which he determined the boiling point of water at the summit to an accuracy of 0.1 per cent.[50]

46. Daumas in Crombie, *Sci. Change* (1963), 421–3, and in Singer, *Hist.*, IV (1958), 382–4.

47. Middleton, *Hist. Barom.* (1964), 178–9, 188, 197, 243–5; Daumas, *Instruments* (1953), 273.

48. Shuckburgh-Evelyn, *PT*, 67:2 (1777), 524, 557n; [Fontana], *JP*, 9 (1777), 105.

49. Middleton, *Hist. Therm.* (1966), 80, 119, 127–8, 133; Daumas, *Instruments* (1953), 280–1; Cavendish, et al., *PT*, 67:2 (1777), 831. Cf. G. Turner in Forbes, *Marum*, IV, 257.

50. Crommelin, Ned. aardr. gen., *Tijds.*, 66 (1949), 327–31; Charles le géomètre, *MAS* (1787), 574–82; cf. Fontana's thermometer for measuring the heat of moonlight, *JP*, 9 (1777), 107.

The electrometer, about which there will be much to say later, came into existence about 1750, crudely made, without standards or standardization, and without much agreement on the part of its makers about what it measured. The progress of theory, the improvement of technique, and, above all, the need to standardize measurement—of the ratings of machines for trade,[51] of the shocks given in medical treatments, of the leakage of charge into the atmosphere—produced a strong demand for reliable instruments. It was a job for a Deluc, or so van Swinden told him, encouraging him to do for electricity what he had done for the atmosphere. 'Although electricity has been treated by a great many physicists, it has not yet been considered with the precision that physicists who care about mathematical precision could wish.'[52] In the event others developed the instruments, which existed by the mid-eighties. These in turn reacted upon the theory, simultaneously embodying and confirming the important relationship $Q = CT$ between the charge, capacity and 'tension' of an electrified conductor.[53]

The third of our three types of instrument is represented by the air pump and the electrical machine and their many accoutrements. These instruments, found in every respectable cabinet of the late eighteenth century, could be used for research as well as for demonstration. The power of both instruments increased dramatically after 1750. The usual pump of the mid-century, built according to the designs of Hauksbee, 'sGravesande and Musschenbroek, probably reached 1/40 or at best 1/50 atmosphere. About the same time Smeaton obtained an exhaustion of perhaps 1/80 at. by soaking the leather fittings of his pump in a mixture of alcohol and water.[54] The common pump of the 1770s, still considered an 'excellent machine' in France a decade later, attained 1/165 at. In the same period Nairne advertised an improvement of Smeaton's pump that could provide a vacuum of from 1/300 to 1/600 at. in six minutes' working.[55] These improvements enabled physicists to investigate, among other matters, the vexed question whether vacuum conducts or insulates.

As for the electrical machine, it was capable of generating about 10,000 volt when first introduced in the 1740s. Van Marum's white elephant of 1785, which employed a glass plate 65 inches in diameter, probably gave over 100,000 volt. The earlier machines when used to charge the favorite capacitor of the period, a boy or a gun barrel hung from the ceiling by insulating cords

51. The effect of competition appears clearly in Cuthbertson's search for better measures of electrical output; Hackmann, *Cuthbertson* (1973), 33, and in Forbes, *Marum*, III, 349–51.

52. Van Swinden to Deluc, 16 May 1783 and 23 April 1784, Deluc Papers (Yale): 'L'électricité, quoiqu'elle ait été traitée par un si grand nombre de physiciens, n'a pas encore été considérée avec cette précision que des physiciens, qui font quelque cas de la précision mathématique, pouvoient désirer.'

53. *Infra*, iii.4; cf. Daumas in Crombie, *Sci. Change* (1963), 428–30; Kühn, *Neu. Ent.* (1796), 218–83.

54. Smeaton, *PT*, 47 (1751–2), 420, estimating 10^{-3} at. with a pear gauge later shown to be inaccurate; Nairne, *PT*, 67:2 (1777), 619, 622–5, 634.

55. *Ibid.*, 635–6; *JP*, 30 (1787), 434; Daumas, *Instruments* (1953), 287.

(fig. 3.3), could accumulate an electrical energy of some 0.0008 joule. Van Marum's engine could supply his battery of 100 Leyden jars, put in service in 1790, with perhaps 3000 joule. The increasing power of these instruments will appear from Table 1.1.

TABLE 1.1

Energy Available in 18th-Century Sparks[a]

Date	*Instrument*	*Spark length (inches)*	*Voltage (volt × 10^4)*	*Capacitance (farad × 10^{-9})*	*Energy (joule)*
1747	Glass rod (Franklin's)	1	0.5	0.0005	.000006
1750	Globe machine (Franklin's)	2	1		
	with gun barrel			0.015	.0008
	with Leyden jar			2.0	0.1
1773	Cylinder machine (Nairne's)	14	3		
	with prime conductor			0.05	0.2
	with Leyden jar			2.0	0.9
	with 64 jars			130	58
1785/90	Plate machine (v. Marum's)	24	8		
	with prime conductor			0.2	0.6
	with Leyden jar			5.6	20
	with 100 jars			560	2000
	Storm cloud				5000000

[a] Spark lengths and capacitances from Finn, *Brit. J. Hist. Sci.*, 5 (1971), 290; potentials from Prinz, *Museosci.*, 11:5 (1971), 26, and 12:5 (1972), 14.

With the large installations of the end of the century one could electrocute small animals, melt several meters of wire, electrolyze water, magnetize needles, and study the chemical effects of electricity in motion.

SOME EIGHTEENTH-CENTURY QUANTITATIVE PHYSICS

The availability of good measuring instruments helped supply pressure for the establishment of quantitative relations between physical parameters. Numerical tables, some worked out to crowds of illusory decimals, were duly filled up by experimentalists. To obtain significant relations, however, a theory was required, and often an instrumentalist one. Examples from theories of heat and magnetism will illustrate the fitting together of improved measures, refined theory, instrumentalism and quantification. Each example shares features with the more difficult case we shall examine at large, the quantification of electrostatics.

Heat At the beginning of the eighteenth century there was no standard and reliable measure of any heat phenomenon, and no agreement about the nature of heat. There did exist a few quantitative rules, such as Newton's law of cooling and an unconfirmed theorem, an old saw among medical writers, about the final degree of heat of a mixture of bodies of unequal heats. The medical man assimilated the problem to that of the average price of a number of goods at different prices, whence $H = \Sigma m_i h_i / \Sigma m_i$, H being the final heat, m_i and h_i the masses and initial heats of the several bodies.[56] The question reopened when good thermometers became available and physicists, without much justification, took temperature as a measure of heat. In fact over a large range the quantity of free heat in a body is closely proportional to its temperature as given by a mercury thermometer, for mercury expands nearly linearly with heat. There seems to have been little persuasive evidence for this proposition before the experiments of Black and Deluc.[57]

In 1744, G. W. Krafft, professor of mathematics and physics at the Petersburg Academy of Sciences, attacked the old problem of mixing with thermometers probably obtained directly from Fahrenheit's successor Prins. With them he got for the final temperature T of the mixture of two masses of water m_1 and m_2 at initial temperatures t_1 and t_2, $T = (11m_1t_1 + 8m_2t_2) / (11m_1 + 8m_2)$. Recent experiments have shown that this peculiar formula is precisely what one obtains by performing as crudely as possible, ignoring heat absorbed by the mixing vessel and thermometer, or lost to the atmosphere. Krafft apparently regarded his task as a search for the numerical relation holding in a specific mixing operation; he omitted from consideration complications that theory might suggest, and he did not aim at—or even suggest the desirability of—a general law correctible according to circumstance.[58]

Not so Krafft's former student and, in 1745, his successor, G. W. Richmann. Richmann had an excellent physical intuition, which assured him of the linearity of Fahrenheit's thermometer and of the parochiality of Krafft's numbers. Richmann conceived his task to be the discovery of general quantitative relations, which, in the case before him, could not fail to be the old medical average, $T = (m_1t_1 + m_2t_2) / (m_1 + m_2)$. Careful measurements, taking into account what we would call the water equivalent of the instruments, confirmed Richmann's universal equation. 'The whole business,' he then wrote, 'shows clearly that physics should avoid mathematical abstractions with all diligence whenever possible, and attend to every circumstance in individual cases.'[59] In fact it more nearly shows the opposite: it was Richmann who began with an abstraction.[60]

56. Zubov in *Mélanges* (1964), I, 654–61, and Ak. nauk, Inst. ist. est. tek., *Trudy*, 5 (1955), 69–93; McKie and Heathcote, *Discovery* (1935), 54–9.

57. Deluc, *Recherches* (1772), §§418m–422rr; McKie and Heathcote, *Discovery* (1935), 125, citing Black's experiments of 1760; cf. Polvani, *Volta* (1942), 208–9. There had been earlier trials, e.g., Brook Taylor's on the linseed-oil thermometer (*PT*, 32 [1723], 291), and Richmann's on Fahrenheit's (*NCAS*, 4 [1752–3], 277–300). Cf. Middleton, *Hist. Therm.* (1966), 109, 124–6.

58. Krafft, *CAS*, 14 (1744–6), 218–39; McKie and Heathcote, *Discovery* (1935), 55–63.

59. Richmann, *NCAS*, 1 (1747–8), 152–67, 168–73; McKie and Heathcote, *Discovery* (1935), 65–76; *DSB*, XI, 432–4.

60. As to the nature of heat, Richmann, *NCAS*, 4 (1752–3), 278, held it to be 'a certain motion of certain corporeal particles,' a view Krafft probably shared.

Richmann's law answered an old problem in new language, and provided the opportunity for a splendid discovery through recognition of its limitations. As Richmann observed, 'only if we have accurately determined the properties of bodies can we legitimately infer other truths with certainty.'[61] In 1769 J. C. Wilcke, mathematician and experimental physicist at the Swedish Academy of Sciences, made the 'paradoxical' observation that water cooled below 0°C warms on freezing. With this paradox in the back of his mind he immediately grasped the significance of an observation made early in 1772.[62] Wishing to wash snow from a courtyard, he was surprised to find that hot water did not melt nearly so much as it should according to Richmann's law. He thereupon sought a new rule, by a new method. Having mixed hot water at temperature t_1 with melting snow and measured the resultant temperature T, he computed the difference between T and R, the final temperature to be expected from Richmann's law if water at 0°C had been used in place of snow. In the simplest case, all masses being equal, $R-T = 36$ and 3/28 degrees. Hence, as Wilcke concluded, it required somewhat more than 72° of heat to melt unit mass of snow at 0°C. He observed that these 72° must disappear, or as we would say become latent, in liquefying the ice, and that liquefaction occurs without change of temperature.[63]

Unknown to Wilcke, some ten years earlier (c. 1761) his discovery had been made by Joseph Black, then professor of chemistry at the University of Glasgow.[64] Black measured latent heat, as he called it, by exposing diverse mixtures of ice and water in a lecture room maintained at the comfortable Scots level of 47°F. He measured the time, and so the relative quantity of heat, that each mixture took to climb from its initial temperature of 32°F to that of the ambient air. Subtracting out the water equivalent of the containers, he calculated that the amount of heat required to melt unit mass of ice would heat the same mass of water from 32° to 173°F, or by some 78° on Wilcke's centigrade scale.[65] It is noteworthy that Black's line of thought was probably much the same as Wilcke's: he began with Fahrenheit's observation, as recorded by Boerhaave, that supercooled water warms on freezing, and inferred that, in solidifying, ice-cold water gives up heat without change of temperature.[66] In his lectures he represented this conclusion as obvious and commonsensical; for if, he said, ice immediately becomes fluid when the temperature rises above 32°F, we should have great spring torrents and floods, which would 'tear up and sweep away everything, and that so suddenly that mankind should have great difficulty to escape from their ravages.'[67] Whatever the merits of this

61. *NCAS,* 4 (1752–3), 241–2.

62. Wilcke, *AKSA,* 31 (1769), 87–108; Oseen, *Wilcke* (1939), 156, 174–8.

63. Wilcke, *AKSA,* 34 (1772), 93–116. Cf. McKie and Heathcote, *Discovery* (1935), 78–94.

64. Black's doctrine dates from 1757–8, his experiments from 1761, but he published nothing about either. McKie and Heathcote, *Discovery* (1935), 35; Black to Watt, 15 March 1780, in Robinson and McKie, *Partners* (1970), 83–4.

65. McKie and Heathcote, *Discovery* (1935), 17–20.

66. Guerlac, *DSB,* II, 177.

67. Quoted by McKie and Heathcote, *Discovery* (1935), 16.

argument, it was after the fact: both Black and Wilcke came to discover latent heat not by meditating about geophysical catastrophes, but by following up a quantitative discrepancy detected through the promiscuous use of the mercury thermometer. It may also be pertinent that, in contrast to the kineticists Krafft and Richmann, Black and Wilcke took heat to be an expansive substance capable of combining with matter, and thereby altering its state.[68]

From latent heats Black moved to specific ones, again by following up quantitative data from Fahrenheit and Boerhaave, who had satisfied themselves of what amounts to Richmann's law in the case of a mixture of equal volumes of the same substance.[69] (They worked with volumes rather than masses owing to Boerhaave's theory of the uniform distribution of heat.) In the case of mercury and water, however, the temperature of the mixture always fell out closer to the initial temperature of the water than the law allowed; to save the law it was necessary to mix three parts of mercury to two of water. Now Boerhaave held that this experiment confirmed his theory of heat distribution against the more common view that bodies hold heat in proportion to their mass; and indeed Fahrenheit's finding (2:3) was more favorable to Boerhaave than to his opponents, who should have expected about 1:13.[70] But Black refused to consider 2/3 equal to one; and when he compared Fahrenheit's result with measurements by the physician George Martine (1740), which showed that mercury both heated and cooled more quickly than an equal bulk of water, he concluded that bodies have different capacities for heat.[71] That solution may be regarded as instrumentalist. It saved a quantitative discrepancy at the price of implying unspecified connections between the matter of heat and the internal arrangements or chemical composition of bodies.[72]

Once again the parallel to Wilcke is striking. Boerhaave's theory of heat distribution appears in Musschenbroek's *Elementa physicae,* which came out in Swedish in 1747, together with notes by the translator, Samuel Klingenstierna, the first professor of physics at the University of Uppsala. Klingenstierna criticized the distribution law; Wilcke, who had been Klingenstierna's student,

68. Wilcke, *AKSA*, 34 (1772), 101, 107, and 2 (1781), 53–4. Black thought the fluid theory more probable than the kinetic because, e.g., it easily assimilated latent heat to chemical combination, which Lavoisier and Laplace (*MAS* [1780], 359) also took to be its strongest feature. One suspects that Black obtained more guidance from the material theory of heat than he allowed; cf. his famous researches on 'fixed air' (1753), a springy fluid of Hales' type, which alters the state of the bodies with which it combines, which can be recovered by heating, etc.

69. Boerhaave appears to say (*New Meth.* [1741], I, 290) that the final temperature T is $(t_1 - t_2)/2$; he obviously means that T lies $(t_1 - t_2)/2$ above t_2, not above zero. But cf. McKie and Heathcote, *Discovery* (1935), 76–7.

70. Boerhaave, *New Meth.* (1741), I, 290–1. The ratio should have been about 0.45 rather than 0.67.

71. McKie and Heathcote, *Discovery* (1935), 12–15; Guerlac, *DSB*, II, 178–9.

72. Richmann, *NCAS*, 3 (1750–1), 309, 323, 332–3, reports the rapid heating and cooling of mercury, points out the conflict with accepted theory, and tries to relate heat capacity (his term) to the configuration of particles and pores. But it is plain from *NCAS*, 4 (1752–3), 241–2, that he has not fully grasped the concept of specific heat.

probably picked up the subject from him; and in the 1770s Wilcke demonstrated that bodies hold heat in proportion neither to volume nor to mass. He hit upon the idea of specific heat capacity, though not the name, independently of Black, and found its measure in the following characteristic way.

Immerse a mass of metal at temperature t_1 in an equal mass of ice-cold water; record the temperature of the mixture T; calculate from Richmann's law the amount of water w at temperature t_1 which, when mixed with the same quantity of ice-cold water, will yield the same resultant T; then $w = T/(t_1-T)$ is the specific heat of the metal relative to that of water.[73] Wilcke probably had obtained w for gold and for lead before 1780, when he first learned of Black's work; he then measured it for ten other substances. Since, as in his measurements of the latent heat of fusion, he ignored the heat capacity of the calorimeter, his numerical results were not very good.[74] But the conceptual work had been done, and others more painstaking improved the measurements.[75]

These improvements in the theory of heat illustrate conceptual advance inspired by introduction of instruments and the consequent discovery of quantitative discrepancies between the results of measurement and the predictions of theory. Our second example, the quantification of magnetic force, offers improvement in measurement as a consequence of a previous clarification of concepts.

Magnetism Among pressing unfinished Newtonian business in the early eighteenth century was the establishment of the 'law' of magnetic force. Several important early disciples of Newton, particularly his assistants Francis Hauksbee and Brook Taylor, and the ever-inquisitive Musschenbroek, accordingly undertook to obtain by experiment a magnetic analog to the law of gravitation. Their procedure is instructive, and perhaps a little comforting to those who still fear physics; for it reveals that these Newtonian hierophants had failed to understand the foundations of their doctrine.

Hauksbee and Taylor placed a large lodestone so that its poles sat on an east-west line directed towards the center of a small compass needle; they measured the angle $\phi(d)$ between the needle and the magnetic meridian for several values of d, the distance between the needle's center and the closest pole of the lodestone; and they tried to find a value of n such that ϕ decreased as d^{-n}. They hoped that the procedure would give 'the Proportion of the Power of the Lode-

73. Oseen, *Wilcke* (1939), 232–4, 247–8.

74. Wilcke, *AKSA*, 2 (1781), 48–79; McKie and Heathcote, *Discovery* (1935), 95–108. Wilcke learned of Black in Magellan's *Nouvelle théorie du feu* (1780), which ascribes the discovery of latent heat to Wilcke on the ground of first publication. See the letters from Watt to Magellan in Robinson and McKie, *Partners* (1970), 80–1, 85–8.

75. E.g., Lavoisier and Laplace, *MAS* (1780), 373, who however obtained a poorer value for the latent heat of fusion with their famous ice calorimeter than had Black (75 as against 78); and Johan Gadolin, who got good values for the specific heats of some metals by scrupulously including the water equivalents of his apparatus and by reading temperatures to tenths of a degree. McKie and Heathcote, *Discovery* (1935), 108–15.

stone at different Distances' (Hauksbee) or 'the Law of Magnetical Attraction' (Taylor).[76] But it did not succeed: $\phi(d)$ went as d^{-2} at short distances and as d^{-3} further out.

And why should one expect any simple relationship between d and ϕ ? Why choose them and not, say, a trigonometric function of ϕ and the distance between centers, as variables? Compare Newton and the gravitational attraction between two bodies. Posit first an undetectable reciprocal r^{-2} attraction between every pair of *elementary* particles in the bodies. Derive thence the mathematical consequence that, should the bodies have spherically symmetric distributions of mass or be separated by distances very large compared to their diameters, each would experience a macroscopic force acting at its center, along the common line of centers, and decreasing as the square of the distance between centers. Hauksbee and Taylor did not start with an elementary magnetic force, and had no way to specify measurable variables of interest.

Taylor thought it his business to find, by experiment, 'what point within the Stone, and what point in the Needle, are the Centers of the Magnetical power.' In a word, he sought 'force' as an effect, and indeed as an effect of certain needles and lodestones, as contemporary Newtonian apologetics appeared to recommend. Had it succeeded, the hunt for centers would have allowed a formulation of the 'law of magnetism' that excluded or avoided postulating force-causes, or hypothetical interactions between microscopic entities. Unfortunately the best place for the lodestone's suppositious center fell outside its figure. 'From Whence it seems to appear, that the power of Magnetism does not alter according to any particular power of the distances.' Despite Taylor's failure, Daniel and Jean Bernoulli, in their prize-winning essay on magnetic theory, later advocated precisely the same search for 'what can be called in some sense centers of force.' They assumed that the law of distance was that of inverse squares, and proposed to find the two centers by experiment. The self-consciously Cartesian Bernoullis could propose an experiment identical in purpose to that of the Newtonian Taylor because both parties understood force to mean macroscopic effect.[77]

Consider next the flailings of Musschenbroek, who first took on magnetism ex cathedra in an elaborate dissertation published in 1729. He opens apologizing: 'You may well be amazed at an author who writes about a phenomenon [magnetic attraction] of whose cause he confesses himself ignorant.' But to do otherwise would be to run to hypotheses, which is to say fables, or mere opinion; like Newton, who rejected hypotheses, Musschenbroek will limit himself to careful description.[78]

The first property of the magnet needing attention is the 'proportion of the attractive forces at different distances.' On Christmas eve, 1724, Musschenbroek suspended one spherical magnet above another and measured the weight

76. Hauksbee, *PT*, 27 (1710–12), 506–11; B. Taylor, *PT*, 29 (1714–6), 294–5.
77. B. Taylor, *PT*, 31 (1720–1), 204–5; D. and J. Bernoulli, AS, *Pièces*, 5 (1748), 140.
78. *Dissertatio* (1754^2), 7–8.

W required to counterbalance the attraction (fig. 1.5) as a function of the interval *d* between the magnets, not the distance between their centers. When collected into tables these numbers—as well as others obtained with other pairs of magnets—revealed nothing, or rather, as Taylor found, they showed that 'nulliam dari proportionem virium in diversis distantiis.' The most promising result, that the attractions seemed to diminish as the curvilinear volume between the balls, had no obvious interpretation. Krafft, after confirming Musschenbroek's measurements, threw up his hands: 'I cannot divine to what cause this proportion could owe its origin.' The possibility that the method might not meet the task did not occur to either of them. To be sure, Musschenbroek saw that the forces he measured were compounds of several attractions and repulsions, but he could not contrive a means to unscramble them. 'In such darkness it is best to suspend judgment, and to relate our observations [for the use of] a wiser and more serious age.'[79]

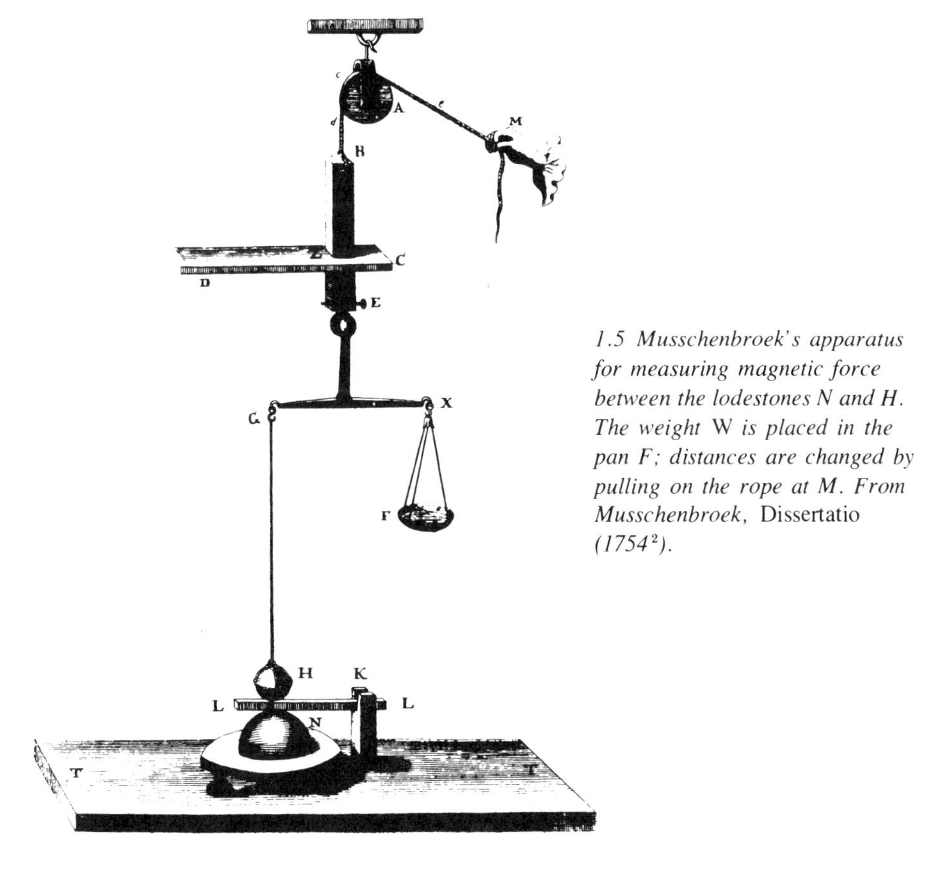

1.5 Musschenbroek's apparatus for measuring magnetic force between the lodestones N and H. The weight W *is placed in the pan F; distances are changed by pulling on the rope at M. From Musschenbroek,* Dissertatio *(1754²).*

79. *Ibid.*, 17, 26–7, 37–8; Musschenbroek, *PT*, 33 (1724–5), 372; Krafft to Wolff, 29 April 1740, in Wolff, *Briefe* (1860), 214. Cf. W. S. Harris, *Rud. Mag.* (1872²), 190–5.

Musschenbroek knew that Newton had found magnetic attraction to decrease roughly as the cube of the distance. 'Would that the experiments from which Newton gathered this result had been recorded! For perhaps that man of stupendous subtlety in mathematics found a way to segregate attractions and repulsions, the proportion of which he found to decrease as the third power of the distance.'[80] Even here Musschenbroek did not grasp the point; for he took Newton's achievement to have been the unscrambling of the *macroscopic* attraction exercised by one magnet on another from the simultaneous repulsion. He never did recommend the search for a law between magnetic elements, although Newton had pointed the way clearly enough.[81]

In his later texts Musschenbroek limited himself to reporting $W(d)$ for many pairs of magnets, and to working out spurious 'laws of attraction' for each arrangement. As for the macroscopic repulsive force, it proved more elusive than the attractive. Musschenbroek measured it by the weight $W'(d)$ required on the side G of his balance to counter the force between magnets with like poles facing. He found W' to be less than W and even negative (needed on side X of the balance) at very small d, when the repulsion sometimes became attraction.[82] It was of course hopeless to look for a 'law of repulsion' without distinguishing this last effect, a change in the strength of the magnetic elements (magnetic induction), from the property sought, the dimunition with distance of the force of a magnet of fixed strength.

The cleansing of the subject was begun by continental mathematical physicists eager to master and to extend the methods of the *Principia*. Between 1739 and 1742 two French Minims, Thomas Le Seur and François Jacquier, both professors at the Sapientia in Rome, issued Newton's masterpiece enriched with many notes, some theirs, others contributed by J. L. Calendrini, professor of mathematics and philosophy at the Académie de Calvin (University of Geneva). The book has an odd pedigree: an anti-trinitarian author, Franciscan editors, Calvinist collaborators, and, as protector and dedicatee, the Anglican Royal Society of London; a miscegenation later compounded by the Presbyterian Scots, who reissued the book and attributed it to the Jesuits. The note on magnetism, probably Calendrini's, was intended to clear Newton's 'rough' result (the force of a magnet diminishes as r^{-3}) from the doubts raised by Musschenbroek.[83] Calendrini did not use a balance, but a needle (of length $2a$ = SN in fig. 1.6); and he developed his theory for the simple case that s = CM, the distance from the center of the magnet to that of the needle, is very large in comparison with a.

Let the 'force' of the magnet — the total force compounded of the actions of its poles — exerted upon an element dm of the needle a distance cM = r away

80. Musschenbroek, *Dissertatio* (1754²), 39. Newton, *Math. Princ.* (1934), 414.

81. *Ibid.*, 415: 'Magnetic and electric attractions afford us example of this [an integrated force, like gravity]; for all attraction towards the whole arises from the attractions towards the several parts.'

82. Musschenbroek, *Essai* (1739), I, 279–82; *Cours* (1769), I, 433–6.

83. Newton, *Phil. nat.*, III (1742), 39–42, tr. in Palter, *Isis*, 63 (1972), 552–8.

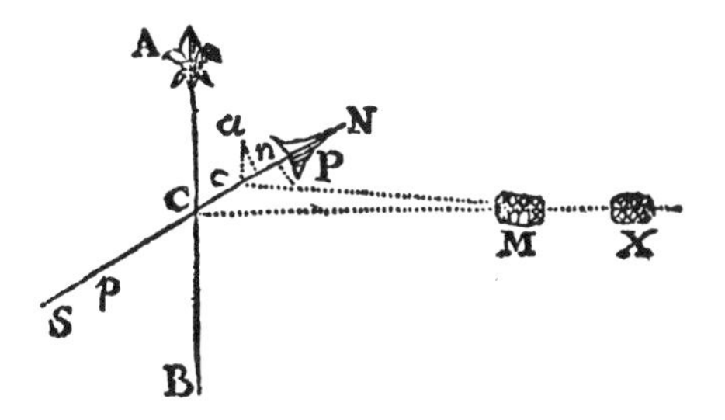

1.6 Calendrini's measurement. SN is the needle, M and X positions of the magnet, AB the magnetic meridian. From Palter, Isis, *63 (1972), 544–58.*

be $Mf(r)dm$. If x is the distance Cc of dm from the center C, the turning moment of this force about C is $dT = Mf(r)dm \cdot x\sin\theta$, where $\theta = \angle$McN; and the total moment, $\int_{-a}^{a} dT$, must equal the total amount exerted by the earth's field H, $\int_{-a}^{a} Hdm \cdot x\sin\phi$, where $\phi = \angle$ACN is the departure of the needle from the magnetic meridian. In the extreme case $s >> a$, $s \simeq r$ and $\theta \simeq \pi/2 - \phi$, whence $f(r)$, the quantity sought, is proportioned to $\tan\phi$. Calendrini found $f(r) \sim r^{-3}$, which he took to be the 'law' of magnetic action.[84] Note that he did not have the elementary law, but a compounded form appropriate only to a magnetic dipole at large distances. Calendrini apparently did not recognize these limitations.

Good steel bar magnets became available soon after Calendrini found the law of dipoles. These 'artificial magnets,' stronger and more uniform than lodestones, lent themselves to investigations of the elementary law of magnetic interaction. The first person to assert that magnetic poles interact according to the law of squares seems to have been John Michell, a Cambridge mathematician and one of the inventors of the method of making artificial magnets. Michell did not demonstrate how he deduced the law or that it saved the phenomena.[85] Nor did Tobias Mayer, professor of applied mathematics at the University of Göttingen, when, in 1760, ten years after Michell, he announced the same result.

The report of Mayer's work shows that he understood the error of the Musschenbroeks. Magnetism, he said, should be approached just as Newton did gravity: admit the 'force of a single part' of the magnet; do not worry about its cause, whether vortical 'or something worse;' measure its macroscopic effects, and secure laws valid for all magnetic bodies. Proceeding thus, we are told, Mayer accounted for all the phenomena 'with mathematical precision.'

84. Calendrini does not proceed quite so neatly, for his experimental arrangement admits values of $s \approx 3a$. He handles such cases by supposing that $dm = kxdx$ and by neglecting the change in θ with x, which allows him to define a magnetic center at P (CP = $2a/3$); the balancing of moments now gives $f(r) \sim \sin\phi/\sin\theta$, θ = angle MPN.

85. Michell, *Treatise* (1750), 17–19; *Traités* (1752), cxix–cxx. Cf. Hardin, *Ann. Sci.*, 22 (1966), 27–9.

He required only an inverse-square force between elements and the assumption that, in a bar magnet, the intensity of magnetism of each element is proportional to its distance from the geometrical center.[86]

Mayer's manuscripts on magnetism, once thought lost, have recently been published. They confirm the public announcement of his work and reveal its technique. Mayer cut through the difficulty of obtaining the elementary law of distance $f(r)$ and of magnetic strength $\nu(x)$ from composite measurements by postulating that $f \sim r^{-2}$ and that $\nu \sim x$, x being the distance from the center of an artificial bar magnet to any cross-section. On these assumptions he computed the angles $\theta(d)$ at which two magnets m and M, aligned as in fig. 1.7, would just part from one another. The measured values of θ serve as an indirect check of the conjectures behind the computations. Mayer found a good fit when he assumed that only 'symmetrical' parts of the magnets interacted, by which he meant parts distant from their respective centers by x_m and x_M, where $x_m/x_M = a_m/a_M$, a representing the length of a magnet.[87] His treatment was ruthlessly instrumentalist; his 'criterion of truth,' a successful mathematical theory. The phenomena of universal gravitation, he said, 'would not be better or more simply explained even if their ultimate causes were known.' The old Cartesian vortex hypothesis may or may not be false: what is clear is that it is 'useless and inept.' 'Nature would seem to have hidden these causes for the very reason that knowledge of them serves no useful purpose, thus reminding us once again of the truth of the dictum that Nature does nothing in vain.'[88]

The announcement of Mayer's 'law' called forth an instructive criticism from Aepinus, a gifted mathematical physicist who had just published a theory of electricity and magnetism of the first importance. He warmly endorsed Mayer's Newtonian goal, the search for quantitative laws. The search for causes had proved vain, he said; 'those who devote themselves to it, taking a cloud from Juno, concoct systems of dreams in the name of theories.' Furthermore, accord-

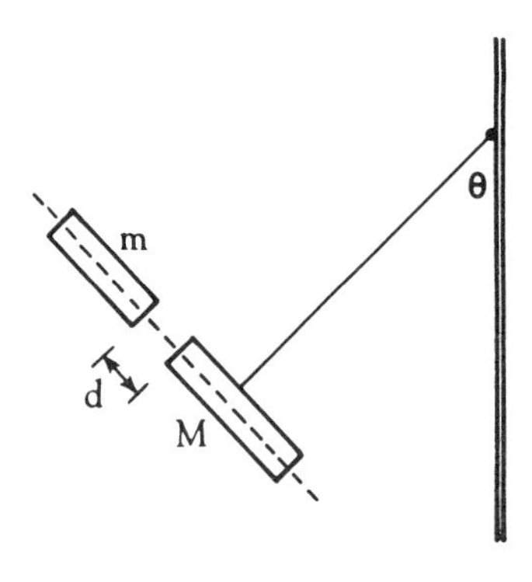

1.7 Mayer's technique for obtaining a law of magnetic force.

86. *GGA* (1760:1), 633–6; *GGA* (1762:1), 377–9; Kästner to Lambert, 13 Dec. 1769, in Bopp, Ak. Wiss., Heidelberg, *Sitzb.* (1928:18), 19, 21–3.
87. Mayer, *Unp. Writ.* (1972), III, 68–79, 83, 86.
88. *Ibid.*, 64–7.

ing to Aepinus, Mayer proceeded correctly in looking for laws of interaction between magnetic elements. But his laws were false: the magnetic center of a bar magnet does not always coincide with its geometrical middle; the intensity of magnetism of an element is not proportional to its distance from either center; and the force between elements cannot be of the gravitational form, which does not allow for induction.[89] Aepinus, who knew exactly what to look for, had no idea how to find it.

Not so Lambert, who admired Mayer as a 'genius of the first rank,' and who despised those who experimented at random, generating useless data, 'like most of that of a celebrated Leyden professor.'[90] Lambert, like Aepinus, advocated Newtonian instrumentalism: 'The example of gravity clearly shows that mathematical knowledge of the objects and processes of Nature depends only slightly on physical knowledge, and that the former can be advanced quickly and expanded wonderfully if the latter is always kept within narrow bounds.'[91] Magnetism has escaped Newtonianization because the macroscopic force between magnets is an integral of the elementary forces we seek but cannot isolate. We must therefore work with composites.

Lambert uses Calendrini's apparatus arranged so that the axis of the magnet need not be perpendicular to the magnetic meridian (fig. 1.8). He assumes that the 'force' of the magnet and of the earth on the needle are $Mf(r)g(\phi)$ and $Hg(\omega)$, respectively, M being the magnet's strength, H the earth's, f and g unknown functions, r an unspecified distance dependent upon SC, the separation of the south pole of the magnet from the center of the needle. Lambert's deduction that g = sine shows him at his best. He found ω and ϕ for two different positions of the magnet at the same distance SC (for example at d and L in fig. 1.8); since $Hg(\omega) = Mf(r)g(\phi)$, g could be deduced from $g(\phi_1)/g(\phi_2) = g(\omega_1)/g(\omega_2)$, the subscripts indicating the two positions of equal r. Lambert recognized that the happy result was an average, since the obliquity of the magnet's action differed from point to point along the needle. As for f, Lambert, like the Bernoullis, *assumed* it to be inverse-square (surely a blemish in his method), whence $r = \text{const.}(\sin\phi/\sin\omega)^{1/2}$. But what to take for r? Lambert plotted the measured distances d from the south pole of the magnet to the north pole of the needle against $(\sin\phi/\sin\omega)^{1/2}$, and gathered that, to within 7 percent, r differed from d by more than the length of the needle! Hence the 'confirma-

89. Aepinus, *NCAS*, 12 (1766–7), 325–40. Mayer is defended by Hansteen, *Untersuchungen* (1819), 290–3, and, on the basis of the Mss., which show that Mayer realized the limited applicability of his work, by Forbes in Mayer, *Unp. Writ.* (1972), III, 9–11. For the problem of $v(x)$ see W. S. Harris, *Rud. Mag.* (1872^2), 231–6.

90. Lambert to Mayer, 2 March 1772, in Lambert, *Deut. gel. Briefw.* (1781), II, 433; Lambert to Euler, 4 April 1760, in Bopp, Ak. Wiss. Berlin, Phys.-Math. Kl., *Abh.* (1924:2), 13. Cf. Lambert, *MAS*/Ber (1766), 26, and *Deut. gel. Briefw.*, II, 25, recording Musschenbroek's patronizing of Lambert.

91. Lambert, *Photometrie* (1760), §5. Cf. Lambert to Holland, 25 Sept. 1769, in Lambert, *Deut. gel. Briefw.* (1781), I, 325.

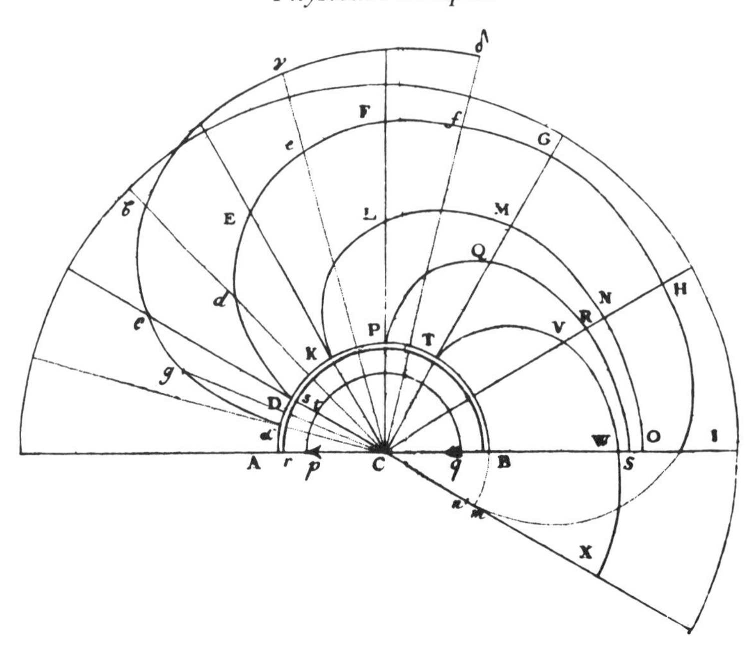

1.8 Lambert's method. The magnet is placed at points in the plane with its south pole S facing C and its axis on the line SC. Each curve is the locus of points S for which the needle takes a constant declination ω. For example, DdEdFfGHI shows the positions of the south pole of the magnet for which ω = 30°. Since the axis of the magnet lies along DC, dC, etc., one can read from the diagram the angular separation ϕ of the magnet's axis and the needle for the various configurations. From Lambert, MAS/Ber (1776), 22–48.

tion' of the inverse-square implied that the effective pole of the needle lay outside it.[92]

The 'force' Lambert sought, the integrated effect of the magnet on the needle, follows r^{-3}, not r^{-2}; and r is the distance between the centers CC′ of the magnet and the needle, not a function of d. In fact his numbers satisfy $f = CC'^{-3}$ as well as his own law.[93] Ever on his guard against arbitrary assumptions, always developing his mathematical accounts in the widest generality, Lambert was nonetheless traduced by a vulgar physical theory. He preferred

92. Lambert, *MAS*/Ber (1766), 22–48. Note that the result, $f = f(d)$, undercuts the deduction of g, which depended on pairs of points at equal distances SP. Harris, *Rud. Mag.* (1872^{2}), 196–209, gives a fair account of Lambert's work.

93. Hansteen, *Untersuchungen* (1819), 155–6, successfully obtains an integrated r^{-3} for Lambert's case using an elementary r^{-2}, and extravagantly praises his work ('Meisterstücke des Scharfsinns,' the only guide from the labyrinth [p. 294]), while admitting the handling of f is 'less satisfying' than that of g (p. 299).

Euler's plenum to Newton's void, attributed gravity to pressure, and conceived magnetic force to arise in the Cartesian manner, from a flow of magnetic aether.[94] He thought of the integrated magnetic force as a material current the density and power of which diminished, by geometrical necessity, as the square of the distance from its source.[95] As for a law for magnetic elements, Lambert rightly observed that it would be very difficult to deduce one from his integrated form.[96]

The foregoing will give a measure of the achievement of Coulomb, a Newtonian applied mathematician and an implacable foe of Cartesian explanations.[97] He had three advantages over his predecessors: a clear if crude representation of magnetism, a new method of measuring force, and, above all, long thin artificial magnets with well-defined poles. These magnets, two feet in length, of excellent steel, magnetized according to Michell's method as improved by Aepinus, acted as if their magnetism was concentrated about 5/6 inch from either extremity. Coulomb explained that their magnetic fluids, the cause or carrier of magnetic force, were confined to small regions; and he made the goal of his measurements the hypothetical force between elements of the fluids. The length and concentrated power of his magnets gave him a good approximation to isolated poles of strengths proportional to the quantity of the presumed magnetic fluid.

Coulomb suspended one such magnet horizontally in its magnetic meridian by a wire of known torsion and placed another vertically in the meridian with its north pole occupying the position from which it repelled that of the first (fig. 1.9). He had found that the force with which a twisted wire strives to unwind is proportional to its angle of twist, or torsion. In the case shown (fig. 1.10), the repulsion of the north poles balances the earth's field $H\sin\phi$ and the torsion $b\phi$, or, for small angles, a total force $\phi(H+b)$. If one now twists the dial clockwise, by, say, θ, the hanging magnet will rest at a new angle, ϕ, urged toward the meridian by a force $(H+b)\phi + b\theta$. Suppose the magnetic repulsive force to be c/r^2, where $r = 2a\sin\phi/2$ is the distance between the north poles. Then, for small ϕ,

$$c/a^2\phi^2 = (H + b)\phi + b\theta,$$

an equation Coulomb found to hold very nearly. The assumed law also gave the observed orientation of a compass needle exposed to the action of the long

94. Cf. *Photometrie* (1760), §18, promising to treat the Euler (wave) and Newtonian (particle) theories of light equally, and *MAS*/Ber (1766), 50–1, on the nature of lines of magnetic force.

95. Lambert to Euler, 15 Jan. 1760 and 6 Feb. 1761, in Bopp, Ak. Wiss., Berlin, Phys.-Math. Kl., *Abh.* (1924:2), 10, 18; *MAS*/Ber (1766), 32–5, 37, 54, 67. Cf. Daujat, *Origines* (1945), III, 482–7.

96. *MAS*/Ber (1766), 48; cf. *MAS*/Ber (1766), 73–7.

97. Coulomb, AS, *Mém. par div. sav.*, 9 (1780), §4 (Coulomb, *Mémoires* [1884], 9): 'Pour les [magnetic phenomena] expliquer, il faut nécessairement recourir à des forces attractives et répulsives de la nature de celles dont on est obligé de se servir pour expliquer la pesanteur des corps et la physique céleste.' Cf. Gillmor, *Coulomb* (1971), 176–9, 193–4.

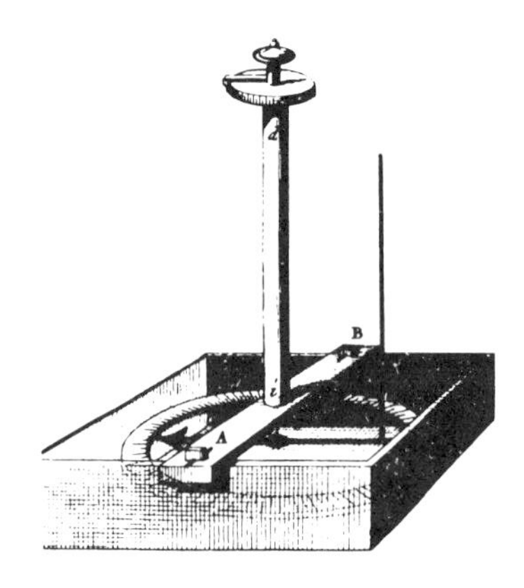

1.9 Coulomb's magnetic torsion balance. The wire of suspension, which runs in the housing di, can be twisted by the graduated knob; the vertical rod is the long thin magnet. From Coulomb, MAS *(1785), 578–611.*

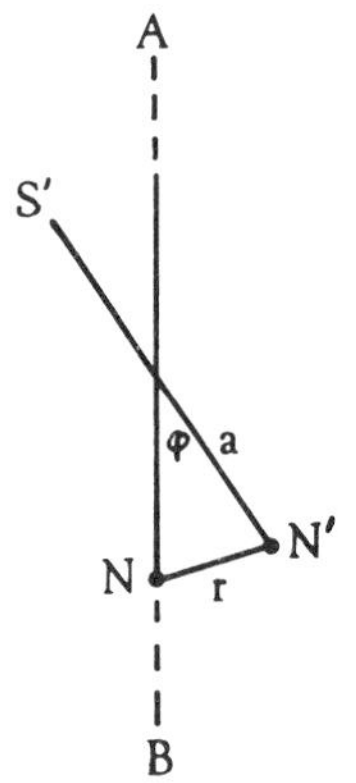

1.10 Coulomb's measurement. AB is the magnetic meridian, N the north pole of the vertical magnet, N'S' the suspended needle.

magnet by direct computation and summation of the *four* elementary interactions between pairs of poles.[98]

The results of the hunt for magnetic force are summarized in Table 1.2. Through most of the eighteenth century 'law of force' was an ambiguous expression, and the means of measuring any given force far from obvious. The complexities and false starts sketched for the magnetic case recur, redoubled, in the electrical; for it took longer to divine the appropriate conceptions in the apparently capricious phenomena of electricity.

98. Coulomb, *MAS* (1785), 589–90, 601–11 (*Mémoires* [1884], 138–45). Hansteen, *Untersuchungen* (1819), 306, criticizes Coulomb for not estimating the error incurred by ignoring the south poles; which, however, is easily done, and amounts to less than one percent in the least favorable case.

TABLE 1.2

Efforts to Obtain a 'Law of Magnetic Force,' 1710–1785

Date	*Investigator*	*Method*	*'Law' as measured*	*Understood as integrated?*	*Sees need for elementary law?*	*Explicitly instrumentalist?*	*Uses artificial magnets?*	*Gets elementary r^{-2}?*
1710/15	Hauksbee/ Taylor	needle	found none	no	no	no	no	no
1729	Musschenbroek	balance	found none	barely	no	no	no	no
1742	Calendrini	needle	r^{-3}	yes	no	no	no	no
1750	Michell	—	—	yes	yes	no	yes	as guess
1760	Mayer	gravity	r^{-2}	yes	yes	yes	yes	yes
1766	Lambert	needle	r^{-2} (wrong)	yes	yes	yes	no	no
1785	Coulomb	torsion	r^{-2} (right)	yes	yes	yes	yes	yes

CHAPTER II

The Physicists

We shall later examine the work of some 210 electricians active between 1600 and 1790. They make up two-thirds of all those then writing on electricity whose publications are noticed in the catalogs of the world's great collections.[1] They include everyone who made significant contributions to the understanding of electrical phenomena.

These early electricians may be divided into five groups according to their chief means of support: members of religious orders; paid academicians; professors; public lecturers; and 'others,' primarily artisans, practicers of professions (doctors, lawyers, ministers), and the independently wealthy. The results appear in Tables 2.1 and 2.2. Numbers in parentheses refer to our 210 electricians; the others, to our electricians plus additions from the catalogs. (Since people active in more than one time period are counted more than once, the total of the numbers in parentheses exceeds 210.) The sum S of the first four groups, A, B, C, D, is always greater than the number of 'others,' E: S/E ranges from 1.2 to 3.0, and, in each time period, is about the same whether calculated for all electricians or for our 210. We may therefore be confident that most early writers on electricity fall into groups A through D. Hence the rationale for taking these groups as 'the physicists:' they are identifiable, roughly homogenous, and they predominated, at least in the study of electricity.

This dominance is imperfectly indicated by the numbers. Except for certain unsalaried Fellows of the Royal Society of London, members of groups A through D were always the leaders, and those in E the followers. Paid academicians in particular made substantive contributions much out of proportion to their numbers. That the British did not conform to the general pattern is easily explained. Britain had neither Jesuits nor paid academicians, and interest in experimental physics declined at her universities during the middle of the eighteenth century, when study of electricity flourished at Continental institutions. A consequence and measure of this disparity are the strong showing of the British in Table 2.2.

The gross temporal variation in the numbers in the tables may be roughly accounted for as follows. Electricity did not claim the attention of many physi-

1. Compiled from Ekelöf, *Cat.* (1964–6), Frost, *Cat.* (1880), Gartrell, *Elect.* (1975), Rossetti and Cantoni, *Bibl.* (1881), and Weaver, *Cat.* (1909), omitting those who (a) wrote only on medical electricity or on the installation of lightning rods, or (b) are known for but a single article less than five pages in length, or (c) could not be identified. Members of this last group are represented in the catalogs by only one publication each.

TABLE 2.1

Electricians by Profession, 1600–1789

	1600–99	*1700–39*	*1740–49*	*1750–59*	*1760–69*	*1770–79*	*1780–89*
A. Jesuits	7(7)		6(2)	6(4)	6(2)	6(2)	1(1)
Univ. profs.	4(4)		4(1)	1(1)	1(1)	4(2)	1(1)
College profs.	1(1)		2(1)	4(2)	4(1)	2(0)	
B. Academicians[a]	3(3)	1(1)	4(4)	10(10)	4(4)	5(5)	10(7)
Big three[b]	1(1)	1(1)	4(4)	10(10)	3(3)	3(3)	5(4)
Others	2(2)				1(1)	2(2)	5(3)
C. Professors[c]	4(4)	10(10)	31(24)	22(16)	16(7)	29(15)	25(17)
Universities	4(4)	9(9)	24(20)	18(14)	11(5)	23(11)	15(12)
Colleges		1(1)	7(4)	4(2)	5(2)	6(4)	10(5)
D. Lecturers	3(3)	4(4)	4(4)	2(2)	2(2)	3(2)	3(3)
E. Others	10(10)	5(4)	43(32)	32(23)	19(17)	28(15)	40(26)
Britain	5(5)	5(4)	22(18)	14(13)	11(10)	15(12)	13(12)
Elsewhere	5(5)		21(14)	18(10)	8(7)	13(3)	27(14)
TOTALS	27(27)	20(19)	88(66)	72(55)	47(32)	71(39)	79(54)

[a] Salaried only, except for associés at AS.

[b] Paris, Berlin, Petersburg.

[c] Exclusive of Jesuits.

TABLE 2.2

Breakdown of 'Others' from Table 2.1

	1600–99	*1700–39*	*1740–49*	*1750–59*	*1760–69*	*1770–79*	*1780–89*
Artisans[a]		1(1)	12(9)	5(4)	4(3)	7(5)	7(4)
Britain[b]		1(1)	10(8)	4(4)	4(3)	6(5)	3(3)
Elsewhere			2(1)	1(0)		1(0)	4(1)
Law, Govt. or Military Service	1(1)		9(7)	9(4)		5(2)	9(4)
Practicing M.D.	3(3)	1(1)	8(5)	4(4)	3(2)	6(2)	8(5)
Regular Clergy[c]		1(1)	6(4)	5(3)	5(5)	2(1)	5(4)
Britain[b]		1(1)	4(3)	1(1)	3(3)	1(1)	2(2)
Elsewhere			2(1)	4(2)	2(2)	1(0)	3(2)
Independent	4(4)	1(0)	6(5)	8(7)	6(6)	5(5)	4(4)
Britain[b]	2(2)	1(0)	3(2)	6(6)	4(4)	4(4)	2(2)
Elsewhere	2(2)		3(3)	2(1)	1(1)	1(1)	2(2)
Others	2(2)	1(1)	2(2)	1(1)	1(1)	3(0)	7(5)
TOTALS	10(10)	5(4)	43(32)	32(23)	19(17)	28(15)	40(26)

[a]Includes instrument makers, apothecaries, engineers.
[b]Includes American colonies.
[c]Includes dissenting clergy.

cists until the invention of the electrical flare and the Leyden jar in the 1740s: these novelties brought five times as many people to study or play with electricity in the decade 1740–9 as had done so in the preceding thirty years. A decline in numbers occurred in the 1750s as the novelty wore off; that they fell by only 20 percent is owing to the restimulation of interest by the demonstration of the identity of lightning and electricity. The great drop in the 1760s corresponds to a general slump in the cultivation of the arts and sciences, caused in part by the Seven Years' War. The revival in the 1770s and 1780s owed much to the inventions of Volta.

1. JESUITS

Knowledge about electricity was kept alive during the seventeenth century by Jesuit polymaths. They also enriched the subject with valuable observations. In the eighteenth century the relative significance of the direct contributions of the Society sharply declined. It continued to be important indirectly, however, as Schoolmaster to Catholic Europe. One indication of its effectiveness is that it educated at least 20 percent of the 195 members of the Paris Academy of Sciences honored by éloges in its *Histoire* during the Ancien Régime.[2]

THEIR MAGISTERIUM

The reputation of the Jesuits as teachers dates from the foundation of their order. In the 1550s they received more invitations to establish schools than they could handle.[3] 'It is better for a town to found a college of Jesuits than to build highways or harbors,' the French proverb runs. These colleges taught poor children along with rich, opposed the spread of heresy, attracted and produced civilized and educated people, and—not to be ignored—brought in fresh money in the form of student expenditures for food and lodging.[4]

The Society's pedagogical principles, as set forth in its *Ratio studiorum* of 1599, guided the organization and conduct of all Jesuit schools in the seventeenth and eighteenth centuries. The system had many strengths: a steady curriculum, proof against educational faddists but adaptable to local needs; well-trained and educated teachers; dependability, regularity, punctuality and, above all, experience.[5] Both friends and foes of the order praised its schools. Ranke: 'Students learned more from the Jesuits in six months than from others in two

2. Information from Charles Paul, who is preparing a study of the éloges.

3. One example, that of the province of Austria (including Bohemia) will illustrate the multiplication of Jesuit colleges. The first foundation was at Vienna in 1551; 100 years later (1663) Austria had 15 colleges; another century (1767) and it had 38. Paulsen, *Gesch.*, I (1896), 403. There were 113 colleges in France in 1762 (Morney, *Origines* [1947^2], 171), and 107 in Austria and Germany in 1750 (Pachtler, *Ratio,* I, ix–xx).

4. Chossat, *Jésuites* (1896), I, 103, 112–13.

5. H. Weber, *Gesch.* (1879), 87–91.

years.' Bacon: 'As for what pertains to pedagogy, it will be most briefly said: consult the schools of the Jesuits, nothing in use is better . . . If only they were ours.'[6]

The entire course of liberal studies took eight years, of which the first five were devoted to ancient languages and literature. Then came three years of 'philosophy:' logic and divisions of science in the first year; 'physics' (in the large Aristotelian sense) and mathematics in the second; and metaphysics, ethics and psychology in the third.[7] The scheme varied from place to place. At some of the larger colleges, for example at La Flèche beginning in 1626, metaphysics was studied with physics and mathematics became the main subject of the third year.[8] At other, smaller colleges, the philosophy course was reduced to two years, 'to save time,' or, as became necessary in German schools in the eighteenth century, to drop outmoded material, particularly metaphysics, from the curriculum.[9]

The 'physicists' (the second-year philosophy students) read Aristotle's *Physica,* the first book of *Generation and Corruption,* the *Meteorologica,* and, very briefly, *De Caelo,* treating only 'a few questions about the elements, and about the heavens only regarding its substance and influence.' The closer study of astronomy was the province of the professor of mathematics, who taught Euclid and 'something of geography or of the sphere or other matters that students like to listen to.'[10] The *Ratio* further provided that private instruction in 'mathematics,' which could include much of physical science, should be available to students with talent for the subject.

This important and forward-looking provision came at the urging of Christopher Clavius, the distinguished mathematician at the central Jesuit university, the Gregorian College in Rome.[11] It took almost a century to implement. The order had first to seek outside help[12] while Jesuit novices tried to develop a taste for mathematics. A crash program, planned at Rome, had Clavius giving a three-year course to ten students, who would go forth, teach, and populate the earth, or at least Jesuit colleges, with mathematicians.[13] The larger Jesuit

6. Ranke, *Popes* (1847³), 379; Bacon, *Advancement* (1915), 17; *Works,* I (1863), 445, 709. Cf. Schimberg, *Education* (1913), 521.

7. Farrell, *Jes. Code* (1938), 233; Fitzpatrick, *St. Ignatius* (1933), 131–4, 169–71. The philosophy course was not available at the smallest colleges, those with a staff of 30 and an income of 10,000# (according to regulations of 1603); philosophy first entered in 'middle' colleges, those with at least 60 religious and 15,000#. Rochemonteix, *Collège* (1889), I, 89–94. In fact the middle colleges were frequently smaller than the plan required, e.g., Montpellier in 1668 (26 religious, 7400#). Faucillon, *Collège* (1857), 59–61.

8. Rochemonteix, *Collège* (1889), IV, 27–50. At Louis-le-Grand, mathematics and metaphysics occupied the third year, physics and ethics the second; Dupont-Ferrier, *Vie,* I (1921), 181.

9. Chossat, *Jésuites* (1896), 234–6, 248–9; Ziggelaar, *Pardies* (1971), 49, 69.

10. Fitzpatrick, *St. Ignatius* (1933), 170–1, 175.

11. *Ibid.*, 130; Cosentino, *Physis*, 13 (1971), 207–8.

12. E.g., Maurolico, who taught at the Jesuit college in Messina; in 1573–4 a group of Jesuits, including Clavius, was deputed to assist Maurolico to put his papers in order, and to produce a text needed by the Society. Scaduto, *Arch. hist. Soc. Jesu*, 18 (1949), 133–40.

13. Pachtler, *Ratio,* II, 142–3, quoting a document of 1586.

schools of Italy and France made provision for the teaching of mathematics;[14] Germany remained behind, especially after the Thirty Years' War. Towards the end of the century, despite repeated requests from Rome for improvement, the German provinces 'scarcely had one professor who could teach the subject creditably at a large university.'[15]

The Jesuits had several reasons for emphasizing mathematics. According to a draft of the *Ratio* of 1586, it was necessary for other studies, including poetry (astronomical allusions), history (geography), politics (military technology), and ecclesiastical law (exact chronology), but above all for astronomy, navigation, architecture, and surveying.[16] These last subjects were important in the training both of Jesuits bound for foreign missions and of young aristocrats destined for high military or government service. Most 'noble youths,' says Schott, are interested in practical applications of mathematics, and come great distances to study with such Jesuit masters as himself and Athanasius Kircher.[17]

This instruction required instruments, of which the larger colleges often had a good stock: instruments for astronomy, geodesy, dialing, drawing, and, perhaps, for the demonstration or investigation of physical principles.[18] Since this apparatus catered to the interests of an influential portion of the student body, it was not considered a frill; when Schott took up a chair at the college at Würzburg he went far out of his way expressly to trade for mathematical books and instruments to replace those the college had lost during the Thirty Years' War, and to bring it in line with its better-stocked sister institution at Mainz.[19] The English knew how to appreciate what they lacked in this respect. 'This I must always affirm for the honor of my mother the University of Oxford, that if her children had the good utensils, which adorn the colleges of the Jesuits abroad, the world would not long want good proof of their ingenuity.'[20] Description of these instruments, some built to original designs, was a staple in the technical books that the Jesuits produced in profusion in the seventeenth century.[21] They were frequently asked for advice about the procurement of

14. Cf. Chossat, *Jésuites* (1896), 439; Ziggelaar, *Pardies* (1971), 76–7. Note the letter of the General to the Paris Provincial, 11 Sept. 1656, in Dainville, *RHS*, 7 (1954), 15: 'It is not without incredible sorrow that I have learned that mathematics and Hebrew are so neglected by our [priests] that there would be almost no one to teach them if we had to replace the older professors.'

15. Duhr, *Gesch.* (1921), III, 413. Cf. the efforts of the General, Tamburini, in 1724; Dainville in Taton, *Enseignement* (1964), 31–2.

16. Pachtler, *Ratio*, II, 141–2.

17. Schott, *Magia* (1671), sig.++001v; *Pantometrum* (1660), sig. 0003r. Cf. the insistence of the city of Avignon that the Jesuits bring a mathematician, who turned out to be Kircher, into their college there in 1628; Chossat, *Jésuites* (1896), 234–6.

18. Dainville, *RHS*, 7 (1954), 6–21, 109–23. Note the special room for physics in the new college at Montpellier; Faucillon, *Collège* (1857), 88.

19. Schott to Kircher, 12 Sept. 1655, in CK, XIII, f. 49. Cf. Duhr, *Gesch.* (1921), III, 414, for the apparatus at Ingolstadt, a tourist attraction in 1675.

20. E. Bernard to J. Collins, 3 April 1671, in Rigaud, *Corresp.* (1846), I, 159.

21. E.g., Schott, *Pantometrum* (1660); Zucchi, *Nova* (1649). Cf. the résumé of Schott's gadgets in Mercier, *Notice* (1785), 5–14.

instruments, and sometimes acted as intermediaries between purchasers and makers.[22]

A second reason for Jesuit emphasis on mathematics was that it offered a strong ground for confronting, or rather avoiding, the new physics. The mathematician, according to a dodge used by the ancient astronomers, aimed not at discovering the true principles of things, but at 'saving the phenomena,' at concocting adequate quantitative descriptions in the easiest possible way.[23] The Jesuits found this subterfuge comfortable, and taught optics, mechanics and astronomy without troubling about truth.[24] In gnomonics, or dialing, a subject they emphasized, they could reasonably adopt a geocentric point of view. When considering the solar system, they might use Copernicus' theory, treated as an hypothesis;[25] in the last quarter of the seventeenth century both the Gregorian College (Rome) and Louis-le-Grand (Paris) used texts that represented heliocentrism as a useful mathematical fiction.[26] The strength of this ploy decreased in time. In the eighteenth century, in the face of Cartesian and Newtonian physics, it no longer sufficed to protect the natural philosophy of Aristotle.[27]

These organizational generalities apply to Jesuit universities as well as to Jesuit colleges. In most cases the 'university' undergraduate or arts course was nothing other than the three years of philosophy prescribed by the *Ratio*. Humanities were taught at associated Jesuit grammar schools. The separate Jesuit university occurred primarily in Germany and to a lesser extent in Italy, where the Society took over existing institutions;[28] in France, the larger Jesuit colleges—such as Bordeaux, La Flèche, Lyon—were both grammar schools and provincial faculties of arts. In general, the Society could exercise stricter control over colleges than over universities; while the colleges were usually endowed, the universities had at least a part of their operating expenses from states or municipalities, which thereby retained the power to dabble in educational reform. The forced modernization during the eighteenth century of the outmoded curricula at the conservative Jesuit universities under Austrian control is a dramatic instance of government intervention.[29]

22. J. Doddington to Kircher, 17 and 31 Jan., 21 March and 24 Oct., 1671, in CK, V, ff. 21, 50, 97, 104; and the correspondence of Kircher with Leibniz, 1673 (Friedländer, Pont. acc. rom. arch., *Rend.*, 13 [1937], 229–47), and with G. A. Kinn, 1672 (J. E. Fletcher, *Janus*, 56 [1969], 267).

23. Duhem, *To Save the Phen.* (1969), *passim.*

24. Cosentino, *Physis*, 13 (1971), 205. Cf. Descartes' annoyance at the superficiality of his teachers' use of mathematics; Descartes, *Discourse* (1965), 8.

25. E.g., Schott, *Cursus mathematicus* (1661), and Dechales, *Cursus mathematicus* (1674). Note the titles.

26. Eschinardi, *Cursus physico-mathematicus* (1687); Dechales, *Cursus seu mundus mathematicus* (1690[3]), who puts the matter thus: 'si l'on ne se rapporte qu'aux arguments de la science, sans égard à l'autorité de l'Ecriture, il n'y a aucun qui tranche définitivement pour ou contre le système de Copernic.' See Ziggelaar, *Pardies* (1971), 150–6.

27. Cf. Cosentino, *Physis*, 13 (1971), 206–7.

28. Paulsen, *Gesch.*, I (1897), 383–407.

29. *Infra*, ii.3.

The teachers of philosophy and mathematics could hold several sorts of positions within the Society. The most junior taught at the smaller schools, and took their students through all three years of philosophy. They had little time, and usually little talent, for independent scientific work; many, perhaps most, did not continue in teaching careers.[30] Those deemed sufficiently conscientious and able advanced to the bigger schools, where they could specialize in one or another branch of philosophy. There were fifty positions in 'physics' in France in 1700, and 62 at the expulsion of the Society in 1761.[31] These positions often functioned as academic chairs, with the peculiarity, however, that their less distinguished incumbents frequently moved from college to college.

The professor of mathematics followed a different path. He was usually chosen from among the few who had shown mathematical ability, and received special lessons, during study of philosophy. Since their specialty occupied only a part of the physics course, they usually taught something else as well, chiefly languages or physics, or kept books for the college. During the early seventeenth century they changed colleges frequently, but after 1660 their positions stabilized, owing, perhaps, to an increasing demand and higher appreciation for their services.[32] The demand came not only from students, but also from the general public and especially from government, which often used Jesuit mathematicians as consulting engineers. In France the value of the order as a source of technicians was recognized by the creation of seven royal chairs of hydrography in Jesuit institutions in or near seaports. By 1700 the French Jesuits had 21 other chairs for mathematics at their several colleges.[33]

During term the conscientious Jesuit professor had little opportunity to advance, or even to keep up with his subject. Typically he lectured two hours a day, conducted reviews, arranged for and presided over disputations, wrote theses for his abler students, offered lectures to the general public;[34] in addition he said masses, heard confessions, and attended to his own religious exercises. Recognizing that its savants needed occasional relief from these burdens, the Society set up positions free from teaching for distinguished specialists. During their leisure, which might last from two to six years, these scriptors wrote monographs or texts, or coordinated astronomical and geographical data sent from missionary or provincial colleagues. Clavius was alternately professor and scriptor from 1565 to 1612.[35] The Collège Louis-le-Grand, which the order wished to make a showplace, 'where science was done as well as disseminated,' sheltered some 90 scriptors between 1606 and 1672, including physicists and mathematicians such as Jacques Grandami and Noel Regnault.[36]

30. Cf. Dainville, *RHS*, 7 (1954), 14.
31. Dainville in Taton, *Enseignement* (1964), 33–4, 40–1.
32. Dainville, *RHS*, 7 (1954), 15; Huter, *Fächer* (1971), 7.
33. Dainville in Taton, *Enseignement* (1964), 33–4; cf. Ziggelaar, *Pardies* (1971), 28.
34. Chossat, *Jésuites* (1896), 100–3; Guitton, *Jésuites* (1954), 48; Ziggelaar, *Pardies*, 125; Schott, *Pantometrum* (1660), proem.
35. Phillips, *Arch. hist. soc. Jesu*, 8 (1939), 191.
36. Dupont-Ferrier, *Vie*, I (1921), 60–2, 81, 121, 150–1, and III (1925), App. A.

The official scriptors made up only a portion of the personnel supported by the Society in research and writing. Frequent sabbatical could be arranged for distinguished mathematicians. An interesting case is Gregory de St. Vincent, who had not only leave but assistants, furnished by his General in the expectation that many fine texts for the improvement of novices, and many fine monographs for the glory of the Society, would result. Gregory spent much of his time trying to square the circle; when he announced success he was called to Rome to be examined, lest his 'proof' embarrass the order in precisely the field it had made its own. The Roman mathematicians could not certify Gregory's circle-squaring. They nonetheless thought his other work worthy of support, which he continued to enjoy for forty years.[37]

Probably the most important result of all this writing was the preparation of standard texts and reference works, which, especially in the seventeenth century, served those outside the order as well. One thinks, for example, of Fabri's *Synopsis geometrica,* an introductory manual used by Leibniz, by Flamsteed, and, perhaps, by some of Newton's students;[38] and of Pardies' *Elémens de géometrie,* 'the plainest, shortest, and yet easiest Geometry' ever published, according to the preface to the eighth English edition.[39] Texts on the design and application of simple instruments, like Schott's on the pantograph and Zucchi's on elementary machines, also circulated widely.[40] Among works of reference Riccioli's *Almagestum novum* played a unique role: although belligerently anti-Copernican, it became a standard source for astronomical data even among adherents of the new cosmology. The Jesuit literature on the magnet well served students of the lodestone, again without regard to doctrine. And so the elder Huygens, who was no friend of the Order, exhorted the reformer of modern philosophy to rely upon the data of the Jesuits: 'For these scribblers,' he wrote Descartes, 'can serve you in matters *quae facti sunt, non juris.* They have more leisure than you to provide themselves with experiments.'[41]

The Jesuits gathered and disseminated much of their information through correspondence. One knows the important part played by the Minim monk, Marin Mersenne, in effecting communication among physicists and mathematicians in the 1630s and 1640s. But Mersenne was an individual, the Jesuits an organized and disciplined society; their world-wide missions housed a network of informants able to identify natural and artificial novelties, and to observe astronomical phenomena invisible from Europe. Requests for and dissemination of this information, and for the supply of rara naturalia, constituted much of the

37. Bosmans, *Biog. nat. Belg.*, XXI, cols. 145–51, 156, 168.

38. Fellmann, *Physis,* 1 (1959), 6–25; Leibniz, *Phil. Schr.* (1875), IV, 245; Flamsteed to Collins, 20 May 1672, and Collins to Newton, 30 April 1672, in Rigaud, *Corresp.* (1846), II, 146, 320.

39. Pardies, *Elémens* (1671); *Short, but yet Plaine Elements of Geometry* (1746^8); Ziggelaar, *Pardies* (1971), 64–8; Wieleitner, *Arch. Gesch. Naturw. Tech.*, 1 (1908–9), 438.

40. Cf. Borelli to Collins, 10 April 1671, in Rigaud, *Corresp.* (1846), I, 165.

41. Huygens to Descartes, 7 Jan. 1643, in Descartes, *Corresp.* (1926), 186; Riccioli, *Almagestum* (1651); Koyré, *Metaphysics* (1968), 89–117; Flamsteed to Collins, 29 Sept. 1673, in Rigaud, *Corresp.* (1846), II, 168. For Jesuit magnetism see Daujat, *Origines* (1945), II, *passim.*

learned correspondence of the order. Kircher, for example, channeled foreign observations of eclipses of the moon to his colleague Riccioli, who rewarded him by naming a lunar crater in his honor.[42] Schott exchanged letters with much of learned Germany, including Otto von Guericke, the Lutheran mayor of Magdeburg, who found it convenient to announce his discoveries in Schott's encyclopedic works. The fine museum of the Settalas in Milan was filled with specimens furnished by Jesuit missionaries.[43] The order's connections also assisted the circulation of books at a time when no regular international trade in them existed. Jean Bertet, 'the true Mersennus of France,' a Jesuit mathematician, a student of Fabri's and friend of Pardies', labored for years to procure continental books on his subject for the use of English philosophers.[44]

The very reputation of the successful Jesuit savant might constitute a subtle but important aspect of his magisterium: since touring intellectuals tended to seek him out, he often functioned as the nucleus of a tiny international congress, an ever changing polyglot academy with shared interests, similar educations, and a common learned language. No study trip to Rome was complete without a visit to Fabri and Kircher; Arriaga, according to an old slogan, was a chief attraction of the capital of Bohemia, 'Pragam videre, Arriagam audire;' while the Minims Mersenne and Maignan drew even royal visitors to their cells in Paris and Toulouse.[45]

The powerful Jesuit educational system, its celebrated pedagogues and part-time researchers, made the Society the leading patron of physical and mathematical sciences during the seventeenth century. According to Leibniz, all the chief scientific men in Italy in the 1670s were Jesuits; the discipline and cooperation of the Society's savants served as the model, and perhaps the inspiration, for Leibniz' plans for a German academy of sciences.[46] Perhaps Bacon too took hints from them: 'partly in themselves [he observed] and partly by the emulation and the provocation of their example, they have much quickened and strengthened the state of learning.'[47]

Jesuit work provided starting points for investigations in many branches of physics and mathematics, especially optics, mechanics, magnetism and electricity; the Royal Society of London routinely reviewed Jesuit books and sought to

42. J. E. Fletcher, *Manuscr.*, 13:3 (1969), 157–8. Kircher was consulted on all respectable subjects (J. E. Fletcher, *Janus*, 56 [1969], 259–77) and many questionable ones; in a single year he examined and rejected ten designs for perpetual-motion machines (Gutmann, *Kircher* [1938], 14).

43. Fogolari, *Arch. stor. lomb.*, 14 (1900), 59, 113, 119; Rota Ghibaudi, *Ricerche* (1959), 46–7.

44. Rigaud, *Corresp.* (1846), I, 139–40, 151, 162, and II, 22–3, 530.

45. Oldenburg, *Corresp.*, V, 294, 423, 439; Eschweiler, *Sp. Forsch.*, 3 (1931), 253–85; Ceñal, *Rev. est. pol.*, 46 (1952), 111–49. Monconys, *Voyages* (1665–6), gives a lively picture of the international intellectual life of the day, when it was possible for one man to know 'most of ye Intelligent and Curious men in the world'; Oldenburg to Boyle, 24 Dec. 1667, re Carcavi, in Oldenburg, *Corresp.*, V, 79.

46. Friedländer, Pont. acc. rom. arch., *Rend.*, 13 (1937), 242–5. Cf. Grosier, *Mémoires* (1792), I, xxvi–xxviii.

47. Bacon, *Advancement* (1915), 41; *Works*, I (1863), 469. Cf. Reilly, *Line* (1969), 75.

open correspondence with the leading members of the Society.[48] As the national academies of science grew in size, resources and importance, and science returned to the secular universities, Jesuit patronage of mathematics and physics became less significant and appropriate. Moreover, the Society lost ground to the moderns by first opposing and then belatedly assimilating Cartesian philosophy. In Germany especially their defense of the old ways dropped their universities behind the leading Protestant schools during the eighteenth century. The German provinces took with peculiar literalness the Society's reaffirmation, at its general congregation of 1730–1, of the fundamental principles of Aristotelian philosophy.[49] But the Jesuits of France and Italy continued to train productive scientists until their expulsion and the dissolution of the Society (1773). 'It was the most beautiful work of man,' lamented a former student, J. J. de Lalande, astronomer, pensionary of the Paris Academy, professor at the Collège Royal, etc.; 'no human establishment will ever approach it.'[50]

THEIR ECLECTICISM

Although St. Ignatius had counselled his Society to follow St. Thomas in theology and Aristotle in philosophy, it did not long remain subservient to either. On the theological side the Jesuits produced their own doctor, Suarez, who did not hesitate to disagree with Aquinas.[51] Suarez' *Disputationes metaphysicae* (1597), freer and more orderly than the traditional Aristotelian commentary, have a vigorous, eclectic air;[52] they were used both in Jesuit colleges and in the Protestant universities of Holland and Germany.[53] During his student years Leibniz eagerly read the *Disputationes* and Descartes, it is said, carried them with him in his travels.[54]

The free, eclectic style of Suarez' popular metaphysics exercised an important though subtle influence on physics. Suarez' colleagues at Coimbra wrote their authoritative commentaries on Aristotle's natural philosophy in a style

48. Duhem, *Rev. met. mor.*, 23 (1916), 58–9; Reilly, *Arch. hist. Soc. Jesu.* 27 (1958), 340–4, and *Line* (1969), 89–90, 93–6; Oldenburg, *Corresp.*, VI, 119, 274, 317, and V, 315, 564; Boyle to Oldenburg, 3 April 1668 (*ibid.*, IV, 299): 'I am glad you are like to settle a correspondence with *Rome,* that being the chief center of intelligence.' In a quaint and ambitious reciprocation, the Jesuit college at Liège announced in its *Prospectus* of 1685 that it had perfected all the Royal Society's discoveries; Reilly, *Line* (1969), 14.

49. Hammermayer, *Gründ.-Frügesch.* (1959), 237–8; Prandtl, *Gesch.* (1872), I, 539; Paulsen, *Gesch.*, I, 425, and II, 103–4.

50. Quoted by Delattre, *Etablissements*, II, 1554n. Cf. Costa, *Rev. stor. it.*, 79 (1967), 849 f.; Varićak, Jug. akad. znan. umjet., *RAD,* 193 (1912), 248; Grosier, *Mémoires* (1792), I, vi–vii.

51. Mahieu, *Suarez* (1921), 522. Other commentators have tried to minimize the disparity; see Riedl in Smith, *Jes. Think.* (1939), 18n. A useful guide to the technical differences between Suarez and St. Thomas is Werner, *Suarez* (1889^2).

52. Grabmann, *Mitt. Geist.*, I, 528–35; Jansen, *Phil. Jahrb.*, 50 (1937), 418–23.

53. Eschweiler, *Sp. Forsch.*, 1 (1928), 251–325, esp. 288–91; cf. Petersen, *Gesch.* (1921), 283–338.

54. Riedl in Smith, *Jes. Think.* (1939), 5; Eschweiler, *Sp. Forsch.*, 1 (1928), 254, 259; Gilen, *Schol.*, 32 (1957) 47.

between the old commentary and the new, freer synthesis; they did not fear to leave doubtful points unsettled, or to add essays that made possible 'a somewhat freer consideration of the topics than their original context provided.'[55] A symbiosis similar to that of Suarez and the Coimbra commentators established itself later at Prague, where Arriaga, an adept in Suarez' methods, presided over a group of physicists that included Gregory de St. Vincent, Juan Caramuel, Marcus Marci, and Balthasar Conrad.[56] It cannot be mere coincidence that several of those who advanced the study of electricity in the seventeenth century also played a part in the dissemination of Suarez' work.

Jesuit writings on natural philosophy contained much that conflicted with the Aristotelian doctrines that they purported to transmit, and that their authors were obliged to teach.[57] In the last third of the sixteenth century the Society hoped to establish a standard interpretation of peripatetic physics for use in its academies. It began by turning from the superficial summulae of the late Renaissance to Aristotle's texts, as in the products of Coimbra, which gave the original Greek, a Latin translation, and a running commentary, as well as the free expositions. The exigencies of the philosophical course also required shorter compendia or *cursus*; and these, despite the Society's best intentions, could never be made uniform. The general councils of the 1590s recommended the adoption of official textbooks, the characteristics of which are set forth in the early drafts of the *Ratio studiorum*; but the practicing pedagogues consulted, and especially the Spanish school, urged that the official doctrine be kept to a minimum.[58] 'In scientiis,' St. Thomas had said, 'auctoritas minime valet.' Early in the seventeenth century the Society formally abandoned the project for an official textbook of natural philosophy.[59] Such a text, by freezing the curriculum just as the new science began to develop, would seriously have handicapped Jesuit efforts to educate Catholic Europe for survival in the modern world.

As novelties unknown to Aristotle accumulated, the books of conscientious Jesuit natural philosophers departed more and more from their ancient model.

55. Coll. conim., *Comm. de Caelo* (1603²), 487. Among doubtful points: whether stars act by 'influence' or only by light and heat, *ibid.*, 200; whether qualities alone among accidents can act ('non videtur nobis absolutè pronunciandum'), *Comm. in phys.*, I (1602), col. 393; the number of categories ('suas habeat difficultates'), *Comm. in univ. dial.*, I (1607), 340.

56. Caramuel (of whom more below) was a Cistercian; St. Vincent and Conrad, Jesuits; Marcus Marci, a physician educated by the Jesuits, was only prevented by ill-health from becoming one himself. Eschweiler, *Sp. Forsch.*, 3 (1931), 253–85; Werner, *Suarez* (1889²), II, *passim;* Marek, *RHS,* 21 (1968), 109–30, and *Bohemia,* 16 (1975), 98–109.

57. 'In matters of any consequence let him [the professor of philosophy] not depart from Aristotle unless something occurs which is foreign to the doctrine that academies everywhere approve of.' *Ratio* (1599), in Fitzpatrick, *St. Ignatius* (1933), 168. Cf. *ibid.,* 151.

58. The Spanish Jesuits also opposed a definitive catalog of forbidden propositions; the first drafts of the *Ratio* had about 600; that of 1590, 200, that of 1599, none. Fichter, *Suarez* (1940), 137–44.

59. Jansen, *Phil. Jahrb.*, 51 (1938), 187; *Ratio* (1599), in Fitzpatrick, ed., *St. Ignatius* (1933), 119–21, 170–1; Hilgers, *Index* (1904), 194–206.

An early stage is represented by Niccolò Cabeo's commentary on Aristotle's *Meteorologica* (1646). Despite its title, it is not a classical commentary, but almost an independent *philosophia universalis,* a label under which it reappeared, unaltered, in 1686.[60] Cabeo no doubt chose the *Meteorologica* as his vehicle because, as the most practical and specialized of Aristotle's treatises on the inorganic world, it gave the 'commentator' the best purchase for an exposition of terrestrial physics.[61] Not that Cabeo professed to follow Aristotle closely: the true task of the peripatetic commentator, he holds, is first to clarify the writings, and then to judge their truth, 'not on his authority, because *ipse dixit,* but by plain reasoning.' 'If you never question Aristotle's doctrines your commentary will not be that of a philosopher but that of a grammarian.' And Cabeo does take shocking liberties, like interpreting the form of volatile substances as a material emanation.[62]

Jesuits soon dropped the commentary form in favor of comprehensive treatises, such as the *Physica* of Honoré Fabri. Though Fabri maintained the pious fraud of holding to hylomorphism 'most religiously,' and pretended to deduce all of physics from six peripatetic principles, viz., heat, impetus and the four elements, he in fact fashioned an adroit compromise between Aristotle and the corpuscularians.[63] Where he thought Aristotle erred, as in the theories of natural motion and celestial composition, Fabri did not pretend to follow: on the contrary, he admitted all the gains of early modern physics but the Copernican geometry. Professing to eschew hidden virtues, antiperistasis, sympathies, antipathies and atoms, he nonetheless invoked an occult cause (in the guise of 'magnetic particles') and exploited corpuscles so freely that contemporaries classified him as Cartesian or Gassendist. Moderns understandably find him eclectic.[64] Fabri always denied that he subscribed to corpuscular philosophies; his eclecticism, he insisted, was only in the eye of the beholder, misled by the commentaries of 'impious Arabs' and of 'modern scholastics, who have never understood Aristotle's thought.'[65]

More overtly eclectic is Fabri's younger colleague, Francesco Lana, a student of Kircher's. 'It is a most vulgar error,' he says, 'to believe that our mental images of the truth are unique, as if many likenesses cannot be made of

60. The title of this edition (Rome, 1686) is given by Ferretti-Torricelli, At. di Brescia, *Comm.* (1931), 384, but does not appear in Sommervogel. Lana Terzi, *Prodromo* (1670), 13, singles out Cabeo and Gassendi as the best modern writers on physics.

61. Descartes saw as much when he offered his *Météores,* which amounts to a free commentary on the *Meteorologica*, as an example of the application of his new method to physics. Descartes, *Oeuvres,* IX:2, 15.

62. Cabeo, *Met. Ar. comm.* (1646), I, 'Ad lect.,' and p. 253.

63. Fabri, *Physica,* I (1669), 'Auctor-Lect.,' vii–viii, xxxviii; *Epistolae* (1674), 34–5; Morhof, *Polyhistor* (1747[4]), II, 267, 334–5; Cromaziano, *Restaurazione,* I (1785), 131.

64. Daujat, *Origines* (1945), II, 350–75 (Fabri's magnetism); Lasswitz, *Gesch.* (1890), II, 460, 486, 490 (Fabri's eclectic corpuscularism); Morhof, *Polyhistor* (1747[4]), II, 218; Pernetti, *Recherches* (1757), II, 122.

65. Fabri, *Epistolae* (1674), 56.

the same statue. Accordingly I will use many models [*imagines*] to express clearly the truth to which the science of natural things can attain: the more representations of an object are present to the intellect, the more intimately and clearly the object is known; what one model lacks, another will supply.' Lana supplements Aristotle's hylomorphic and elemental doctrines with the *tria prima* of the Paracelsians; with purposeful Democritean atoms; with sympathies, antipathies, powers, principles, the fixed, the volatile, the acid, the alkaline.[66] The primary obstacles to the advance of science, according to Lana, are excessive attachment to antiquity and, beyond that, to pet subjects, as Aristotle to logic, Plato to theology, Proclus to mathematics, Gilbert to magnetism, and chemists to fire.[67]

The knowledgeable natural philosopher changes his principles to suit his problems; when Aristotle falters he must take another guide. This program, which might be called 'moderate peripateticism,' was not peculiar to Lana or to the Jesuits. Juan Caramuel, for example, the friend of Arriaga, correspondent of Kircher's, mathematician, papal envoy and bishop, had earlier endorsed the same method. 'In the dead of winter, in a failing light, you come to a difficult crossing you do not dare to attempt; Peter approaches from the opposite direction, carrying a bright lantern. You cross, he crosses: but you do not follow him because you have used his light; you continue your journey in your original direction. It often happens thus in our philosophizing. We come to a difficult and obscure place; we hale Zeno, Plato, Aristotle; their doctrines are a torch that dissipates the darkness. We cross in the greatest security; and yet we do not therefore follow them, but proceed as we had begun.'[68]

One of the most original directions was that chosen by Emmanuel Maignan. Schooled by the Jesuits of Toulouse, Maignan determined to follow the religious life among the neighboring Minims when his masters awarded a classmate a literary prize he thought he had earned. Such intellectual sensitivity was then not characteristic of the *minimi minimorum,* an austere order of Franciscans, some of whom had not troubled to learn to read; but the few savants the order did produce and encourage in the seventeenth century were aggressive and independent thinkers, untied to medieval doctors, opposed to frilly argumentation, and interested in matters accessible to experiment. These qualities appear in Maignan as they do in his colleagues Mersenne and J. F. Niceron.[69]

Maignan's chief work is a *Cursus philosophicus* (1653), which, despite its global and pedestrian title, is an original treatise primarily devoted to natural

66. Lana Terzi, *Magisterium,* I (1684), 'Auctor-Lect.' Cf. Morhof, *Polyhistor* (1747[4]), II, 266, who holds that the eclectic method is the soundest for physics, 'ut non omnia omnes videant.'

67. Lana Terzi, *Prodromo* (1670), 5–6; cf. the Idols in Bacon's *Novum organum.*

68. Caramuel, *Rat. real. phil.* (1642), 62. Cf. Ceñal, *Rev. fil.,* 12 (1953), 101–47; Fernández Diéguez, *Rev. mat. hisp.-amer.,* 1 (1919), 121–7, 178–89, 203–12; Glick, *Isis,* 62 (1971), 279–81.

69. Bayle, *Dictionnaire* (1740[5]), III, 280–3; Rochot, *Corresp.* (1966), 7; Whitmore, *Order* (1967), 112–19; Ceñal, *Rev. est. pol.,* 46 (1952), 111–49; Nicéron, ed., *Mémoires,* XXXI (1735), 346–53.

philosophy.[70] He has no patience with received authority in science. 'I would think myself exceedingly stupid,' he writes, 'were I to deny the truth of what I see and touch in my experiments, and credulously to embrace what someone else has concluded from some abstract and fanciful little arguments.' The Aristotelians are his main targets. He rejects hylomorphism, 'the solemn daily teaching of the peripatetic schools;' the *horror vacui;* and the distinction between potency and act, which he calls 'voces sine re,' or claptrap. Yet he is no corpuscularian. He conceives that there are many kinds of matter, the magnetic and the non-magnetic, for example, whose different properties cannot be reduced to differences in the sizes, shapes and motions of constituent particles. Maignan is a qualitative atomist: he holds that bodies act upon one another only during contact, but denies that the act is purely mechanical. Among the non-mechanical qualities his corpuscles possess are principles of self-movement and sympathetic powers competent to activate them: a magnet, for example, emits streams of characteristic particles that trigger and direct the self-movements of neighboring pieces of iron.[71]

Maignan's qualitative atomism, with its sympathies and principles, was a fair compromise between radical corpuscularism and the traditional philosophy.[72] His *Cursus* well illustrates the breadth of options open to moderate peripatetics of the seventeenth century, to those who wished to develop the received philosophy according to their own lights, and to come to grips with the new physics of Galileo and Descartes.

During the eighteenth century the Jesuits stopped struggling to reconcile physics and mathematics with the doctrines of Aristotle and Ptolemy. As noticed earlier, they were teaching Copernicus 'hypothetically' before the end of the seventeenth century. Later Descartes, and still later Newton, entered their textbooks. These liberties were acknowledged after the fact, in an edict of 1751, which freed Jesuit physicists of Aristotle except for 'the first principles of natural body. . . , namely of matter and form in the peripatetic sense, of prod-duction *de novo,* and of the existence of some absolute accidents.'[73] There is no difficulty in clothing any physical theory in the language of hylomorphism.

THEIR CONSTRAINTS

To balance our rude portrait of Jesuit physicists we should notice two obstacles they met with in their work. The first was the problem of support. The gluttonous intellect of a Kircher consumed an amount of treasure little consistent with the vow of poverty, or with the funds made available by superiors often hard

70. The second edition (1672), 742 pages in-folio, devotes 75 pages to logic, 50 to metaphysics and the rest to 'philosophia naturae, seu physica.' Cf. Ceñal, *Rev. fil.*, 13 (1954), 15–68.

71. Maignan, *Cursus* (1673²), 127, 143, 195, 226, 590–6, 608; Ceñal, *Rev. fil.*, 13 (1954), 15–68; Lasswitz, *Gesch.* (1890), II, 492–3.

72. Cf. Jansen, *Phil. Jahrb.*, 50 (1937), 433–6; Sander, *Auffassungen* (1934), esp. 9–10.

73. Pachtler, *Ratio,* III, 435f. Cf. Specht, *Gesch.* (1902), 199n; Dainville in Taton, *Enseignement* (1964), 49n; Bednarski, *Arch. hist. Soc. Jesu,* 2 (1933), 213.

pressed to make ends meet. Kircher found his Maecenases among Catholic and Protestant princes, who sometimes contributed so handsomely to his treasury that he was able to assist the work of others. The celebrated museum grew from a private bequest.[74] Occasionally proceeds from the sale of books might be made available for scholarly purposes: Schott secured special permission to handle money in connection with his publications, and any surplus probably ended in new instruments.[75] To do original work in the sciences the moderate peripatetic probably needed outside support and certainly required the cooperation of his colleagues and superiors, the use of the order's instruments and facilities, and free time. These advantages were not always easy to procure. Would-be physicists assigned to remote, impoverished or understaffed colleges, where the customary teaching apparatus might be incomplete, experienced considerable, and perhaps insurmountable difficulties.

Opposition from colleagues posed another obstacle. Simplicios did exist, and exercised their influence both unofficially, through local cabals, and institutionally, through the censorship of the press. The irregular practices were probably the more effective: conservative rectors or prefects of studies might burden their modernizing professors with routine assignments or secure their transfer to non-academic positions. A cabal of conservatives at the college of Lyon, alarmed at Fabri's aggressive taste for novelties, is said to have engineered his reassignment from France to Italy, and from teaching to the papal bureaucracy. It is difficult to estimate the harm done the eclectics and moderate peripatetics by such maneuvers. In Fabri's case it backfired; he found Rome congenial, enjoyed the support of the Jesuits' General, and became more productive, innovative and combative than before. Similarly Pardies, under attack from his colleagues at Bordeaux for his weakness for Cartesian physics, was transferred by the General to Louis-le-Grand, where he could do even greater mischief.[76]

The practical working of the censorship is also difficult to determine from its formal acts. Few scientific works ever earned a place on the *Index of Prohibited Books*. The Church directed its vigilance against errors in faith, morals and canon law;[77] its sortie against the Copernicans carried it into territory it had previously avoided and could not hope to hold. After the blunder over Galileo (1633), the Church again avoided indexing physics books,[78] and sought to sway their authors during review. The Roman censorship was particularly officious. Schott, who knew that its dilatory, frightened and often ignorant functionaries had annoyed even Kircher, refused an invitation to return from Germany to the Collegio Romano in 1664 primarily because he wished to

74. Brischar, *Kircher* (1877), 43–59; Villoslada, *Storia* (1954), 183–4; Schott to Kircher, 16 June 1657, in CK, XIII, 45: 'Ringratio molto a V^{a} R^{a} per li 6 scudi offertimi di nuovo per li miei studij.' Maignan, *Cursus* (1673^{2}), Praef., acknowledges the financial support of Cardinal Spada.

75. Duhr, *Gesch.* (1921), III, 590–3.

76. Vregille, *Bull. Soc. Gorini,* 9 (1906), 5–15; Ziggelaar, *Pardies* (1971), 69–78.

77. Putnam, *Censorship* (1906), I, 182–93, and II, 127–9.

78. Galileo first appeared upon the *Index* in 1664; Mendham, *Lit. Pol.* (1830^{2}), 175–7.

write, and at Würzburg the censors gave him little trouble.[79] But even in Italy few if any scientific books were suppressed. On balance, the censorship seems rather to have harassed than guided; a physicist willing to suffer delay and indignity would probably see his book approved much as he had written it, provided, of course, that he did not openly espouse heliocentrism. To pick a late example, the first volume of the proceedings of the Bologna Academy of Sciences was delayed for several years because the unenlightened authorities—Bologna then (1730) being a Papal State—demanded that whenever the proceedings mentioned Copernicus, or discussed the moderns 'respectfully and charitably,' the editor characterize the novelties as ingenious hypotheses. He refused, saying that it would make the Academy ridiculous. That was in 1729. Two years later, when the enlightened Lambertini, later Benedict XIV, became Archbishop of Bologna, the proceedings appeared as originally planned.[80]

The differential workings of the censorship, as applied to doctrinal and scientific matters, may be illustrated by the experiences of Maignan and Fabri, each of whom has managed to secure a permanent place on the *Index*. Maignan had vigorously attacked the natural philosophy of the schools, and ended closer to Descartes than to Aristotle. He was indexed not for assaulting peripatetics, however, but for trying to justify the practice of usury.[81] Fabri published a *Physica* in 1670 that strayed close to Descartes, praised Galileo, and took a soft line towards Copernicus. It was not for this that he then found himself in the jails of the Inquisition, but for an untimely attack upon the Jansenists, and an excessive defense of the slippery doctrine of probabilism.[82] 'As for his books on philosophy, don't worry about them,' writes the General, Oliva, to the Provincial of Lyon. 'One can indulge a man of such parts.'[83] The point is important, as Fabri's contretemps has given rise to the idea that his brethren sacrificed him because he had 'busied himself with science.'[84] In fact, Fabri's scientific connections helped extricate him from the clutches of the Inquisition. Cardinal Leopold dei Medici, who had known Fabri as a correspondent of the Florentine Accademia del Cimento, intervened on his behalf and helped restore him to a position of honor and influence.[85] We may take it that the Congregation of the Index had no interest in electricity.

79. Schott to Kircher, 21 Aug. 1664, in CK, VIII, 110: 'Hactenus non multum desideravi redire Romam, quoniam existimo me hic habere meliorem occasionem scribendi et imprimendi libros meos, quam ibi.'

80. Bortolotti, *Storia* (1947), 157–8.

81. Maignan, *De usu licito pecuniae* (1673), prohibited 24 Oct. 1674; *Index* (1948), 293.

82. Fabri, *Apologeticus doctrinae moralls societatis Jesu* (1670), prohibited 27 Jan. and 22 March 1672; *Index* (1948), 168. For Fabri's favorable view of Copernicus see Thorndike, *Hist.*, VII, 665–70.

83. Guitton, *Jésuites* (1954), 55–6. The Jesuits were never very enthusiastic about the *Index;* cf. Hilgers, *Index* (1904), 205.

84. Middleton, *Experimenters* (1971), 324; Boffito, *Bibl.*, 44 (1942), 176–84; cf. Middleton, *Br. J. Hist. Sci.*, 8 (1975), 144–6, 154, acknowledging that during the last third of the seventeenth century churchmen in Rome were free to study and write about the new physics.

85. Bofitto, *Bibl.*, 44 (1942), 176–84. Ceyssens, *Franz. St.*, 35 (1953), 401–11, does not support his claim that, after 1670, Fabri had trouble gaining official approbation for his books.

2. ACADEMICIANS

Most eighteenth-century natural philosophers managed to enter one or more of the learned academies whose rapid multiplication was a characteristic of the age. The proliferation of these institutions is suggested by Table 2.3. The more important societies are listed separately; all, including those lumped together as 'others,' published memoirs and / or offered prizes for papers on scientific subjects. The list could be extended indefinitely by admitting unproductive societies and intellectual drinking clubs. Cross-national comparisons on the basis of the table should be hedged, since, owing to the state of scholarship, coverage for France is better than that for other countries, especially the Germanies.

Leaving aside differences in quality, we may divide these organizations into two natural classes in two different ways. The first division opposes the special to the general: the scientific academies, like London, Paris, Bologna, Stockholm, Haarlem, to the all-purpose society of 'sciences, arts, et belles lettres,' such as Berlin, Petersburg, Brussels, Edinburgh, and most of the academies of France. Only the activities of the scientific classes of these general societies will concern us. The second, and more important division opposes open academies, with large, coopting, self-supporting memberships, to closed institutions, a few men supported in whole or part by princes who established their salaries, appointed their colleagues, and fixed their number.

CLOSED SOCIETIES

The model of closed institutions was the Académie Royale des Sciences, Paris, set up as a part of the bureaucracy of Louis XIV. In return for instruments, quarters, and salaries that ranged from 1500# to 2000# for ordinary members and 6000# to 9000# for foreign heavyweights like Huygens, the academicians were to advance science and to apply it as government consultants on technological problems, patent applications, and other technical matters. The Academy's procedures were first codified at its reorganization in 1699; after a few further refinements promulgated in 1716, its structure and functioning remained essentially unaltered until 1785.

According to the constitution of 1716, the Academy had 44 regular scientific members, distributed horizontally into six subject classes, three 'mathematical' (geometry, astronomy, mechanics) and three 'physical' (chemistry, anatomy, botany), and vertically into three levels, adjuncts, associates, and pensionaries. Two adjuncts, two associates and three pensionaries constituted a class; the perpetual secretary and treasurer, both pensionaries, completed the company. No one living outside Paris, or 'attached to any religious order,' was eligible for regular membership. There were also several categories of irregular members: occasional supernumerary regulars; twelve *honoraires,* usually unscientific; twelve *associés libres,* unconnected with any class and ineligible for promotion; six (later eight) foreigners, always distinguished scientists; and an indefinite number of 'correspondents,' provincials or foreigners given the right—

TABLE 2.3

Numbers of Academies Founded 1660–1800

	1660/1724	*1725/49*	*1750/74*	*1775/99*
Britain				
RS, London	1662			
RS, Edin.				1783
Manch. Lit. and Phil.				1781
France[a]				
AS, Paris	1660			
AS, Bordeaux	1712			
AS, Lyon	1724			
AS, Montpellier	1706			
AS, Toulouse		1729		
Others	2	6	5+12[b]	2
Germanies				
AS, Berlin	1700	[1744]		
SW, Göttingen			1751	
AW, Munich			1758	
Others[c]		1	4	2
Italy				
AS, Bologna	1714			
AS, Turin			[1757]	1783
Netherlands[d]				
HM, Haarlem			1752	
Teylers Stichting				1778
Others			2+ 1[b]	11
Elsewhere				
AS, St. Petersburg		1725		
AS, Stockholm		1739		

[a]Compiled primarily from Delandine, *Couronnes* (1787); purely literary societies, even those that toward the end of the century occasionally ventured into agriculture or commerce, have been discarded.
[b]'Sociétés d'agriculture.'
[c]Compiled from McClellam, *Int. Org.* (1975), 249–56, 499–502, an admittedly incomplete sample.
[d]Compiled from Rooseboom, *Bijdrage* (1950); Bierens de Haan, *Holl. Maat.* (1752), 35–6; Muntendam in Forbes, *Marum,* I (1969), 5–6.

and the honor—to communicate scientific news and the results of their own researches to a specified regular member of the Academy.[1]

The duties of the academician remained as before: to work at his science and to advise when consulted. To maintain the pace of work, the Academy met twice a week when not on one of its statutory vacations, which amounted to

1. Maindron, *Académie* (1888), 19, 23, 48.

fourteen weeks a year. At each meeting, according to the regulations of 1716, one pensionary and either an associate or an adjunct were to read an original paper; if enforced, the provision would have produced 152 papers a year.[2] A great many excellent papers were written, and published in the Academy's *Mémoires,* which began to appear regularly from 1699, or rather from 1702, when the volume for 1699 was issued. These *Mémoires* and their accompanying *Histoire* (chiefly summaries of the memoirs and stylized éloges of deceased members) set the eighteenth century its highest standards of scientific work and — through the éloges — its ideal of an honorable, dedicated, selfless career in science.

The Parisian academician needed dedication. In principle only the pensionaries of each class received salaries. After 1775 they had 3000#, 1800# and 1200#, in order of seniority; salaries had probably averaged the same (2000#) since 1716. Occasionally an associate might receive up to 500#, and very rarely, as in the case of d'Alembert, an adjunct might receive something too. In addition, one earned a small sum for attendance at meetings.[3] Already in 1716 the academicians complained that they could not live in Paris on 1500# a year; 3000# was probably nearer the minimum needed during most of the century.[4] Consequently only the senior pensionaries had anything approaching a sufficiency from their academic employment, and then usually only after a long wait; promotion generally followed seniority, although the King reserved the right to intervene and the company itself sometimes perpetrated an irregularity.[5]

Nor were provisions for research generous. The Academy's regular budget for general expenses appears to have been about 12,000#. Additional grants might be made for special projects, for example the 12,000# given Réaumur annually to do experiments and to prepare the *Descriptions des arts et métiers.* The fate of this grant is instructive. The government agreed that on Réaumur's death it would go to the Academy. When he and his authority died in 1751 the government punctually honored its commitment: Réaumur's 12,000# went to the Academy, and the Academy's 12,000# went back to the treasury.[6]

To work and to eat the Paris academician had to have a private income, a pension from outside the Academy, or a job. The prestige of his position and his access to influence usually brought him a post, if he needed one, associated with his specialty: teaching at the Collège Royal or at the state military, naval, or technical schools; editing the Academy's almanac, the *Connaissance des temps,* which brought 800# toward the end of the century; special consulting at the mint, mines, or the government porcelain works; drawing maps for the

2. Chapin, *Fr. Hist. St.,* 5 (1968), 385–93.

3. Maindron, *Académie* (1888), 100–1; Chapin, *Fr. Hist. St.,* 5 (1968), 384–6; Hankins, *D'Alembert* (1970), 43.

4. Bioche, *Rev. deux mondes,* 107:2 (1937), 181; R. Hahn, *Min.,* 13 (1975), 501–13. The total for salaries was 50,000# in the 1770s, of which 42,000# went to pensionaries (the secretary and treasurer each had 3000#) and 8000# for doles ('petites pensions').

5. E.g., in the election of Dufay to associate in 1724; cf. Chapin, *Fr. Hist. St.,* 5 (1968), 387, and Hankins, *D'Alembert* (1970), 137.

6. K. Baker, *RHS,* 20 (1967), 248.

Navy; writing for encyclopedias; inspecting industries; serving, semi-retired, in the military.[7] This moonlighting might make ends meet; it also reduced ouput and, perhaps, deflected some from an academic career. 'What kind of work can one expect of savants forced to spend their days on the pavements of Paris instead of in their studies? Is a man who comes home tired and distraught ready for work that demands his full powers? Will he spend his evenings doing experiments? . . . A gifted young man who wishes to follow his scientific bent will find himself opposed by family and friends who do not want to see him enmeshed in studies which, while they might bring him some glory, will certainly lead him to starvation.'[8]

The Academy of Berlin, established in 1700 by the King of Prussia, tried to follow the lead of Paris, including class divisions, salaries, corresponding members, practical applications and instruments furnished by the government. Its income, however, which derived from a monopoly on the sale of almanacs, could not initially maintain more than a president, a secretary, two astronomers and household help. Only a small fraction of the seventy regular members it engaged between 1700 and 1740, when Frederick the Great came to the throne, had had anything but token support; in 1721 only its anatomist received a pension.[9] Moreover the Academy had little international importance; its journal, rightly called *Miscellanea,* appeared at long intervals filled with inhomogeneous and often stale material. These blemishes caused the livelier academicians themselves to crave reform. In 1743, after joining with new spirits drawn to the capital by Frederick's promise, they set up a new, and newly Frenchified, academy.[10]

According to its revised statutes (1746), the Académie Royale de Berlin was a general learned society, consisting of four classes (experimental philosophy, mathematics, speculative philosophy, literature) each made up of three pensionaries and three associates. In addition, there were a top-heavy administration (a president, a secretary, a director for each class), correspondents, and emeriti or 'veterans.' Financial support came from the almanac income, which had grown considerably with the acquisition of Silesia, and which Frederick allowed to reach the Academy intact, a policy his father had not often followed.[11]

Frederick wished his academy to have the best brains in Europe. Shortly after his accession he tried for the most advertised brain in Germany, Wolff's. Wolff preferred to remain a professor and to return in triumph to Halle. As head of his

7. R. Hahn, *Min.,* 13 (1975), 501–13; cf., regarding Clairaut's income, Doublet, *Bull. sci. math.,* 38:1 (1914), 95, 189–90.

8. Maindron, *Académie* (1888), 106, quoting an anonymous Ms. of c. 1720, apparently the work of leading academicians.

9. Harnack, *Gesch.* (1900), I:1, 75, 158–9, 230–1, 240–4; Dunken, *Deut. Ak.* (1960). In 1737 the Academy's mathematical class had six members and its medical-physical class eight, mostly physicians. Hagelgans, *Orbis* (1737), [Pt. ii], 9.

10. Harnack, *Gesch.* (1900), I:1, 248–92.

11. *Ibid.*, 230–1, 299–302; Gottsched, *Hist. Lobs.* (1755), 108–9.

mathematical class Frederick captured Euler from the Petersburg academy. It was as well that Wolff did not come, for Euler detested the monad, which he took in a physical sense and held to violate both mathematics and the Christian faith.[12] The other chief scientific members of the new academy were Newtonians, an orientation urged by Voltaire and Algarotti, who happened to be in Berlin in the early 1740s. For president Frederick recruited Maupertuis, who had the double and unusual advantage of being both attractionist and French. For director of the physics class he chose his physician, J. T. Eller, who had studied in Leyden and even in London. He was a 'passionate Newtonian,' according to Wolff, 'who hotly rejects everything that does not agree with his pitiful principles, which he incorrectly thinks are Newton's.' Frederick had also hoped to have a Dutch Newtonian, but none came.[13]

Although Wolff himself had declined to fight in Berlin, his epigoni J. G. Sulzer and Samuel Formey, members of the Academy's non-scientific classes, were prepared to do battle. The first rounds went to the anti-Wolffians. But after the deaths of Maupertuis (1759) and Eller (1760), the Wolffians gained ground, supported by extreme representatives of the French Enlightenment, whom Frederick delighted to honor. This odd alliance was celebrated in 1766 in a speech on the 'Reconciliation of the Philosophies of Leibniz and Newton' given in the Academy on the occasion of the King's birthday. Later that year Euler sought refuge in Petersburg from rampant rationalism and what he took to be Frederick's disfavor. In the twenty years of his directorship the Berlin Academy had come to rival that of Paris. The old *Miscellanea* were replaced by regular *Mémoires,* printed in French; as Maupertuis explained, his native tongue was perfect, and also rich in words for describing the advances of science.[14] To be entirely Parisian the *Mémoires* wanted only an *Histoire,* which, according to the Academy's secretary Formey, Maupertuis refused to supply out of antagonism to the inventor of the genre, Fontenelle.[15] This imperfection was corrected in 1770.

Salaries at Berlin ran a little higher than at Paris. The average seems to have been about the same, namely 500 RT; but the better emoluments, like the 1600 RT (some 6400#) paid Euler or the 12,000# given Maupertuis, greatly exceeded even the best Paris pensions.[16] Moreover Berlin was cheaper than Paris.[17] It also appears that Frederick wished to raise the average salary of his

12. Winter, *Registres* (1957), 16, 32, 37–8.

13. Brunet, *Maupertuis* (1929), I, 77–8; Stieda, Ak. Wiss., Leipzig, Phil.-Hist. Kl., *Abh.*, 83:3 (1931), 7, 13; Wolff to Manteuffel, 29 Aug. 1747, in Ostertag, *Phil. Geh.* (1910), 117, and Manteuffel's answer, *ibid.*, 118. For Eller, *DSB*, IV, 352–3; *MAS*/Ber (1761), 498–510.

14. Winter, *Registres* (1957), 45–8, 52–6, 68, 84, 89, 91; Brunet, *Maupertuis* (1929), I, 127; Stieda, Ak. Wiss., Leipzig, Phil.-Hist. Kl., *Abh.*, 83:3 (1931), 30–7; Calinger, *Ann. Sci.*, 24 (1968), 239–49.

15. Formey, *Souvenirs* (1797^2), II, 254.

16. Harnack, *Gesch.* (1900), II:1, 487–91; Bartholmess, *Hist.*, I (1850), 179; Winter, *Registres* (1957), 14, 17, 19; Clairaut to Cramer, 27 April 1744, in Speziali, *RHS*, 8 (1955), 221.

17. Von Freyberg, *Lehmann* (1955), 52.

academicians, for in trying to woo Mayer from Göttingen in 1758 Euler offered 550 RT (later 700 RT) and the assurance of a raise at the next vacancy, 'for the king intends to reduce the number of members and contrariwise to increase their salaries.'[18] As at Paris, at Berlin academicians had access to further technical jobs, teaching or advising; an unusually successful example is F. K. Achard, once the Academy's leading electrician and director of its physics class, who received an additional pension of 500 RT for assistance to the tobacco industry.[19]

The third major European foundation on the Paris model was the Academy of Sciences of St. Petersburg, consisting of three classes (mathematical, physical, rhetorical) of salaried members, whose number fluctuated according to supply and the policy of the regime. The Academy was the last project of Peter the Great, who conceived that Russia might reach the eighteenth century within a generation or two if the natives could be trained, and the government advised, by Western experts. His academicians were to advance and popularize their sciences, train talented Russians in an 'academic university,' and direct the translation of Western texts to reach those enticed by the popularizations. At the head of this pedagogical pyramid was to sit the inevitable Wolff, with a salary of 2000 rubles (2500 RT). But in 1724, after protracted negotiations, Wolff rejected the offer and stayed in Marburg, where he had just found refuge from the theologians of Halle.

The following year Peter died, and although his successors maintained his academy, they did not implement his general instructional scheme.[20] This failure by no means inconvenienced the academicians. They could do their own work, as long as their relatively high salaries were paid (from 600 to 1000 rubles [750 to 1250 RT] or more[21]). And, if they could endure the weather, the constant changes of government, the dictatorship of the bureaucrats placed over them, the jealousy of the Russian members, and their own bickering, they might find St. Petersburg very 'agreeable.' Or so Euler said in transmitting an offer of 1000 rubles plus moving expenses to Mayer, adding that no one had yet complained of the life and forgetting the reasons that had prompted him to accept a call to Berlin.[22]

Many of the first academicians were Germans or Swiss Germans recruited by

18. Euler to Mayer, 15 May 1753, in Forbes and Kopelevich, *Ist.-astr. issl.*, 10 (1969), 404. In fact, however, salaries were rarely increased, 'however good or bad they are'; Lagrange, *Oeuvres*, XIII, 258.

19. Harnack, *Gesch.* (1900), I:2, 481, 512; Stieda, Ak. Wiss., Leipzig, Phil.-Hist. Kl., *Abh.* 39:3 (1928), 16.

20. Kunik in Wolff, *Briefe* (1860), xiv–xv, xxii–xxv, xxix–xxxiv; Blumentrost to Wolff, 23 May and 27 Dec. 1723, *ibid.*, 167, 171.

21. *Ibid.*, 167, offering between 700 and 800 roubles; cf. *ibid.*, 177–82, and Lipski, *Isis*, 44 (1953), 349–54. The academy's income derived from customs dues and a monopoly on calendars; Kunik in Wolff, *Briefe* (1860), xxix; Vucinich, *Science* (1963), I, 71, 88, 96.

22. Euler to Mayer, 11 June 1754, in Kopelevich, *Ist.-astr. issl.*, 5 (1959), 391; Vucinich, *Science* (1963), I, 72–82.

Wolff, whom the Russians retained at 300 RT a year to suggest candidates and negotiate contracts.[23] Not unnaturally these pioneers inclined toward Wolffian philosophy. The contract of Christian Martini, professor of natural philosophy, bound him to 'expound physics according to Wolff's principles.' His successor, G. B. Bilfinger, 'who [as Lambert put it] in many respects rendered greater service to Wolff's philosophy than did Wolff himself,' had already infected the University of Tübingen with his master's teachings. Euler and Daniel Bernouilli, although they despised Wolff's metaphysics, did not reject his physics. They would have agreed with the spirit of Bilfinger's effort to save the theory of vortices, for which he won a prize from the Paris academicians in 1728: 'Nothing is simpler than the Cartesian vortices; hence I think everything should be tried before they are given up; and if they will not work properly, I should wish them to be changed as little as possible.'[24]

Mechanistic physics had a prosperous career in Russia. Euler dominated mathematics and physics in the Academy until 1741, and continued to influence their development from Berlin. His colleagues Krafft and Richmann shared his views, as did the first important Russian member of the Academy, Lomonosov, who had been trained by Wolff at Marburg. The Academy's first Newtonian physicist, Aepinus, arrived in 1757. Aepinus had moved into the government bureaucracy by the time Euler returned in 1766 to reinvigorate the tradition of Cartesian physics, which, in the easily digested form of his *Letters to a German Princess,* enjoyed a vogue in Russia in the late 1760s and 1770s.[25]

The Kings of France and Prussia, and the Czar of Russia, could afford to spend upwards of 15,000 RT a year on their academies. Lesser potentates contented themselves with lesser academies and fewer academicians. George II of England, as Elector of Hanover, refused to spend more than 450 RT on his Societät der Wissenschaften in Göttingen; it opened in 1751 with five members whose academic salaries summed to 400 RT. They came cheaply because they were professors, and hence already on the royal payroll. The Society improved its income and science by publishing an almanac, a few volumes of memoirs and an important literary review, the *Göttingische Zeitungen* (later the *Göttingische Gelehrte Anzeigen*), which played a role in the history of electricity. But learned publications are not a dependable source of revenue, and the Society remained financially, intellectually, and, by the coopting of professors, personally tied to the university.[26]

23. Blumentrost to Wolff, 19 March 1725, Wolff, *Briefe* (1860), 185; cf. *ibid.*, 43–6.

24. Boss, *Newton* (1972), 105, 110n; Winter, *Registres* (1957), 8–9. For the bitter squabble between Bilfinger and Bernoulli see Ak. nauk, *Materialy*, I, 501–67; Lambert, *Neues Organon*, §632, quoted by Wahl, *Zs. Phil. phil. Kritik,* 85 (1884), 68.

25. Boss, *Newton* (1972), 139, 146–51, 162, 169, 211, 215; *infra,* xvi.3. The *Letters* were written in 1760–1 but not published until 1768. On the negotiations for Euler's return to Petersburg (final settlement: his salary, 3000 roubles; his son's, 1000; and moving expenses for 14 people), see Stieda, Ak. Wiss., Leipzig, Phil.-Hist. Kl., *Abh.*, 83:3 (1931), 26–35.

26. Joachim, *Anfänge* (1936), 29, 36–7, 68–9; Ak. Gott., Wiss., *Sekul.*, 40–53. The membership increased to a maximum of about 15 in the 1770s.

A similar arrangement had existed since 1714 in Bologna. The Accademia delle Scienze dell'Istituto di Bologna had an endowment, library and instruments from a wealthy and well-travelled townsman, Luigi Ferdinando Marsigli, and a building and operating expenses from the city, with the approval of the Pope. Professors were appointed to the Institute to teach and to study subjects outside the university curriculum. Although it was not, as Fontenelle fancied, 'Chancellor Bacon's Atlantis in actuality, and the dream of a savant come true,' the Bologna Academy, like that at Göttingen, was a valuable adjunct to its university.[27]

Among the smaller proprietary academies not tied to universities those of Turin and Munich deserve notice. In 1783 Victor Amadeurs III, King of Sardinia, granted a royal charter to a group that had existed in Turin since 1757; it thereby became the Académie Royale des Sciences, at a cost to the king of a building and a 'generous allotment' for its upkeep, for instruments, prizes and medals.[28] The original group, although small, had 'provide[d] models in everything,' according to Lambert, and not least in electricity. The larger Academy strove to retain its quality by admitting only those who had already acquired a reputation by their published works.[29]

The Akademie der Wissenschaften of Munich, established in 1758, received its charter, meeting place, instruments and operating expenses from the Elector of Bavaria and from the usual source of academic funds, the calendar monopoly. In the 1760s it had an income of over 8000 guldens (21,000#) annually, of which about 3000 went to pay its secretary, the directors of its two classes (historical, philosophical), and a few 'professors,' whose emoluments ranged between 600 and 800 guldens (1600 and 2100#) plus lodging, fuel and light. Two-thirds of the members elected between 1759 and 1769 were bureaucrats or priests, in equal measure, and consequently already salaried by state or church. Lay professors made up less than a tenth of the company.[30] In contrast to the Protestant Göttingen Society, the Catholic Munich Academy was set up in opposition to the local university, run by the Jesuits at Ingolstadt. One of the Academy's founders, who wished to make it a 'lodge of Wolffians,' urged the exclusion of Jesuits, 'because they are scholastics and Jesuits,' and supported the appointment of Lambert, a Protestant, to a paid professorship. 'What has orthodoxy to do with mathematics, physics, chronology and calendar making?'[31] Several resourceful experimental physicists were to hold the position

27. Bortolotti, *Storia* (1947), 149–53.

28. Anon., Acc. Sci., Turin, *Mém.*, 1 (1784–5), ii–iii, xiii, xix, xxxii–xxxiii.

29. *Ibid.*, xix; Lambert to Euler, in Bopp, Ak. Wiss., Berlin, Phys.-Math. Kl., *Abh.* (1924:2), 29: 'cette société donne en tout des modèles.'

30. Doeberl, *Entwicklungsgeschichte,* II (1928), 321–2; Hammermayer, *Gründ.-Frühgesch.* (1959), 106, 164, 193, 298, 368.

31. *Ibid.*, 239–40, 248; Lambert thought the pension of 600 guldens (1600#), for which he obliged himself to some administrative work and the provision of three original memoirs a year, sufficient. Lambert to Kästner, 24 March 1761, in Bopp, Ak. Wiss., Heidelberg, *Sitzb.* (1928:18), 14–15.

briefly occupied by Lambert, and one or two made contributions to the study of electricity.[32]

OPEN SOCIETIES

The home of the Newtonians, the Royal Society of London, had obtained its charter, a silver mace, and a property worth £1300 from Charles II in the 1660s; its operating expenses at first came from the pockets of its Fellows, who agreed to pay two pounds for admission and a shilling a week thereafter.[33] New members were first elected upon the proposal of a Fellow, and two-thirds of the votes of those present; the recruit entered on the same footing as the other regular members, there being none of those 'disagreeable distinctions' (as Voltaire called them) that ordered the Paris Academy. 'The Royal Society of London lacks two things most necessary to mankind,' he says, 'rewards [pensions] and rules. It is worth a small fortune in Paris for a geometer or a chemist to be a member of the Academy; it costs one in London to be a Fellow of the Royal Society.'[34] The Fellowship was free and poor, and long poorer than it needed to be; its unpaid dues mounted steadily, to almost £2000 in 1673. Newton, who had been excused his dues, moved energetically against malingerers on assuming the Society's presidency in 1703.[35] Further tough measures and gifts of land and stock secured the Society's finances without raising the dues. After 1752 the admission fee was five guineas, the increase owing to the Society's assumption of financial responsibility for the *Philosophical Transactions* in that year. Most new Fellows preferred a composition fee or life membership, which rose from 20 guineas to 26 after 1752.[36] Although by no means a small sum, it probably worked no hardship on those otherwise eligible for election.

At first recruitment proceeded apace, and the Society numbered 199 in 1671. It declined to about 150 in 1700, and picked up again with Newton's presidency. The annual recruitment almost doubled, from about nine a year in 1700 to fifteen anually between 1701 and 1720; for the rest of the century it averaged 23. The membership reached 303 in 1741, 545 in 1800. In the same period its income rose from £232 to £1652.[37] More members meant more money; for a time admission came very easily, particularly for foreign candidates. Voltaire had the idea that anyone who declared his love of science, and deposited his fee, was immediately received a member, and d'Alembert is said to have boasted that he could arrange the election of any traveller bound for England,

32. Ildefons Kennedy (Benedictine), F. X. Epp (ex-Jesuit), M. Imhof (Augustinian), Coelestin Steiglehner (Benedictine). Cf. Westenrieder, *Gesch.*, II (1807), 110–11, 390, 415.

33. RS, *Record* (1940⁴), 9, 11, 24, 93; Weld, *Hist.* (1848), I, 100, 145.

34. RS, *Record* (1940⁴), 10–11; Voltaire, *Lettres* [1734] (1964), II, 170–1.

35. RS, *Record* (1940⁴), 44–5; Weld, *Hist.* (1848), I, 231, 250; Hunter, RS, *Not. Rec.* 31(1976), 17–18, 24–5, 49–57.

36. RS, *Record* (1940⁴), 36, 94–5; Weld, *Hist.* (1848), I, 181–2, 523, II, 43.

37. RS, *Record* (1940⁴), 49; Weld, *Hist.* (1848), I, 473, II, 51, 227–8; Hunter, RS, *Not. Rec.*, 31(1976), 26–32.

'should he think it an honor.'[38] The society countered such censure by requiring each nomination to be put forward, in writing, by at least three Fellows, and by admitting no more than two foreigners a year, 'till [their] number be reduced to eighty.'[39]

Since the ordinary income of the Society seldom exceeded £1000 before 1790, and in its first years did not reach half that, it could not afford to engage much help. Its earliest paid staff, exclusive of household help, consisted of the principal Secretary and a Curator. The former, Oldenburg, was to have £40 a year, the latter, Hooke, £30 and an apartment in the Society's rooms. But it appears that Oldenburg received no regular salary until 1669, and that Hooke agreed in 1662 to 'furnish the Society every day they meet [once a week], with three or four considerable experiments, expecting no recompense until the Society get a stock enabling them to give it.'[40] That happened in 1664. Thereafter special curators of experiments were engaged from time to time, the last of whom, J. T. Desaguliers, held the post from 1714 to 1743. His pay varied from about £10 to £40, depending upon the number of experiments he furnished. After Desaguliers' time the post lapsed, as the Society's statutes of 1775 explain, because the Fellows themselves had become 'so well acquainted with the mode of making experiments, that such accomplished curators have not been found necessary.'[41] Meanwhile the burden of the Secretary had been divided, and a clerk-librarian engaged. In 1780 the two chief secretaries, both members of the Society, received £70.5, a little more than junior pensionaries of the Paris Academy, and the Clerk, a full-time employee ineligible simultaneously to be a Fellow, had £220. These emoluments were raised to £105 and £280 respectively, in 1800, in acknowledgment of the inflation at the end of the century.[42]

Salaries and household expenses left little for research. The Society gradually built up a good library and collection of curiosities, mainly by donation, but it could not treat members to the services of mechanics or instrument makers or commission special apparatus. Very rarely it paid for a research project, as when it gave a man a guinea for permission to transfuse twelve ounces of sheep blood into him. (Why sheep's blood? Because, according to the transfusee, who survived the trial, 'the blood of a lamb has a certain symbolic relation to the blood of Chirst, since Christ is the agnus Dei.'[43]) A worthier project was measuring the gravity of a mountain, a work of Newtonian piety that cost the

38. Voltaire, *Lettres* (1964), II, 171; d'Alembert as quoted in Weld, *Hist.* (1848), II, 152.

39. These measures were introduced in 1728–30 and 1761, respectively. RS, *Record* (1940[4]), 49–51; Weld., *Hist.* (1848), I, 459–61.

40. *Ibid.*, 137–8, 173, 204, 360; RS, *Record* (1940[4]), 11. According to Magalotti, who attended meetings in the 1660s, Hooke performed experiments chosen by the Secretary from among those suggested by the Society. Crino, *Fatti* (1957), 159.

41. RS, *Record* (1940[4]), 30; Weld, *Hist.* (1848), I, 286–7, II, 87–8.

42. *Ibid.*, I, 302–3, II, 228–9. Halley resigned his fellowship in 1686 to become the Society's first Clerk, at a salary of £50. F. Hauksbee, Jr. received about the same (£47.35) in 1763. Henderson, *Ferguson* (1867), 273–4.

43. Weld, *Hist.* (1848), I, 220–1.

Society almost £600 in the early 1770s. More consistent with its means was appeal to the government for grants for meritorious projects, such as expeditions to observe the transits of Venus in the 1760s. The budget of one observer, the same man who measured the attraction of the mountain, Neville Maskelyne, may be of interest: estimated expenses for eighteen months came to £290, of which £140 were needed for 'liquors.'[44]

Since the Fellows undertook almost all their researches at their own charge and initiative, and since their scientific attainments varied greatly, their activities ranged from inexpensive observations of two-headed cows to elaborate calculations in the style of Newton. The *Philosophical Transactions* were of unequal quality. 'It is not astonishing that the memoirs of our Academy are superior to theirs,' says Voltaire; 'well disciplined and well trained soldiers must in the end overpower volunteers.'[45] But volunteers, who outnumber regulars ten or twenty to one, can afford frequent misfires, provided some among them are clever or lucky enough occasionally to hit the mark. Even during the presidency of Martin Folkes, when, according to the Society's nineteenth-century historian, the *Philosophical Transactions* had more than their usual quantity of puerile and trifling papers,[46] they also carried reports of important and original experiments, particularly on electricity.

The Royal Society had a few domestic descendants late in the eighteenth century, notably its namesake at Edinburgh (1783) and the Manchester Literary and Philosophical Society (1781), the latter organized as a corrective to indolence: 'Science, like fire, is put in motion by collision,' its founders said, confidently looking forward to productive bumps among its members.[47] Both societies supported themselves by dues, a guinea per member per year; neither could afford even part-time support for a research or teaching position. The wealthiest and, for our subject, the most important of the learned academies inspired by English example was created not in Britain, but in Sweden, as a result of the pushing of, among others, Mårtin Triewald, who had lived for many years in England, knew the leading Newtonians there, and was himself a member of the Royal Society.[48]

True to its model, the Swedish Academy of Sciences at Stockholm at first received nothing from the crown but the right to call itself royal. It had its operating expenses from its members, a third of whom were aristocrats or high civil servants, and a fifth professors. Their contributions, from a ducat (about 0.5 guinea) to 300 Dkmt (100#) could scarcely have caused them hardship.[49] Help soon came, however: bequests that amounted to well over 70,000# in

44. *Ibid.*, II, 79–82, 14–15.

45. Voltaire, *Lettres* (1964), II, 171–2.

46. See Weld, *Hist.* (1848), I, 483–5.

47. Manch. Lit. Phil. Soc., *Mem.*, 1 (1781–3), vi; RS, Edinb., *Trans.*, 1 (1788), 1–15. For the elaborate bumping behind the RSE see Shapin, *Br. J. Hist. Sci.*, 7 (1974), 1–41; for bibliography of provincial English academies, Schofield, *Hist. Sci.*, 2 (1963), 76–7.

48. Lindroth, *Hist.* (1967), I, 2–4, 14; B. Hildebrand, *Förhistoria* (1939), 140–2, 257, 281–2.

49. Lindroth, *Hist.* (1967), I, 28–32, 102–3; the membership reached 64 in 1742, and then rose to, and remained about, 100 (*ibid.*, 12, 15).

35 years, and, most important, control and profit of the almanac business, acquired in 1747. By 1765 the almanacs brought between 15,000# and 20,000#; in the sixties and seventies the total income of the Stockholm Academy—including income on its endowments, often lent to members at 6 percent—was about half that of the Paris Academy.[50] Most of this money went to support research or travel. Only a little over a fourth of the total went to salaries. The most important office of the Society, the secretaryship, worth only 1800 Dkmt (600#) in the 1740s, brought a good salary, 3000#, in 1771. The emolument for the only other significant paid post, a lectureship (later professorship) in experimental physics, also rose from below subsistence level (2000 Dkmt when founded in 1759) to adequate (6000 Dkmt or about 2000# in 1776).[51]

The lecturer in physics had to show experiments to general audiences and to report his researches for publication in the Academy's *Handlingar* or *Transactions*. Accordingly the Academy made available substantial funds for the purchase of instruments. The first full-time lecturer, Johan Carl Wilcke, who held the post for almost forty years, made capital contributions to the study of electricity. After accepting his position, he published his work, as anticipated, in the *Handlingar*. Fortunately their value was so great, particularly for technology and physics, that it proved financially profitable to translate them into German in their entirety. The indefatigable Kästner saw the opportunity and seized it: he taught himself Swedish and issued 53 volumes of faithful translations between 1740 and 1790.[52] These translations, a vigorous correspondence, a capable membership and an excellent income made the Stockholm Academy one of the leading learned societies of the Enlightenment.

A less distinguished form of the Academy of the London type bred promiscuously in the provinces of France. To be sure these societies often aped the Paris Academy. They divided themselves into classes (mathematical, physical, historical, literary) and orders (regulars, associates, correspondents, honoraries), fixed the number of regulars (from eight, as at Agen, to forty, as in the final specifications for Lyon and Bordeaux), and set up requirements of residence and attendance.[53] But they followed the Royal Society of London in most important respects: meagre resources; miscellaneous memberships, drawn mainly from nobles, lawyers, and high clerics;[54] and the freedom to study what they pleased, so far as their poverty and competence allowed. Usually they

50. *Ibid.*, I, 104–9, 147, 150, 153.

51. *Ibid.*, 45–6, 51–2, 462–70.

52. *Ibid.*, 185–98, 208. On the practical orientation of the *Handlingar*, *ibid.*, 114–15.

53. Delandine, *Couronnes* (1787), *passim;* J. Bernoulli, *Letters*, II (1777), 131–2. The most Parisian in organization was Montpellier, which enjoyed the privilege of submitting one memoir a year for publication by the Paris Academy. Dulieu, *RHS*, 11 (1958), 232–3.

54. A good example is the general academy (science, literature, technology) of Besançon, established in 1752, with 40 members, almost all 'gens d'Eglise, de robe et d'épée'; the 'scientific' members were a physician, an engineer, and a surgeon (Cousin, *RHS*, 12 [1959], 327). Cf. the legal-financial domination at Pau (Desplat, *Milieu* [1971], 41), Bordeaux, Dijon, and Châlons-sur-Marne (Roche, *Livre* [1965], 176).

chose to study the natural and human history of their region, its curiosities, agriculture, commerce and industry. Often enough they did nothing at all. 'They make many promises, but their ardor soon cools; few are willing to force themselves to compose thorough and thoughtful works.'[55] The pretentions, incompetence, and jealousies of provincial academicians often made them ridiculous. 'Ci git qui ne fut rien pas même academicien,' reads a gravestone in Dijon.[56] Members of the Paris Academy, particularly Condorcet, tried unsuccessfully to improve provincial performance by encouraging the association of smaller societies with larger ones,[57] such as Bordeaux, Lyon, Montpellier or Toulouse, which contributed their mites to science, and provided a few opportunities for scientific careers.

The history of the Bordeaux Academy, almost the earliest and certainly the most active of these societies, will illustrate their vicissitudes and opportunities. Founded in 1712 primarily by the parliamentary aristocracy under the protection of a local magnate, the Duc de la Force, the Academy at first struggled to meet its housekeeping expenses. Most of its income came from dues, which were set so high (300#) that few paid them regularly; they were dropped altogether towards 1732 when income from 60,000# given by de la Force became available. Other wealthy citizens made gifts or bequests: the Academy soon had its own buildings, a library, a natural-history collection, a cabinet de physique, and an average income, from 1739 to 1771, of about 3000# a year. It began to overextend itself. In the 1740s it bought 4500# of physical instruments; it hired a curator and a librarian, and sponsored public lectures; by 1761 it could no longer meet its debts or pay its employees. Further gifts and better administration renewed its income, which averaged 7000# a year from 1772 to 1780.[58] The same financial difficulties, and, though less often, the same accomplishments, are met with elsewhere.[59]

Bordeaux was an academy of 'sciences, literature and arts.' At first it emphasized physics, and, in the 1740s, experimental physics. After the mid-century it moved towards technology, agriculture and commerce. These subjects then were fashionable; a dozen provincial agricultural societies were created in the 1760s. Even academies that had been entirely literary, such as Arras, Caen, Grenoble, Nîmes and Pau, began to cultivate agronomy.[60] 'In our

55. Sequier (Nîmes) to Condorcet, 1774; quoted by Barrière, *Académie* (1951), 349.

56. 'Here lies one who was nothing, not even an academician'; the gravestone is exhibited in the Musée des Beaux Arts, Dijon. Cf. Roche, *Livre* (1965), 105.

57. Baker, *RHS*, 20 (1967), 258, 262, 266–77. Cf. Priestley, *Hist.* (1775^3), I, xviii–xx, recommending progress by subdivision of labor, by funnelling money from large academies to small.

58. Barrière, *Académie* (1951), 20, 25–7, 31–3, 39, 97–8.

59. For example Montpellier, too poor to publish its memoirs regularly, nonetheless supported a lecturer in physics from 1780, with the help of the Estates of Languedoc (Dulieu, *RHS*, 11 [1958], 234–7). There was a similar position at La Rochelle from 1785 (Torlais, *RHS*, 12 [1959], 111–25) and one in chemistry at Dijon from 1783 (Delandine, *Couronnes* [1787], I, 259–63).

60. Barrière, *Académie* (1951), 350–1, 354–5; Delandine *Couronnes* (1787), I, 190–1, 244–5, 276, II, 54–7, 64–9. Cf. Roche, *Livre* (1965), 163–8.

time,' wrote a correspondent of Voltaire's 'all the women [he might have said provincial academicians] had their beau esprit, then their geometer, then their abbé Nollet; nowadays [1760s] it is said that they all have their statesman, their politician, their duc de Sully.'[61] It was in the first half of the century that the provincial academies contributed to the study of electricity. Thereafter they occasionally helped indirectly, as the Montpellier Academy encouraged Coulomb, but the subject no longer answered the interests, as it eluded the competence, of most provincial academicians.[62]

The establishment of scientific societies came late to Holland, among other reasons because the universities and independent lecturers satisfied much of the interest in experimental philosophy. In 1756, however, domestic pride and foreign example encouraged the formation of the Hollandsche Maatschappij der Wetenschappen at Haarlem. The organization of this body, unique at the time, anticipated that of the Kaiser Wilhelm Gesellschaft in our century: funds were supplied by 'directors,' who paid an admission fee of 60 florins (130#) and dues of *f.*50 (later *f.*100); the work was done by 'scientific members,' who had access to the society's books, instruments and collections, but no salary. It began with 23 members, all local, including several professors from Leyden; foreign members were elected beginning in 1758 and, as we have seen, so promiscuously as to bring discredit on the Academy, and measures to decrease their number.[63]

To minister to its collections and correspondence, the Academy employed a secretary at *f.*700 (from 1777 *f.*1000) and, from 1777, a curator at *f.*300 and fringe benefits that included housing among the specimens. This curator, who held the job throughout the century, was Martinus van Marum, the leading Dutch electrician of his time.[64] His work on electricity was supported by Teyler's Tweede Genootschap, a small general-purpose learned society set up at Haarlem under the will of a Peter Teyler in 1778. As a member of this society and, from 1784, the director of its library and collections, van Marum had *f.*1500 a year and the confidence of its officers, who supplied the money for his great electrical machine.[65]

ACADEMIC FUNCTIONS

The purpose of the academies was the advancement of science and technology. The Royal Society's charter of 1663 specifies its goal as 'promoting Naturall Knowledge,' which the members took to mean useful information (as opposed to scholastic speculation) about the physical world (as opposed to the super-

61. Pierres de Bernis to Voltaire, 26 July 1762, in Voltaire, *Corresp.*, XLIX, 139–40. Nollet stands for experimental physics; Sully for finance, commerce, transportation.

62. Cf. the prize questions, *infra,* ii.2.

63. Bierens de Haan, *Holl. Maat.* (1752), 3, 8–11, 273–4; *supra,* i.7.

64. Bierens de Haan, *Holl. Maat.* (1752), 43–4; Muntendam in Forbes, *Marum,* I, (1971), 17–18, 41. Van Marum became secretary of the Society in 1794 while retaining the curatorship; the Society then could not afford to pay its secretary anything (*ibid.*, 33–4).

65. *Ibid.*, 5, 20–1.

natural).[66] The same concept recurs in the names of many later foundations: the Hollandsche Maatschappij tot Voortsetting en Aanmoediging van Nuttigge Konsten en Wetenschappen, the American Philosophical Society for Promoting Useful Knowledge, the Kungl. Svenska Vetenskapsakademie [for improving] Vetenskaper och Konster 'som tiena til en almän nytta.'[67] The program of the New Atlantis, of perfecting the arts by perfecting the sciences, of simultaneously seeking fruit and light, or the useful and the true, also informed the plans made by Leibniz for the Berlin Academy, and the hopes of the provincial academicians of France.

The balance between the two parts of the program was struck in different ways by different institutions. Stockholm excepted, the more prestigious the society the closer it stood to pure science; in 1798 Frederick William II of Prussia reprimanded his academicians for having moved too far into 'speculative investigations' at the expense of 'works of general utility.'[68] There is no doubt that the academies' implied promise to improve man's estate, as well as the services they rendered as technological consultants and patent officers, gave rationale for much of their support. To us, however, the chief feature of the academies is not their promise to be useful but their explicit dedication to the 'cultivation,' 'advancement,' or 'promotion' of their sciences, or, in a word, to research. 'Ein Academiste muss erfinden und verbessern oder seine Blösse unvermeidlich verrahten.'[69]

It was commitment to research that distinguished the academician. Not until the end of the eighteenth century, and then only in the leading universities, did the cultivation of science—as opposed to its preservation and dissemination—begin to be a responsibility of the professoriate. In theory the Academy complemented the University; the one taught the known, the other explored the unknown. The putative duties of the Petersburg academicians, to 'cultivate their sciences and give a course of lectures once a year,' implied a new sort of institution, 'not a complete university, not an academy of sciences, but rather a combination of both.'[70] This oddity exacerbated the problems of recruitment: should one seek able and ambitious men, eager to join an academy where they might win reputations as savants, or should one settle for a more common and docile type, 'those who aspire only to be professors?'[71] As it happened the brilliance of the first recruits and the failure of the 'academic university' reduced the planned hybrid institution to a first-class academy.

The best example of the complementary character of academy and university

66. Weld, *Hist.* (1848), I, 126, 138.

67. The elucidation of the purpose of the Stockholm Academy comes from the introduction to the first number of its *Handlingar;* B. Hildebrand, *Förhistoria* (1939), 374.

68. Quoted by Westenrieder, *Gesch.*, II (1807), Vor.

69. Haller (1751), quoted by Joachim, *Anfänge* (1936), 52.

70. Blumentrost to Wolff, 23 May 1723 and Feb. 1724, in Wolff, *Briefe* (1860), 167, 173. Cf. *Encyclopédie,* art. 'Académie': 'Une académie n'est point destinée à enseigner ou professer aucun art, quel qu'il soit, mais à en procurer la perfection.'

71. Memo by A. Golovkin, 1724, in Wolff, *Briefe* (1860), 181.

125548

is Göttingen. The founders of the university wished to create it and its research arm, the Societät der Wissenschaften, together. The Society would provide a supplementary salary to the ablest professors to enable them to reduce their large teaching loads; on their 'released time,' as we might say, they were to meet once a week, 'improve and elucidate' their sciences, and produce an annual volume of memoirs.[72] The regime decided to create the university first. The Society followed fifteen years later, with the modest numbers and stipends already mentioned. Its ambitions were anything but modest: 'to increase the realm of knowledge with new and important discoveries, encourage professors both to write solid works and to apply themselves to their lectures, and to spur students to praiseworthy zeal for science and good morals.'[73] If we disregard Bologna, the arrangements at Göttingen were unique in Europe, and proved useful in argument against those who objected when 'learned university professors or those capable of making discoveries become academicians.'[74]

As this argument suggests, universities did not always welcome the establishment of academies in their neighborhoods. The academic spirit of free enquiry opposed the professoriate's commitment to established learning. The conflict was most serious in Southern Germany and Austria; the Jesuits there helped to defeat plans for establishing an academy in Vienna in 1749–50, and the one in Munich was set up against their protests and, as we have seen, to their exclusion.[75] Another ground for professorial opposition to academies was jealousy, expressed as a fear that the new institutions would diminish the lustre of the old. An example is the campaign mounted by the Senate of the University of Leyden against the Hollandsche Maatschappij. 'The lustre of the University [they said] does not derive entirely from the merit and importance of its professors, but also from its authority [which would be greatly impaired] if a second society of letters were established in the province.' Look at France and England. Oxford and Cambridge have declined since the establishment of the Royal Society, and the University of Paris, formerly so famous, 'has scarcely been heard from since the Royal Academy has been made to flourish there under the particular protection of the King.' This interested argument could not arrest the progress of the Maatschappij, but it did result in introducing two restrictions in its charter: the new society should not sponsor public lectures and —to hide its light as much as possible—its publications were to be entirely in Dutch.[76]

72. Joachim, *Anfänge* (1936), 5–8; Smend, Ak. Wiss., Gött, *Festschrift* (1951), vi.

73. *Ibid.*, 15–16.

74. F. A. Wolff, 'Berliner Universitätsdenkschrift' (1807), quoted by Smend in Ak. Wiss., Gött., *Festschrift* (1951), vii.

75. Huter, *Fächer* (1971), 6; Huber, *Parn. Boic.* (1868), 3, 14; Westenrieder, *Gesch.*, II (1807), 93. The Jesuits were excluded 'non par les loix mais par voie de fait': Lambert to Kästner, 24 Jan. 1764, in Bopp, Ak. Wiss., Heidelberg, *Sitzb.* (1928:18), 16. Universities and neighboring academies sometimes worked together; besides Göttingen, Uppsala and Stockholm (Lindroth, *Hist.* [1967], I, 33), and Halle and Berlin.

76. Bierens de Haan, *Holl. Maat.* (1752), 32–5; the victorious society later joined with the University in unsuccessfully opposing the establishment of a rival academy in Rotterdam, the

The argument of the Leyden Senate had some merit; at least as regards natural philosophy, the University suffered the decline it anticipated. The academies did attract men who, had they not had such opportunities, might have increased the stock of research-oriented members of the universities. Or, to put the point the other way, the academies very quickly lost their importance as scientific institutions, though perhaps not as pressure groups, once research became an expected, and supported, professorial activity.

The eighteenth-century academy promoted science in three ways. First, intramurally, by the mutual encouragement of its members at its frequent meetings, which, as a rough rule, took place weekly for the most distinguished societies, fortnightly for the less, and monthly for the least. To this must be added salaries and instruments that supported the work of members of academies rich enough to provide them. A second service was the publication of the results of the researches of members and associates. The memoirs of the chief academies of the Paris type carried only the work of its members; those of more open societies, such as the *Philosophical Transactions* of the Royal Society, printed papers by unaffiliated people when submitted through a Fellow. The majority of scientific work published in the eighteenth century appeared in the periodicals of learned societies; both the big book and the independent scholarly journal, like the *Journal des sçavans* and the *Acta eruditorum,* became rapidly outmoded as outlets for research results after 1700. The connection of the decline of the *Acta* with the multiplication of academies was recognized at the time.[77]

There is balance in all things. The proliferation of academies created quantities of publications often difficult to procure[78] and impossible to survey. Moreover, they did not always appear regularly, and even those that did came out a year or more after the memoirs they contain were first presented. The delay of the *Philosophical Transactions* averaged eighteen months; of the *Mémoires* of the Berlin and Paris academies, two and three and a half years, respectively; of the Petersburg *Commentarii,* almost five years. Hence the dates printed on the memoirs and cited in our notes should not be interpreted as the time at which they became generally available. Taking into account the delay caused by difficulties of travel and the inefficiencies of the book trade, one should allow on the average a minimum of three years from the date of presentation of a memoir (usually also the date of the volume containing it) to the time of its arrival, printed and bound, on the library shelves. To improve the system review journals were started, such as the *Commentarii de rebus in scientia naturali et medicina gestis* (Leipzig, 1752–98), and, more valuable yet, periodicals that both excerpted academic publications and provided quick and

---

Bataafsch Genootschap der Proefondervindelijke Wijsgebeerte (*ibid.*, 37–9), which countered by pointing out the number of competing academies in France.

77. Cf. Kästner to Maupertuis, 15 April 1750, in Kästner, *Briefe* (1912), 8–9.

78. Much of the learned correspondence of the time concerns the procurement of books and journals. Even Wolff at Marburg had trouble obtaining the *Commentarii* of the Petersburg Academy. Wolff to Schumaker, 1727, in Wolff, *Briefe* (1860), 97.

accessible publication of research results. Of these the most important in the early history of electricity was the *Journal de physique* (Paris, from 1773).[79]

The third and characteristic promotional activity of the academies was the prize competition. A society too poor to award an occasional prize for an essay on a subject of its choosing scarcely qualified as an academy.[80] The major institutions offered one and sometimes more prizes a year; and they were prizes worth having, ranging from the 2500# Rouille prize at Paris through the 50 or 60 ducats (550# to 660#) of Munich and Berlin down to about 300#, as at Bordeaux, Dijon, Marseille, Montpellier and Rouen.[81] The poverty-stricken society at Pau could manage only a little over 100#. Since a respectable award amounted to a sizeable fraction of the income of the smaller academies,[82] they could offer them only when a donor could be found. That happened surprisingly often. Delandine, writing in 1787, lists 1037 competitions proposed in France alone since the foundation of the Paris Academy, and warns that the number of jousts has begun to exceed the supply of knights: 'Already there are no longer enough men of letters and of science to compete successfully for the number of prizes annually proposed.'[83] Many of the competitions, as he says, were slight and parochial, and their subjects and winners soon forgotten; but those of the national academies often brought intense competition, advanced the careers of the victors, and influenced the course of science.

In theory academies chose prize questions for their timeliness, and judges did not know who had written the essays presented in competition. In practice distinguished savants might be invited to compete and the question set to attract them,[84] while the judges could frequently identify the authors of the anonymous papers before rendering their verdict. This foreknowledge came most easily in mathematical tourneys, in which the same men were alternately the judges and the judged, and knew one another's handwriting. One complained that Parisian academicians favored contenders from Berlin over those from St. Petersburg; that d'Alembert could not win a competition of which Euler was a judge, or Daniel Bernouilli one over which Maupertuis had influence.[85] The ethical level of the business may be gauged from Lagrange's remark that, had

79. For the chief rationale of the *Journal,* the proliferation of journals, see *JP,* 1 (1773), iii–iv; K. Baker, *RHS,* 20 (1967), 262–7; *supra,* i.1.

80. Cf., Roche, *Livre,* 157–8, and Lambert's proposal for instant glory for the Munich Academy: offer a big mathematics prize to attract Euler and friends. Lambert to Kästner, 24 Jan. 1764, in Bopp, Ak. Wiss., Heidelberg, *Sitzb.* (1928:18), 16.

81. Maindron, *Académie* (1888), 13–14; Delandine, *Couronnes* (1787), *passim;* Dulieu, *RHS,* 11 (1958), 235; Barrière, *Académie* (1951), 351; Roche, *Livre,* 160.

82. Desplat, *Milieu,* 73–5, 123.

83. Delandine, *Couronnes* (1787), I, vii. Consequently prizes were withheld more and more frequently in France after 1750; Desplat, *Milieu,* 72; Roche, *Livre* (1965), 161–2.

84. Cf. Bouguer to Euler, 8 April 1754, in Lamontagne, *RHS,* 19 (1966), 238.

85. Frisi's complaints in Costa, *Riv. stor. ital.*, 79 (1967), 873–4; d'Alembert to Lagrange, 26 April 1776, in Lagrange, *Oeuvres,* XIII, 316; Lamontagne, *RHS,* 19 (1966), 229; Hankins, *D'Alembert* (1970), 45–6, 49, 59; Winter, *Registres* (1957), 71.

he known that Condorcet was the author of an essay for a certain competition, he 'would have made an effort to have the prize awarded to him.'[86]

The choice and phrasing of the question, and the informed adjudication of the prizes, gave many opportunities to influence the development of a field. The Paris Academy, for example, liked to reward theoreticians of the vortex, as in Bilfinger's victory in 1725 and the essays on magnetism of the 1740s; no doubt the hope of winning 1000# or 2000# strengthened the resolve of wavering Cartesians. The same influence was exerted by the two most important competitions on electricity, the Berlin of 1745 and the Petersburg of 1755; both in the statement of the question and in the awards the academicians approved and confirmed a Cartesian approach. Other prize questions directed attention to the relation between electricity and magnetism (Lyon, 1747), and between electricity and lightning (Bordeaux, 1748).[87] Perhaps the most important contribution of the prize competitions to the study of electricity was Coulomb, who came to his measurements of electrical force by following up work for his winning entry for the Paris prize on magnetism of 1777.

The continental organized prize competition fit neither the permissiveness nor the purse of the Royal Society of London. Rather than stimulate research on a specified subject, the Society preferred to reward, chiefly with nonnegotiable honors, the author of any discovery or invention that its administrators deemed worthy. Two such distinctions encouraged eighteenth-century physicists, the Copley medal, formally established in 1736,[88] and the Bakerian lectureship, initiated in 1775. The endowment of each was £100, making the income, and hence the value of medal or lectureship, about £5 (125#). Many received awards for electricity. Among Copley medallists were Gray, Desaguliers, Watson, Canton, Franklin, Wilson, Priestley, Volta; among Bakerian lecturers, Ingenhousz and Cavallo.[89] The Manchester Literary and Philosophical Society likewise rewarded the best work done, rather than the best answer to a set question.[90] So did the Stockholm Academy until 1760; thereafter it gave prizes in the continental manner, but—perhaps to illustrate its most frequent subject, economy—usually of comparatively little value.[91]

86. Lagrange to d'Alembert, 1 Oct. 1774, in Lagrange, *Oeuvres*, XIII, 292.

87. Delandine lists 11 French prizes on electricity, which divided into two groups: 4 prizes, 1747–9, on physical properties; 7 prizes, 1760–83, on applications to plants and animals (3 prizes not awarded). There were at least 18 competitions on electricity throughout Europe in the Ancien Régime; Barrière, *Académie* (1951), 136.

88. The income from the Copley bequest, received in 1709, initially paid part of the salary of the curator, Desaguliers; the first award of a medal was made in 1731, to an electrician. RS, *Record* (1940[4]), 112, 345; Weld, *Hist.* (1848), I, 385; Wightman, *Physis*, 3 (1961), 346–8.

89. RS, *Record* (1940[4]), 345–6, 364–5.

90. Manch. Lit. Phil. Soc., *Mem.*, 1 (1781–3), xvi, offering a silver medal of about two guineas' value annually to 'encourage the exertions of young men.'

91. From about 50# to 300#; Lindroth, *Hist.* (1967), I, 143, 150; Nordin-Pettersson, Sv. Vet., *Årsbok* (1959), 435–516.

3. PROFESSORS

Between one-quarter and one-half of the electricians active at any time during our period were 'professors' in a university or secondary school (Table 2.4). The rationale for grouping all professors together is that the level of instruction in the university's philosophical faculty, where the physicist usually held forth, did not differ much from what prevailed in the better secondary schools. It was not unusual for a professor to exchange a university for a college post or to hold both simultaneously. Some secondary schools had stronger philosophy courses than some universities; the Jesuit college at Lyon, the Académie de Calvin at Geneva, the Scuole palatine in Milan, the gymnasium in Nuremberg, the dissenting Academy at Warrington, were more distinguished than the 'dwarf universities' (to borrow Eulenburg's phase) of Germany and Italy.[1]

Although in principle universities differed from colleges in possessing schools of law, medicine and theology, they often lacked some, and sometimes all, of these 'higher faculties.' The only reliable mark of a university was a legal one, its right to grant degrees. Our profile of the eighteenth-century professoriate will be drawn primarily from information about universities so defined. For reasons already given, however, it will apply to instructors in senior classes in secondary schools as well.

Almost every European—as opposed to British—university in 1700 had a professor responsible for instruction in physics. There were perhaps 75 such men, and almost all professed a literary, all-inclusive physics. Their number did not increase much during the century. Consequently, to account for the entries in Table 2.5, we must suppose that, by the 1740s, a good fraction of physicists were doing experiments.[2] This new activity, when expressed in alluring demonstrations, brought a wider audience than literary physics could command, and, by its requirements of space and equipment, made the professor a more expensive—and consequently a more valuable—member of the faculty. His prestige and value also rose outside the university, at least among those concerned to modernize instruction; for experimental physics, like economics, history, vernacular instruction and Cameralwissenschaft, breathed the spirit of Enlightenment.

THE SETTING

The professor of experimental physics was not commonly an experimental physicist, or expected to be one. The proposition we have already met—the professor teaches, the academician researches—was emphasized as strongly by the eighteenth-century professoriate as by the spokesmen for learned societies.

1. Eulenburg, *Frequenz* (1804), 2–3. See Borgeaud, *Hist.*, I (1900), 498–9, for the discussion of a defeated proposal to seek university status for the Geneva Academy in 1708.

2. Table 2.5 includes contributions from professors other than physicists, and omits experimental work other than electricity done by the physics professoriate. If the former were subtracted and the latter added, the numbers in the table would doubtless be larger.

TABLE 2.4

Ratio of Professorial to All Electricians

	1600–99	*1700–39*	*1740–49*	*1750–59*	*1760–69*	*1770–79*	*1780–89*
Professorial	9(9)	10(10)	37(26)	27(19)	21(9)	35(17)	26(17)
Table II.1, row C	4(4)	10(10)	31(24)	22(16)	16(7)	29(15)	25(17)
Ibid., row A	5(5)		6(2)	5(3)	5(2)	6(2)	1(1)
Ratio	.33(.33)	.50(.53)	.42(.39)	.38(.35)	.45(.28)	.49(.44)	.33(.31)

TABLE 2.5

Breakdown of 'Professors'[a] from Table 2.1

	1600–99	*1700–39*	*1740–49*	*1750–59*	*1760–69*	*1770–79*	*1780–89*
Catholic Universities[b]	2(2)	2(2)	3(3)	7(5)	4(3)	13(6)	7(5)
Austria					1(0)	1(0)	1(0)
France		2(2)			1(1)		
Germany						3(2)	2(2)
Italy[c]	2(2)		3(3)	7(5)	2(2)	9(4)	4(3)
Protestant Universities	2(2)	7(7)	21(17)	11(9)	7(4)	10(5)	8(7)
Britain	1(0)		1(1)			1(1)	1(1)
Germany	1(0)	7(5)	13(10)	6(5)	2(1)	5(3)	5(4)
Netherlands		2(2)	2(2)	2(1)	1(0)	2(0)	1(1)
Scandinavia			4(3)	2(2)	2(1)	1(0)	
Switzerland			1(1)	1(1)	2(2)	1(1)	1(1)
Secondary Schools		1(1)	7(4)	4(2)	5(2)	6(4)	10(5)
Protestant		1(1)	5(3)	3(1)	2(2)	2(1)	3(0)
Catholic			2(1)	1(1)	3(0)	4(3)	7(5)
TOTALS	4(4)	10(10)	31(24)	22(16)	16(9)	29(15)	25(17)

[a] The author (or respondent) of a doctoral thesis is counted as one-half; all fractions are rounded *up*.
[b] Exclusive of Jesuits, for whom see Table 2.1, row A.
[c] Lombardy is counted as Italy.

One such affirmation occurs in Johann Christian Förster's centennial history of the University of Halle, a context that gives it a special authority, for Förster was a progressive professor of philosophy, an economist, the supervisor of the University's botanical gardens, and Halle was one of Germany's most advanced higher schools. According to him, 'a professor by no means needs to discover new truths or to advance his science. Should he do so, he is in fact more than an academic teacher, he has done opera supererogationis,' he has worked beyond the call of duty.[3] The prescient pro-rector and academic senate of the University of Marburg sounded the same theme in 1786 in opposing the establishment of a Hessian academy of sciences that might tempt and turn professors from their Hauptwerk, teaching.[4] The minister most responsible for the foundation of the University of Göttingen, G. A. von Münchausen, wanted to require the faculty to improve their sciences as well as their students. They soon set him right. 'Whoever justly considers the various duties and offices of professors,' said Albrecht von Haller, speaking at the opening of the Göttingen Society of Sciences, 'will easily see that so great a burden falls upon them that it is entirely unfair to ask them to do any special scientific work, *peculiares singularium inquisitionum labores*.' 'To do more than teach,' says his colleague Michaelis, 'is to do the work of societies of science.'[5]

And yet there is no doubt that by 1790, at the leading universities, calls and promotions came most easily to those who contributed their bits to science. As early as 1750 Tobias Mayer's appointment to Göttingen specified not only that he teach applied mathematics, but also that he devote himself to 'Forschungsarbeiten,' and later, in countering an offer from the Berlin Academy, the Hanoverian government acceded to Mayer's wishes for research facilities as well as for an increase in salary.[6] Even when their contracts did not explicitly require it, Göttingen professors were expected to write, and did so, 'as if the entire empire of letters acknowledged their academic scepter.'[7] Those who did not measure up, who neither wrote nor researched, were as out of place among Göttingen professors as (to quote their colleague Kästner) 'mouse turds among pepper grains.'[8]

Perceiving that research and writing made the Göttingen faculty glorious, the weak University of Vienna thought to improve itself by ordering each of its professors to publish two papers every year.[9] There were other straws in the wind. At Pavia, Volta extracted many improvements for himself and his labora-

3. Förster, *Übersicht* (1799), 2–3; Hamberger and Meusel, *Gel. Teut.* (1786[5]), II, 381. Cf. Turner in Stone, *University* (1974), II, 505–28.

4. Hermelink and Kähler, *Phil.-Univ.* (1927), 458–9; cf. Paulsen, *Gesch.* (1896–7), II, 133–4.

5. Haller (1751) and Michaelis (1768), quoted by Joachim, *Anfänge* (1936), 2.

6. Quoted by E. G. Forbes, *Jahrb. Gesch. oberd. Reichs.*, 16 (1970), 149, 155; Mayer to Euler, 6 Oct. 1754, in Kopelevich, *Ist.-asst. issl.*, 5 (1959), 414. Cf. Schimank, *Rete*, 2 (1974), 207ff.

7. Bose (Wittenberg) to Formey (Berlin), 4 Aug. 1754 (Formey Papers, Deut. Staatsbibl., Berlin).

8. Kästner, *Briefe* (1912), 215–18; Müller, *Abh. Gesch. math. Wiss.*, 18 (1904), 135–6.

9. Kink, *Gesch.* (1854), I, 594.

tory from the Austrian government of Lombardy on the strength of his reputation as a discoverer. His justifications of these benefits show how matters stood. More room, he said, would allow '[me] to busy myself in research and to give private courses on it to capable students;' more money (a raise of 600 lire or 440#) would bring 'all my talents [to bear] on advancing the science I profess, and the instruction of students of it.'[10] Note the order in which he put his obligations.

Just after the turn of the century a Hanoverian minister responsible for the affairs of the University of Göttingen made the new concept of the university explicit. He allowed that faculty had a two-fold obligation: on the one hand, 'to preserve, propagate, and, where possible, to increase the sum of knowledge;' and, on the other, to teach, guide, and inspire. 'Without research, there is a great fear that we shall concentrate only upon the useful, to the ruin of science and, eventually, of teaching itself.'[11]

Now the University of Göttingen had been established and was maintained as Hanover's counterweight to the strong Prussian institution at Halle, itself founded in 1694 as a competitor to the then leading universities, Leipzig (Saxony) and Jena (Weimar).[12] The University of Pavia had been brought from the decadence characteristic of Italian higher schools in an effort to show the benefits of the enlightened despotism of Maria Theresa, who turned her attention to university reform in the 1750s. The happy results of this 'particular attention' greatly impressed the census-taker of Europe's intellectual riches, Jean III Bernoulli, when he visited Pavia in 1775.[13]

Most of the universities of Europe were behind Göttingen and Pavia in 1790. Among the institutions that might stand comparison with them in physics were, in Germany, Halle and perhaps Leipzig;[14] in Italy, Turin and perhaps Bologna; in Switzerland, Geneva. The Dutch universities had by then lost the ascendency in physics that they enjoyed earlier in the century. Neither Utrecht, which Musschenbroek left in 1740 for Leyden, nor Leyden, where he died in 1761, was able to replace him; the lead in research in physics passed to Van Marum, who had no university post, and to Van Swinden, who left Franeker for a post in Amsterdam in 1785.[15] At Paris virtually all the productive professors were also academicians. Great Britain defies generalizations.

The eighteenth century seems not to have been a prosperous time for universities. They were pinched for money by war and inflation and they seldom had

10. Volta to Wilzeck, 15 Jan. 1785 and 3 Feb. 1786, *VE*, III, 283, 330.

11. Brandes, *Betrachtungen* (1808), 16–18, 31; *ADB*, III, 241–2.

12. Selle, *U. Gött.* (1937), 4, 12; Förster, *Übersicht* (1799), 14, 32.

13. Bernoulli, *Lettres*, III (1779), 56–68, 63, 66.

14. Fester, *Gedike* (1905), 13, 21, 78, 87–8; Hermelink and Kähler, *Phil.-Univ.* (1927), 390.

15. Ruestow, *Physics* (1973), 153; Kernkamp, *Utrech. U.* (1936), I, 211; *DSB*, XIII, 183–4; Spiess, *Basel* (1936), 146, giving a student's view of the relative strengths of Leyden and Utrecht in 1760. Cf. Cuthberston, *Prin. Elect.* (1807), v, on the low state of experimental physics in Holland in 1769, and Z. Volta, Ist. lomb., *Rend.*, 15 (1882), 32, on Volta's estimate of Utrecht in 1782 ('the instruments of physics are nothing much').

first claim on the resources of their controlling prince or municipality. Enrollments stagnated or fell. The total annual matriculations in the German universities averaged 4200 from 1700 to 1750, and then declined almost linearly to about 2900 in 1800. Oxford and Cambridge fell from about 300 each in 1700 to some 200 in 1750; Cambridge then stagnated while Oxford recovered half its losses, to about 250.[16]

At its largest the eighteenth-century university was not very large. The biggest in Germany, Halle, never had a faculty greater than forty; its student body fluctuated between about 680 and 1500, and averaged 1000. The next in size, Jena and Leipzig, averaged 930 and 740 between 1700 and 1790; the smallest, Herborn and Duisburg, sixty and eighty, respectively.[17] The range in the Scottish universities was similar: at the end of the century St. Andrews had about 100 and Edinburgh, one of the few older foundations to grow during the period, 1000, up from 200 in 1720.[18] Oxford ranged between 1000 and 1700, at the outside.[19] The Italian universities and the Jesuit colleges doubtless stayed below 1000.[20]

These figures do not imply that professors of physics lacked students. In schools where the old order of learning still held, where philosophy preceded professionalization, he would teach all students who survived into their second year. Such programs characterized the Jesuit universities and colleges. At Würzburg, for example, an average of about forty students annually were 'physicists;' there, and at Dillingen, Freiburg and Fulda, an average of 65 percent of the student body was enrolled in the philosophy faculties.[21] There were also Protestant schools with fixed and frequented curricula in arts, particularly the Scottish universities, which required 'natural philosophy' in the fourth and final year.[22]

In the German Protestant universities, however, enrollment in the philosophy faculty was usually very small, often less than ten. Students went directly into the higher faculties; in a sample of six universities—Duisburg, Erlangen, Göttingen, Halle, Kiel, Strassburg—43 percent of the student body matriculated in theology, 38 percent in law, and 11 percent in medicine.[23] Often these

16. Eulenburg, *Frequenz* (1904), 132; Stone in Stone, ed., *University* (1974), I, 6.

17. Eulenburg, *Frequenz* (1904), 146, 153, 164–5, 319. Eulenburg's figures differ from Gedike's contemporary survey, which makes Leipzig 1200–1300 and Jena and Göttingen each between 800 and 900 in 1790 (against Eulenburg's 670, 780 and 810, respectively). Fester, *Gedike* (1905), 33, 78, 87.

18. Great Britain, *Sess. Papers*, 37 (1837), 248; Dalzel, *Hist.* (1862), II, 307–25.

19. These figures come from the matriculations via the multiplier 5.6 (Stone in Stone, *University* [1974], I, 87), which is probably too high; the table (*ibid.*, 95) suggests one between 3 and 4.

20. E.g., 400 at Parma in 1775 (Bernoulli, *Letters*, III [1779], 184); 'a few hundreds' at Bologna (Simeoni, *Storia* [1940], 89–90); still fewer at Modena (Pietro, *Studio* [1970], 146) and Ferrara (Visconti, *Storia* [1950], 91–2).

21. Computed from Eulenburg, *Frequenz* (1904), 207–9, 310, 312.

22. Morgan, *Scot. U.* (1933), 72–4; Rait, *U. Aberdeen* (1895), 202, 300; Morrell, *Isis*, 62 (1971), 160–1, 207.

23. Eulenburg, *Frequenz* (1904), 207.

students cared only for professional training, for 'jus, jus et nihil plus,' as the lawyers said.[24] Under these circumstances it took a good man to draw audiences to courses in experimental physics; and audiences he needed, for it was their approbation and fees that made possible the purchase of the necessary apparatus.

The successful professor of experimental physics had to be a showman. In Vienna, where he had a captive audience, he was nonetheless directed by statute to strive for '[die] nöthigen Popularität.'[25] Playing to the gallery did not improve the morale of serious savants. We already know that Kästner gave up teaching the standard lecture course because his students came only to be entertained. The expression of the dilemma of his class of pedagogue is best left to one of them, John Robison, professor of natural philosophy at the University of Edinburgh, who had rights to fees from fourth-year students, if he could entice them to stay. 'As I endeavor to conduct my lessons in such a manner that [some students may learn something], I render them less pleasing to the generality of my hearers, who aim at nothing but getting a superficial Knowledge, or, more properly speaking, whose only aim is a frivolous amusement. This renders me a very unpopular teacher, and as I cannot think of becoming a showman, I do not expect to grow rich in the profession.'[26]

Despite the need to perform, despite stagnant enrollments and rising costs, the professors of experimental physics managed to establish their subject firmly in universities and colleges during the eighteenth century. And, despite the consensus that the professor need do nothing more than teach, a few performed supererogatory works, and inspired students to do the same.

THE FIRST COURSES IN EXPERIMENTAL PHYSICS

A very few professors were illustrating their lectures on physics with occasional demonstrations by or just after 1700. Those whose performances had some influence may be counted upon the fingers of one had: Burchard de Volder, a moderate Cartesian at Leyden, the inventor of an improved air pump; J. C. Sturm, who had studied at Leyden and who developed demonstrations based upon the experiments of the Accademia del Cimento, at Altdorf;[27] G. A. Hamberger, a disciple of Sturm's, at Jena; and Pierre Varignon, the follower of Malebranche, at the Collège Mazarin.[28] The introduction of experimental physics into the universities had therefore begun before the advent of Newtonian experimental philosophy. For a time, it proceeded on the continent under

24. Paulsen, *Gesch.* (1896–7), I, 531–2, II, 127.

25. Meister, Ak. Wiss., Vienna, Phil.-Hist. Kl., *Sb.*, 232:2 (1958), 95 (an edict of 1774).

26. Robison to Watt, 22 Oct. 1783, in Robinson and McKie, *Partners* (1970), 130. Cf. Grant, *Story* (1884), I, 241–2; Kästner, 'Verbindung' (1768), in *Verm. Schr.* (1783³), II, 363.

27. Klee, *Gesch.* (1908), 28–30; Crommelin, *Sudh. Arch.*, 28 (1935), 131; Günther, Ver. Gesch. Stadt Nürn., *Mitt.*, 3 (1881), 18; Will, *Nürn. Gel.-Lex.*, III (1757), 800–9; *ADB*, XXXVIII, 39–40.

28. Steinmetz, *Gesch.* (1958), I, 130-2, 206, and Jöcher, *Allg. Gel.-Lex.* (1750–1), II, 1338 (Hamberger); *infra,* ii.4 (Varignon).

Cartesian fellow travellers opposed to Newton's methods and bemused by his apologetics.

One indigenous European pattern, common to the Protestant universities of Germany and Scandinavia, may be illustrated by the career of Christian Wolff. His precocious interest in Cartesian method led him to the study of mathematics and physics, which he chose to pursue at Jena under Hamberger. He went to Halle in 1706 as professor of mathematics. A professor of medicine then taught physics as a second field; Halle's founders had tried to pry Sturm from Altdorf, but he had declined to move. Wolff was soon offering a course in physics based on Sturm. Later he formally took responsibility for instruction in the subject.[29] The course succeeded. Wolff's modernized Cartesian physics, demonstrated in the style of Sturm and Hamberger, spread to many universities of Central Europe between 1720 and 1750, replacing either biological physics taught by a member of the medical faculty or literary physics taught by a professor of philosophy.[30]

Among institutions that followed this pattern were Marburg, to which Wolff himself brought it in 1723;[31] Kiel, where a Wolffian physicist, J. C. Hennings, arrived in 1738, in succession to a Cartesian physician;[32] Leipzig, where in 1750 J. H. Winkler, a Wolffian electrician, took a chair just released by a philosopher and poet, who had had it from a physician;[33] Uppsala, where the first professorship of physics, established in 1750 under external pressure, went to a former student of Wolff's, Samuel Klingenstierna.[34] In these later cases, the new physics came with Newtonian admixtures; for example, Klingenstierna was directed to teach the usual range of mechanics, 'aerometry' (Wolff's specialty), and 'the discoveries of Newton.' Nonetheless the native contribution remained in evidence: the most popular physics texts in Protestant Germany in the 1730s and 1740s were probably Wolff's *Nützliche Versuche* and the *Elementa physices* of G. E. Hamberger,[35] who succeeded his father at Jena in the 1720s.

The little stream of Sturm, Hamberger and Wolff is easily overlooked against the flood of Anglo-Dutch Newtonianism. Just after 1700 Newtonian epigoni, installed in Oxbridge chairs, began to offer courses of experimental philosophy. At Cambridge Newton's successor as Lucasian professor of mathematics,

29. Förster, *Übersicht* (1794), 27, 29, 55, 95

30. Paulsen in Lexis, *Deut. U.* (1893), I, 28–32, and *Gesch.* (1896–7), I, 532; Nauck, *Beit. freib. Wiss.-Univ.gesch.*, no. 4 (1954), 10–11; Helbig, *U. Leipzig* (1961), 55.

31. Hermelink and Kähler, *Phil.-Univ.* (1927), 348, 384n.

32. Schmidt-Schönbeck, *300 Jahre* (1965), 30–5. It appears that Hennings lacked some of the apparatus for the lectures he announced.

33. Leipzig, U., *Festschrift*, IV:2 (1909), 26–9.

34. Anon., Uppsala, U., *Aarssk.* (1910), 35–42; cf. Hildebrandsson, *Klingenstierna* (1919), 12–17.

35. *ADB*, X, 470–1; Börner, *Nachrichten*, I (1749), 52–71. For Hamberger's text, Stieda, *Erf. Univ.* (1934), 27.

William Whiston, and Roger Cotes, named the first Plumian professor of astronomy and natural philosophy in 1706, collaborated on such a course. Their association came to an unnatural end in 1710, when Whiston was unseated for unorthodoxy. Cotes carried on alone until he died in 1716, leaving his professorship and his lectures to his cousin, Robert Smith.[36] At Oxford the rash John Keill, then deputy Sedleian professor of natural philosophy, offered a course in experimental physics with the assistance of Desaguliers; it lasted until 1712, when Keill became Savillian professor of astronomy. The Keeper of the Ashmolean Museum moved into the void, from which he was chased by James Bradley, who succeeded to Keill's professorship in 1721.[37] These professorial lectures represented only a portion of the interest in experimental physics at Oxbridge in the first quarter of the eighteenth century. Many tutors assigned reading in Newton and Keill as well as in Rohault, and some offered experimental demonstrations.[38] The decline of physics at the ancient English universities set in after the death of the first generation of Newtonian professors.

The work of Keill, Cotes and Desaguliers inspired the authoritative texts of 'sGravesande, whose excellent order, convenient size, and copper plates did much to spread the cause of experimental physics on the continent.[39] Among the first institutions to teach physics in the Dutch style were the universities of Duisburg and Utrecht, which shared the services of 'sGravesande's agent Musschenbroek in the 1720s. Musschenbroek was trained as a physician, but taught as a philosopher, within the philosophy faculty.[40] Giessen probably also belongs in this group; in 1729 it set up a chair of 'physica naturalis et experimentalis' for a professor of medicine who had studied and travelled in Holland.[41]

The Catholic universities of Germany and Austria were slower to take up the new physics. The Jesuits who dominated the philosophical faculties there held to the *Ratio studiorum* and the regental system, and stayed suspicious of Descartes. Efforts to introduce Cartesian physics and Sturm's demonstrations into Ingolstadt failed early in the century. Towards 1730 the medical faculty renewed the attempt, and succeeded by threatening to teach experimental physics itself.[42] Responsibility and initiative for developing the subject fell to Joseph Mangold, S. J., a Cartesian of Euler's type. The course of development at Dillingen was similar. In its case the local bishop led the attack against tradi-

36. Hans, *New Trends* (1951), 49–50; *DSB*, III, 430–3.

37. Hans, *New Trends* (1951), 47–8; *DNB*, II, 1074–9; *DSB*, VII, 275–7.

38. Mayor, *Cambridge* (1911), 55, 457; Wordsworth, *Scholae* (1877), 68; Rouse Ball, *Hist.* (1889), 94–5; Gunther, *Early Sci.*, I (1920), 196; Frank, *Hist. Sci.*, 11 (1973), 253–5.

39. Brunet, *Physiciens* (1926), 40–2, 48–54, 61; *supra*, i.1.

40. Kernkamp, *Utrech. U.* (1936), I, 305. Duisburg was closely associated with the Dutch schools; Ring, *Gesch.* (1949²), 179.

41. Jöcher, *Allg. Gel.-Lex.* (1750–1), IV, 1523–4; *ADB*, XXXIX, 615–16; Lorey, Giess. Hochsch., *Nach.*, 14 (1940), 23–31, 15 (1941), 83–7.

42. Schaff, *Gesch.* (1912), 6, 154–6; Prandtl, *Gesch.* (1872), I, 541, 610; Hofmann, *Math.* (1954), 14–15.

tion; in 1745 he recommended the texts of the 'acatholic' Wolff. By the 1750s the Dillingen Jesuits were giving experimental demonstrations and lecturing on physics in German.[43]

The acceptance of experimental physics at the German Catholic universities in the 1750s owed not a little to Maria Theresa's reforms of the Jesuit-run higher schools of Austria: Freiburg i./B., Graz, Innsbruck, Prague and Vienna. A government commission during the reign of her father, Charles VI, had already criticized the philosophical faculty for teaching 'empty subtleties;' its insistence that Descartes' physics be *introduced* into the schools tells plainly enough what was considered up-to-date in Vienna in 1735.[44] The Austrian universities, particularly Vienna and Innsbruck, did acknowledge experimental physics in the 1740s, but Aristotle remained their official guide until 1752, when the Empress insisted on reducing the philosophical course to two years, on using German in instruction, and on eliminating metaphysics, ethics, and 'everything useless.' The old curriculum, she said, had nothing to do with the common concerns of men and states; 'the ungrounded theory (which can not be confirmed by experience) of peripatetic form and matter is henceforth entirely forbidden;' class time 'shall be devoted to true physica experimentalis.'[45]

The progress of the institutionalization of experimental physics accelerated after the Seven Years' War. In the German Protestant universities true Fach-Physiker began to appear, men trained by the experimental physicists who had established themselves in the 1740s and 1750s. At Kiel, for example, the first physics course thoroughly illustrated by experiment was initiated by a Wolffian professor of medicine, J. F. Ackermann, who from 1763 also held a chair in the philosophy faculty; his successor, C. H. Pfaff, M.D., who had finished his studies under Lichtenberg at Göttingen, made his career as a professor of physics.[46] Other Göttingen graduates garnered physics chairs at Giessen (G. G. Schmidt) and Altdorf (J. T. Mayer).[47]

The suppression of the Jesuits in 1773 resulted in a short-run improvement in the former Jesuit universities of Germany and Austria even though—or rather because—instruction continued in the hands of ex-Jesuits. As an Austrian commission charged to consider their replacement reported in 1774: 'We do not have their equal in the mathematical sciences and they are cheaper to maintain than lay professors.'[48] The suppression allowed the concentration of resources and the substitution of modern, specialized, vernacular texts for the Jesuit compendia. For example, Ingolstadt adopted Erxleben's *Anfangsgründe*; Freiburg,

43. Specht, *Gesch.* (1902), 197–200, 317–18; Sommervogel, *Bibliothèque,* V, 481.

44. Paulsen, *Gesch.* (1896–97), II, 107.

45. Schreiber, *Gesch.*, II:3 (1860), 7–11; Fester, *Gedike* (1905), 44; Meister, Ak. Wiss., Vienna, Phil-Hist. Kl., *Sitzb.*, 232:2 (1958), 89–91. For the reforms see Kink, *Fesch.* (1854), I, 424–590.

46. Schmidt-Schönbeck, *300 Jahre* (1965), 30–8, 59.

47. Lorey, Giess. Hochsch., *Nach.*, 15 (1941), 83–7. Mayer soon left for Erlangen, which was well regarded in 1790; Fester, *Gedike* (1905), 70, 74; Kästner, *Briefe* (1912), 244.

48. Halberzettl, *Stell.* (1973), 141, 167, 189; Schreiber, *Gesch.*, II:3 (1860), 50–1. Frederick the Great also kept them on in Silesia; Paulsen, *Gesch.* (1896–7), II, 100–1.

Sigaud's *Anweisungen zur Experimentalphysik*; and Vienna, still reforming, chose Biwald's *Institutiones physicae,* up-to-date in all but language.[49]

In France the Jesuits could not hope to retain their control of education if they opposed novelty as strongly as their German brethren. They had always emphasized applied mathematics; and in the late seventeenth century, as we have seen, they accepted a government charge to provide for instruction in navigation and associated sciences, 'to pray and teach hydrography.'[50] The earliest recorded 'exercises de physique expérimentale' in the French Jesuit system occurred either in colleges associated with hydrographers or in ones where physics was customarily taught by men trained in mathematics: at Aix in 1716, Lyon in 1725, Louis-le-Grand in 1731, Pont-à-Mousson in 1740, Marseille in 1742.[51] Meanwhile the père physicien had identified more closely with his subject: in 1700 30 of the 80 colleges teaching physics used the regental system; in 1761 the figures were 23 and 85.[52]

The higher schools of Paris, the University and the Collège Royal, institionalized the new physics relatively late, partly because the many public teachers of mathematics and physics in the capital allowed them to shirk responsibility. The university's first professor of experimental physics, J. A. Nollet, had been an unusually successful public lecturer; his chair, founded in 1753 at the Collège de Navarre, was a gift of Louis XV.[53] (Other colleges, especially Harcourt, Louis-le-Grand, and Mazarin, which had strong traditions in mathematics, occasionally offered instruction in experimental physics either by the professor of mathematics or by a private lecturer hired for the purpose.[54]) Nollet further broke with tradition by lecturing in French. Although few in the university followed his example—physics in most colleges continued Latin and literary until the Revolution, and the Faculty of Arts did not endorse a vernacular textbook until 1790[55]—opportunities for teaching and learning experimental physics continually improved. For example, in the 1760s Sigaud de Lafond, later of the Academy of Sciences, and his nephew Rouland, instrument makers and public lecturers, advertised themselves as 'demonstrators in experimental physics' at the University, which probably meant that for a fee they brought their own equipment to demonstrate in the class of a professor of physics or philosophy.[56] The arrangement satisfied d'Alembert: 'The University of Paris furnishes convincing proof of the progress of philosophy among us.

49. Schaff, *Gesch.* (1912), 181; Zentgraf, *Gesch.* (1957), 13; Meister, Ak. Wiss., Vienna, Phil.-Hist. Kl., *Sb.,* 232:2 (1958), 93; Paulsen, *Gesch.* (1896–7), II, 110.

50. Dainville, *Géographie* (1940), 435–9; Dainville in Taton, *Enseignement* (1964), 28.

51. *Ibid.*, 40–1; Schimberg, *Education* (1913), 518–19n; Lallamand, *Hist.* (1888), 259.

52. Dainville in Taton, *Enseignement* (1964), 29.

53. Torlais in Taton, *Enseignement* (1964), 627; Jourdain, *Hist.* (1888), II, 274–6, 382–6. For private lecturers, *infra,* ii.4.

54. Lacoarret and Ter-Menassian in Taton, *Enseignement* (1964), 141–5; A. Franklin, *Hist.* (1901[2]), 199, and *Recherches* (1862), 162–75, 109–10; Guerlac, *Isis,* 47 (1956), 212.

55. Lantoine, *Hist.* (1874), 152; Anthiaume, *Collège* (1905), I, 221.

56. Torlais in Taton, *Enseignement* (1964), 633, 637.

Geometry and experimental physics are successfully cultivated there . . . Young masters train truly educated students who leave their [course of] philosophy initiated into the true principles of all the physico-mathematical sciences.'[57]

Outside Paris Nollet's course, and his earlier public lectures, also had significant results. They probably helped the case of experimental physics among the Jesuits, as he claimed that they did among the Oratorians. After the expulsion of the Jesuits, they brought him into demand as consultant to new professors and administrators wishing to set up or to continue instruction in experimental physics.[58] Caen in 1762, Bordeaux in 1763, Pau, Strasbourg, Draguignan, Amiens, all sought his advice.[59]

The Collège Royal, founded in the sixteenth century, had always been more modern than its medieval sister. An example of its precocity was the slow transformation of one of its old chairs of Greek and Latin philosophy into a professorship of physics, a change made permanent with the appointment of Varignon (1694) and his successor Privat de Molières (1722). The chair nonetheless retained its title until 1769. It was then converted into a chair of physics, 'His Majesty having recognized that the two chairs of Greek and Latin philosophy have had little audience since physics has been enhanced by the discoveries of the moderns.' Four years later the Crown combined the chairs of Hebrew and Syriac and established a chair of 'mechanics' (in 1786 changed to 'experimental physics') on the income of the suppressed orientalist.[60]

In Italy the institutionalization of experimental physics had not proceeded very far by mid-century; instruction in philosophy remained literary, and almost exclusively in the hands of clerics, who offered mixtures of Aristotle and Descartes. An exception is Padua, where the Cartesianism of Fardella and the Galilean tradition of applied mathematics met to produce Giovanni Poleni, engineer, philosopher, mathematician and, from 1739, tenant of a new chair 'ad mathematicam et philosophiam experimentalem.'[61] Bologna also made provision for modern instruction in physics and mathematics. In 1737, in association with the moribund University, the Institute established by Marsigli in 1714 set up a professorship of experimental physics. It was later held by Giuseppe Veratti, M.D., among other electricians.[62]

The tempo increased in the 1740s. Benedict XIV encouraged the reform of the Sapienza at Rome 1744–6, which brought a chair of 'rational and ex-

57. D'Alembert in 'Elémens' (1760), *Oeuvres* (1805), II, 460–1.

58. Nollet, *Leçons*, I (1754^{6}), xii–xiv; Jourdain, *Hist.* (1888), II, 175–6, 274–6; Nollet to Dutour 7 Sept. 1768 (Burndy): 'Les nouveaux collèges depuis l'expulsion des Jésuites veulent tout faire de la physique expérimentale.'

59. Torlais in Taton, *Enseignement* (1964), 628; Mornet, *Origines* (1947^{2}), 181; Irsay, *Hist.*, II (1935), 115.

60. Anon., *Rev. int. ens.*, 5 (1883), 406–7; Sédillot, *Bull. bib. stor. sci. mat. fis.*, 2 (1869), 499–510, 3 (1870), 165–6.

61. Vidari, *Educazione* (1930), 132–4, 180–97; Favaro, *U. Padova* (1922), 67–8, 142.

62. Bortolotti, *Storia* (1947), 147–58; Simeoni, *Storia*, II (1944), 116. The institute provided for a professor of experimental physics from the beginning (*ibid.*, 126).

perimental philosophy' (later 'experimental physics') held first by the Minim mathematician François Jacquier. Pisa set up a professorship of experimental physics in 1746, for C. A. Guadagni; in 1748 Turin freed its physics chair from philosophers and entrusted it to G. B. Beccaria, a major figure in the early history of electricity; and similar attempts, apparently not entirely successful, were made at Pavia and Naples.[63] In 1760, if not before, professors of experimental physics existed at Perugia and Modena; in 1764–5, owing to reforms introduced by the King of Sardinia, Charles Emanuel III, the impoverished universities of Cagliari and Sassari had them too.[64] In the 1770s the Austrians made Pavia a showplace; Modena added a second chair of physics; and the pint-sized enlightened despot, Duke Ferdinand I of Parma, enriched his university with the confiscated wealth of the Jesuits and built for it a lecture hall for physics that Volta took as the pattern for his own.[65]

In English universities the cause of experimental physics had several champions at mid-century. Smith and Bradley lived through the 1750s, and found an unlikely colleague in Thomas Rutherford, Regius Professor of Divinity at Cambridge, who wrote an important textbook on natural philosophy. Rutherford's career instances a difficulty in determining the level of academic activity at eighteenth-century Oxbridge: although Rutherford never lectured as university professor of divinity, he taught experimental physics regularly as a member of his college. Similarly, Bradley's successor, singled out by Adam Smith to represent those who had 'given up altogether even the pretence of teaching,' often lectured privately on experimental physics.[66]

There were no doubt many dons and professors sunk in ignorance and indolence, such as Charles Beattie, appointed Sedleian Professor at Oxford in 1720, 'not on account of any skill (for he hath none) in Natural Philosophy, but because he is much in debt to [his] college, occasioned by his Negligence as Bursar.'[67] Yet there was the counterweight of professors who taught well in their or someone else's statutory field, and the dons, particularly at Cambridge, who kept science alive in their colleges. One should not make much of the well-known estimate that the fraction of active English 'scientists' who had been educated at Oxbridge fell from two-thirds in 1650 to one-fifth in 1750.[68] Everything depends upon the definition of scientist. Among those with mathematical training, like the electrician Henry Cavendish, a good fraction

63. Italy, *Monografie*, I (1911), 299–300, 547–55; Scolopio, *Storia*, I (1877), 58.

64. Italy, *Monografie*, I (1911), 78, 166, 429–39, 442.

65. Mor, *Storia* (1953), 93, 175–8; Pietro, *Studio* (1970), 42–95; Italy, *Monografie*, I, (1911), 246–7, 270; Volta to Wilzeck, 4 March 1785, *VE*, III, 295. Bernoulli, *Lettres*, III (1779), 181–3, comments on the excellence of the new facilities for science at Parma.

66. *DNB*, XVII, 499, IX, 1267; Wordsworth, *Scholae* (1877), 72. Cf. the case of Isaac Milner, Jacksonian and later Lucasian professor at Cambridge; *DNB*, XIII, 456–9; Winstanley, *Unr. Camb.* (1935), 131.

67. Godley, *Oxford*, (1908), 82–91; Mallet, *Hist.*, III (1927), 124–6; Winstanley, *Unr. Camb.* (1935), 129–32, 151–2, 179–82.

68. Hans, *New Trends* (1951), 34.

attended Cambridge, where the mathematical tripos called for close study of the *Principia* and treatises on mechanics, optics, and hydrostatics.[69]

Experimental physics came into the Scottish universities when they abolished the regental system and established chairs of natural philosophy. That occurred at Edinburgh in 1708, Glasgow in 1727, St. Andrews in 1747, and Aberdeen in 1753. The fixing of the chairs coincided with large purchases of demonstration apparatus at Edinburgh (1709) and Glasgow (1726); smaller acquisitions by the other schools about 1715 suggest that even before they stabilized their chairs they made some provision for the new science.[70]

In 1703 the Rector of Calvin's Academy at Geneva, praising the 'brief, yet substantial, incomparable, royal,' method of Descartes, insisted on the need for mathematics and experiment.[71] Mathematics prospered first; a chair was set up in 1724 by the municipal authorities against the opposition of the Church, and awarded jointly to two able young men, Calendrini (the collaborator of Jacquier and Le Seur) and Gabriel Cramer. Both then urged the appointment of a professor of experimental physics. That partly came to pass in 1737, when Jean Jallabert received an 'honorary' (unsalaried) post. Jallabert went off to England, France and Holland, procured apparatus, met Musschenbroek and Nollet, became a Fellow of the Royal Society, and returned to lecture to applause. He subsequently followed Cramer, who had followed Calendrini, into one of the Academy's two chairs of philosophy. Their chair passed to Saussure, an excellent physicist and electrician, in 1762; and it became a chair of experimental physics in all but name under him and his hand-picked successor, M. A. Pictet.[72] The rise of Calvin's Academy to a leading center of physics owed not a little to the fact that Calendrini, Cramer, Jallabert, Saussure and Pictet were wealthy members of Geneva's governing class.[73] They could afford to accept half-chairs and honorary posts, and, after election to professorships, to do what they pleased.

BUILDING UP THE CABINET

The professor and his institution shared responsibility for the upkeep and increase of instrument collections according to their relative power and poverty. At Protestant universities, except in Scotland, the professor of experimental physics was expected to furnish some if not all of his equipment; large private collections resulted, such as those of 'sGravesande and Musschenbroek at Leyden, Winkler at Leipzig, Lichtenberg at Göttingen, Bose at Wittenberg, Jallabert and Saussure at Geneva.[74] The extent of the collection, and hence of

69. Rouse Ball, *Hist.* (1889), 191, quoting a source of 1772.

70. Dalzel, *Hist.* (1862), II, 304; Gr. Br., *Sess. Pap.*, 38 (1837), 303; Murray, *Memories* (1927), 110–11; Coutts, *Hist.* (1909), 195; Caut, *College* (1950), 83.

71. Borgeaud, *Hist.*, I (1900), 458–8; cf. Montandon, *Développement* (1975), 45–9.

72. Borgeaud, *Hist.*, I (1900), 483–4, 503–4, 569, 573–7.

73. Montandon, *Développement* (1975), 51–3, 59.

74. Crommelin, *Desc. Cat.* (1951), 13, 21; Lichtenberg, *Briefe* (1901), II, 136, 259–60; Leipzig, U., *Festschrift* (1909), IV:2, 29; Bernoulli, *Lettres*, III (1777), 6; Borgeaud, *Hist.*, I (1900), 569.

instruction, was proportional to the depth of the professor's pocket, 'there being no public subsidies for the improvement of knowledge.'[75] At Kiel, for example, demonstrations sometimes lapsed during the tenure of the Wolffian Hennings, for want of means rather than will; his successor, J. F. Ackermann, had made money at doctoring and provided what was needed.[76] Bose's complaint, that he had to pay 'ready money for all my instruments, without exception, from my air pump to my funnel,' suggests that the professor / demonstrator was not regarded as a good credit risk.[77]

The situation provided an opportunity for wealthy faculty to poach on the preserve of the professor of physics. At Jena, for example, J. C. Stock, M.D., tried to win the physics chair by investing 100 RT (400#) in demonstration apparatus. At Lund Daniel Menlös succeeded in buying his way into a professorship over a man agreed to be his better by promising to acquire for the university the Triewald collection, then (1732) unrivalled in Sweden. At Halle a professor of law, Gottfried Sellius, briefly monopolized instruction in experimental physics. He had married a rich wife, with whose dowry he bought excellent instruments, some made of silver, and all very elegant; but he lived beyond his means and had to flee his creditors, leaving physics to the less ambitious pedagogues from whom he had snatched it.[78]

Despite noteworthy exceptions, such as the annual subvention of 200 RT for 'expensive instruments, books, etc.,' in physics and mathematics given Wolff at Marburg from 1724, the remarkable generosity of the curators of Utrecht during the tenure of Musschenbroek,[79] and the one-time purchase for 6000 Dkmt (2000#) of a 'complete' apparatus to Musschenbroek's specifications for Klingenstierna at Uppsala in 1740,[80] continental Protestant universities did not begin to acquire substantial collections until their professors, who had made the initial investments and suffered the depreciation, started to die off. Then the institution might purchase a working apparatus at a good price, leaving responsibility for further acquisitions to the new professor. Leyden bought 'sGravesande's collection for *f.*3931 (8400#), and several pieces from Musschenbroek's, which fetched *f.*8864 altogether when auctioned in 1761.

The perennially poor University of Duisburg bought half of J. J. Schilling's

75. Kästner to Haller, 1 Aug. 1755, Kästner, *Briefe* (1912), 35.

76. Schmidt-Schönbeck, *300 Jahre* (1965), 35–8.

77. Bose to Formey, 26 July 1750 (Formey Papers, Staatsbibl., Berlin): 'Il me faut payer argent comptant tous mes instrumens, sans en excepter aucun, depuis ma pompe pneumatique jusqu' à l'entonnier, pourtant je suis professeur en physique. A plus forte raison mes téléscopes, la pendule, les micromètres, etc., ne sont acquis qu'à mes propre[s] dépens, n'étant que des opera supererogationis, où je travaille pour ainsi dire par pur[e] magnanimité.'

78. Steinmetz, *Gesch.* (1958), I, 205; Förster, *Übersicht* (1799), 100–1; Leide, *Fys. inst.* (1968), 29–33.

79. Gottsched, *Hist. Lobs.* (1755), Beylage, 34; Kernkamp, *Utrecht. U.* (1936), I, 209–10. The curators bought many instruments when Musschenbroek came in 1723, and again in 1732–3 (almost *f.* 2000 [4300#] worth) when he was pondering accepting a call from Copenhagen.

80. Anon., Uppsala, U., *Aarssk.* (1910), 32–4. One should perhaps also except Lund, which acquired the Menlös-Triewald collection in the 1730s, and Greifswald, which began to build a collection just after 1750. Dähnert, *Sammlung,* II (1767), 828, 889, 985, 999–1000.

collection at his death in 1779.[81] Göttingen bought Lichtenberg's in 1787–8, having meanwhile added nothing to the physical instruments in use there in the 1750s.[82] Leipzig got Winkler's in 1785 for 1064 RT, and brought it up to date in 1808 by acquiring the instruments of his successor, K. F. Hindenberg, for 1000 RT; Altdorf did the same, building upon Sturm's old collection and adding to it, in 1780, that of another of its professors, M. Adelbulner; Marburg and Giessen acted similarly at the turn of the century.[83] Occasionally a philanthropic professor donated his instruments, as Kratzenstein did at Copenhagen (1795), adding an endowment the income of which was still an important part of the economy of the physics institute in 1900.[84]

The case of Halle, that 'garden of free arts and sciences,'[85] is particularly interesting. During his first tenure there, Wolff received a little dole for instruments, which was not continued to his successor in the mathematics chair, J. J. Lange, who gradually built up his own collection. When Wolff died in 1754, leaving Halle for the second time, Segner accepted his post on condition that the University provide adequate teaching apparatus. One thought to buy Wolff's. The heirs wanted a just price. Lange, however, was willing to sell cheaply; and the University could meet its commitment to Segner at little expense by adding a few new machines to those it bought from him. Not until the end of the century did Halle make adequate provision for acquiring physical instruments.[86]

The acquisition of instruments brought with it responsibility for maintenance and modernization. Duisburg, despite its poverty, spent 15 or 20 RT a year on its 'new' physical collection,[87] and Schmidt eventually obtained help from Giessen. A 'mechanic' might be engaged to keep the instruments in order and to help with the demonstrations: Lund employed such a person from the mid-1730s, Leyden from 1752, Utrecht from 1768. All also had to make provision for storage, sometimes, as at Lund, with great difficulty.[88] In 1795 Göttingen obtained a mechanic and other universities did so soon after the turn of the century. No doubt the best known of the tribe is James Watt, who in 1757 took the post of instrument maker set up at Glasgow in 1730.[89]

Catholic universities could not follow the Protestant pattern of acquisition

81. Rooseboom, *Bijdrage* (1950), 15, 107; W. Hesse, *Beiträge* (1875), 90–1.

82. Pütter, *Versuch*, I, 242, II, 267, III, 489; L. Euler to Fred. II, 7 Oct. 1752, in Stieda, Ak. Wiss., Leipzig, Phil.-Hist. Kl., *Ber.*, 83;3 (1931), 54.

83. Leipzig, U., *Festschrift* (1909), IV:2, 28–31; Günther, Ver. Gesch. Stadt Nürn., *Mitt.*, 3 (1881), 9, 30; Lorey, *Nach.*, 15 (1941), 86–7; Hermelink and Kähler, *Phil.-U. Marb.* (1927), 756–7. Cf. Bernoulli, *Lettres* (1771), I, 7n, regarding Frankfurt/Oder.

84. Snorrason, *Kratzenstein* (1967), 55–6; Copenhagen, *Poly. Laere* (1910), 27. The collection was worth at least 20,000# (4000 Rigsdaler).

85. Börner, *Nachrichten* (1749–54), I, 72.

86. Förster, *Übersicht* (1799), 222, 233; Schrader, *Gesch.* (1894), I, 570, 578; Euler to Fred. II, 2 and 20 Nov. 1754, in Stieda, Ak. Wiss., Leipzig, Phil.-Hist. Kl., *Ber.*, 83:3 (1931), 54–6.

87. W. Hesse, *Beiträge* (1875), 91.

88. Rooseboom, *Bijdrage* (1950), 15–18, 32; Kernkamp, *Utrecht. U.* (1936), I, 212–13; Leide, *Fys. inst.* (1968), 35–44.

89. Mackie, *U. Glasgow* (1954), 218.

since many of their professors, as members of religious orders, had very small incomes, and as a rule could not charge fees for courses. Consequently from the beginning instruments were usually provided by the university or by gift. In 1757 Dillingen set up a 'mathematical-physical museum' with the help of 1000 gulden (2600#) from its chancellor;[90] by 1787 it had a very small annual sum, 25 gulden, for instruments. At the same time (1754), Ingolstadt started a 'physical-chemical cabinet' for 1200 gulden.[91]

Both the course and nature of the acquisitions may be illustrated by the purchases of the Jesuit college at Bamberg, of which a full account survives. The first large acquisition was an air pump, the next, in 1747, an electrical machine. Buying increased between 1749 and 1753, and then declined until 1771. In 1789 the college was again looking for a better air pump and an improved electrical machine, for each of which it could pay 650#. By the end of the century it had 125 instruments for general physics, 45 for electricity, 60 for optics, 8 to demonstrate the theory of heat, and 6 for magnetism. These treasures created the usual storage problem; a special cabinet, built in 1755, was much enlarged in the 1790s.[92]

A similar pattern and similar timing occurred in Italy. During the pontificate of Benedict XIV (1740–58), the university of Rome got a 'theater' for experimental physics on the top floor of the Sapienza, a small budget, and instruments, the first six, including an air pump, gifts of Benedict himself. Benedict also enriched the Bologna Institute with a set of instruments made from 'sGravesande's designs. Nollet judged it to be 'assez ample' when he saw it in 1749, but missing a few things, about which he spoke to the Pope, 'who listened favorably,' during an audience in Rome.[93] Nollet also saw and approved the 'rather complete' collection bought for Poleni by the University of Padua in the 1740s; by 1764 it contained 392 items.[94]

Pisa had a cabinet by mid-century, stocked by the private collection of Dutch instruments made by its first professor of experimental physics, C. A. Guadagni; the university put up a small annual sum for apparatus, and, in 1778, partly compensated Guadagni for his total personal outlay with a gift of 100 zecchini (about 1200#). By the end of the century the professor of physics had 500# for the gabinetto and 200# for experiments.[95] Modena's gabinetto was established in 1760, on much the same basis as Pisa's: as nucleus it took the instruments of its professor, the Minim Mariano Moreni, whom it repaid in 1772 with a tiny annuity; from 1777–8 it made available 200# annually for

90. Specht, *Gesch.* (1902), 199–200, 530; Schmid, *Erinnerungen* (1953), 94–5, says that Weber expanded the collection 'modestly,' little money being available.

91. Günther, Ver. Gesch. Stadt Nürn., *Mitt.*, 3 (1881), 10; cf. Schaff, *Gesch.* (1912), 155–6, for earlier purchases.

92. H. Weber, *Gesch.* (1879–82), 338–42.

93. Spano, *Univ.* (1935), 50–1, 253; Nollet, 'Journal' (1749), f. 153v.

94. *Ibid.*, f. 91–2; Favaro, *Univ.* (1922), 142.

95. Italy, *Monografie*, I (1911), 259; Occhialini, *Notizie* (1914), 3, 17–18; Scolopio, *Storia* (1877), I, 79; Nollet, 'Journal' (1749), f. 127v.

improving the collections, and for two Capuchins to look after them.[96] Parma, under its modernizing duke, established a gabinetto in 1770 and furnished it with instruments in the style of Nollet with the help of money and material confiscated from the Jesuits.[97]

Turin acquired the nucleus of its collection in 1739, with the purchase of the many instruments that Nollet had brought to teach physics to the crown prince and with the hiring of a mechanic to keep them in order.[98] The most notable case of government benefaction to Italian physics was Pavia, where Volta established a direct channel to the Governor of Lombardy. He obtained thousands of lire for instruments, grants for foreign travel to select them, salary for a mechanic, a laboratory, storerooms, and a 'teatro fisico,' a large lecture hall where he entertained and instructed the large audiences which, with his reputation as a discoverer, supported his claim upon the treasury.[99]

In France a few Paris colleges—Navarre, Royal, and, to a lesser degree, Louis-le-Grand—had substantial collections by 1790. Navarre's, for example, numbered 235 pieces. Many of the former Jesuit colleges had important instruments, for instance, Dijon, Poitiers, Puy; even the tiny college at Epinal had two electrical machines, an air pump, and devices to illustrate the principles of mechanics, hydraulics and optics.[100] Many of these items found their way into the institutions that divided up the educational empire of the Jesuits in the 1760s.

The Jesuit universities in Austria were furnished with instruments in consequence of the Theresian reforms. Those of 1752 established a cabinet at the University of Vienna; those of 1774, an assistant to set out the apparatus before lecture. The collection excelled in models of machines, which aroused the admiration of Volta.[101] Freiburg had important pieces of apparatus, including an air pump and an electrical machine, by mid-century. It made further acquisitions through a special fee for degrees, introduced in 1752; set up a special cabinet sometime before 1756; and, as at Bamberg, made substantial additions in the 1780s.[102] Innsbruck had built up an impressive 'physical-mathematical cabinet' by 1761, for which a mechanic was engaged in 1774; meanwhile its professor of physics, Ignatius Weinhart, S. J., had assembled a good private

96. Modena, U., *Annuario* (1899–1900), 177; Pietro, *Studio* (1970), 39; Mor, *Storia* (1953), 260–2.

97. Italy, *Monografie*, I (1911), 251; Bernoulli, *Lettres*, III (1779), 181–3.

98. Torlais, *Physicien* (1954), 53–4; *DSB*, X, 145; Nollet, 'Journal' (1749), f. 11v.

99. Correspondence between Volta and Wilzeck, 1785–6, in *VE*, III, 283–4, 295, 311, 401. Volta had a regular budget of 725 lire (560#) annually for instruments around 1780; *VE*, I, 409–10.

100. Torlais in Taton, *Enseignement* (1964), 633; Dainville, *ibid.*, 40–1; Delfour, *Jésuites* (1902), 269.

101. Haberzettl, *Stellung* (1973), 141; *VE*, II, 246; Böhm, *Wiener U.* (1952), 62; Meister, Ak. Wiss., Vienna, Phil.-Hist. Kl., *Sb.*, 232:2 (1958), 36, 95.

102. [Kangro], *Beitr. Freib. Wiss. Univ. Gesch.*, 18 (1957), 10–11; Schreiber, *Gesch.*, II:3 (1860), 109. Bamberg also instituted a fee of from 0.5 to 1 RT in 1785 from each student taking physics to defray the cost of experiments; H. Weber, *Gesch.* (1879–82), 340.

collection, which he wished to be preserved for his order, in whose imminent resurrection he trusted.[103] Good collections were also made at Graz and Tyrnau.[104]

In Scotland the acquisition of instruments depended upon the generosity of friends of the universities. Money for the purchases at Edinburgh and Glasgow previously mentioned came from the town councils; Aberdeen received gifts for instruments from its graduating classes from 1721 to 1756, and from the Society for the Encouragement of Manufactures in Scotland between 1781 and 1785; Edinburgh raised £600 for apparatus during Robison's tenure of its chair of natural philosophy (1773–1797).[105] Oxbridge followed the continental practice. Cotes' instruments seem to have been passed down, probably by purchase, and were still in use in 1776; the collection of his successor once removed, Anthony Shepherd, had a reputation for excellence. Similarly Bradley had bought his apparatus from a previous lecturer, Whiteside the Ashmolean Keeper, for the large sum of £400. No doubt the instruments were sold once again at his death.[106]

PROFESSORIAL FINANCES

Perhaps the chief reason that eighteenth-century professors were not expected to do original work is that they seldom had time for it. Their academic salaries barely answered their needs; to live comfortably they taught more than their contracts required, consulted if they could, wrote textbooks, ran boarding houses, or, if clerics, assumed a share of the chores of parish or monastery.

Taking the simplest case first, the University of Paris established a fixed hierarchy of salaries in 1719, the year in which it abolished fees for courses: 1000# for professors of philosophy, 600# or 800# for regents in the lower forms.[107] The more popular professors suffered from the change, and even those whose income increased had little to celebrate, for 1000# did not support life in Paris. No doubt the fact that most of the professors were clerics made the system work. Salaries climbed slowly by award of supplements; just after the expulsion of the Jesuits the philosophy professors had supplements about equal to their salaries (2000# in all), and in 1783 they reached 2400#, still a very modest income. It is not surprising that Paris professors who were not also academicians contributed little to the progress of science. In the provinces the

103. Huter, *Fächer* (1971), 59–61.

104. According to Bernoulli, *Lettres,* I (1777), 50–4.

105. P. J. Anderson, *Studies* (1906), 151–2; Gr. Br., *Sess. Pap.*, 35 (1837), 132, 169–70; 38 (1837), 303–4; Dalzel, *Hist.* (1862), II, 449–50; *DSB,* XI, 495–7.

106. Winstanley, *Unr. Camb.* (1935), 151; Bernoulli, *Lettres* (1771), 117–18; Hans, *New Trends* (1951), 52; Gunther, *Early Sci.*, I (1920), 200–1.

107. Targe, *Professeurs* (1902), 188, 196–7; Jourdain, *Hist.*, (1888), II, 162–8; a livre of 1719 had 0.83 the silver content of our standard, the livre of 1726.

Jesuit monopoly kept salaries down until the 1760s, when they rose to 1200# in the bigger institutions.[108]

In Italy also, clerical monopoly of the lower faculties—as well as a lower standard of living—kept down salaries throughout the century. At Naples, for example, the physicist got about 900# in 1740, exactly twice the salary of the janitor. At Pavia he recieved some 500#, as against about 800# for the professor of law. There is evidence of improvement in the 1770s and 1780s: Naples was paying about 1300# in 1777; Catania gave about 500# in 1779 and 800# in 1787.[109] Volta's salary reached almost 4000# in 1795, and the senior physics professor at Pisa had about 2700#. These should be regarded as minimum amounts: often a living allowance or perhaps a house was added, and fees for degrees might bring something. At Modena, for example, professors participating in the examinations for the laurea got about 15# each from successful candidates; even the janitors had a share; and since two-thirds of the fee was returned to unsuccessful candidates, the entire university had a stake in preventing failures. The improvement in facilities and salaries helped upgrade Italian physics to the point that, in 1784, Lichtenberg could try to obtain a grant for travel to the peninsula on the ground that 'Italy is now, perhaps more than Britain, the home of true physics, der Sitz der wahren Naturlehre.'[110]

The differences between Oxbridge and the Scottish universities are no better illustrated than in professorial emoluments. The Oxbridge professor had a salary, fixed by statute, in return for which he lectured publicly, that is gratis, on a specified subject. Salaries varied according to the wishes and wealth of the founder of the chair, the prestige of the subject, and the date of foundation. The Lady Margaret Chair of Divinity at Cambridge brought some £1,000 at the end of the eighteenth century, when the Regius professorships in languages, law and medicine still yielded the £40 fixed for them by Henry VIII. The chairs for science, as relatively late foundations, usually carried adequate emoluments, from £100 (the Lucasian) to £300 (the Lowndean); college fellowships paid less, sometimes less than £40, and often about £60. Tutoring brought a pound a student a term. With £100 (2500#) a don might be comfortable, with £200 well off, and with £40 'almost destitute.'[111]

The salaries came whether the professor had few students or many, or indeed whether he lectured or not. Consequently his interest, as Adam Smith remarked, was 'directly in opposition to his duty.' No doubt Smith correctly

108. Targe, *Professeurs* (1902), 308; Lacoarret and Ter-Menassian in Taton, *Enseignement* (1964), 135; Gaullier, *Collège* (1874), 508; Montzey, *Hist.* (1877), II, 158.

109. Amodeo, *Vita*, I (1905), 13, 61, 155; Pavia, U., *Contributi* (1925), 120; Catania, U., *Storia* (1934), 258.

110. Volpati, *Volta* (1927), 42–8; Scolopio, *Storia* (1877), 79; Mor, *Storia* (1953), 110; Schaff, *Gesch.* (1912), 180–1; Lichtenberg to Schernhagen, 30 Sept. 1784, in Lichtenberg, *Briefe* (1901–4), II, 147.

111. Winstanley, *Unr. Camb.* (1935), 97, 101–2, 121, 129, 151–2, 171–3; Mallet, *Hist.*, III (1927), 124; Godley, *Oxford* (1908), 83; Frank, *Hist. Sci.*, 11 (1973), 256.

associated the sinecurism of the Oxbridge professoriate of his time with its inability to exact fees for statutory lectures. It could, however, charge for additional services: Bradley, as Savillian professor of astronomy, asked 3 gns for his course on experimental physics, and drew an average attendance of 57. The Plumian professor at Cambridge also charged for experimental physics; as late as 1802 the incumbent, Samuel Vince, still advertised lectures for the conventional 3 gns.[112]

The exception in England was the rule in Scotland, where professors collected fees of 2 or 3 gns a student. The total incomes of physicists and mathematicians ranged from about £150 to a little over £300. At the turn of the century fees accounted for about half the total at Aberdeen, for a fourth or less at St. Andrew's, for two-thirds to five-sixths at Edinburgh. Robison, for example, had a salary of £52 and an average of £260 a year from students in the late 1790s.[113] The Scottish natural philosopher worked harder to earn more than his English counterpart. No doubt economic incentive helped to make Edinburgh the leading British university in the late eighteenth century.

But the lands of academic opportunity were Holland and Protestant Germany. A man with a reputation might exact a large salary as a price for accepting or refusing a call; he could negotiate important fringe benefits, such as a free dwelling, firewood, bread and beer; and he could complete his happiness by attracting crowds of students to courses for which he was entitled to charge fees. Wolff was particularly successful at this game. He had 200 RT (800#) when he began at Halle in 1706; after calls to Leipzig, Jena and Petersburg, he got 600 RT.[114] Marburg hired him in 1723 at 500 RT with perhaps as much again in fringes; which, as he said, was 'nothing trifling in Germany,' although only half the income of the Italian singers in the Hessian opera.[115] Prussia brought him back to Halle in 1740 at the price of 1000 RT. All the while he took in substantial fees. His 'private' lectures at Marburg had as many as 100 auditors; his total income from university sources—salary plus fringes plus fees—exceeded 2000 RT annually by 1724. He had in addition royalties from his books and rent from student lodgers (1 or 1.5 RT a week).[116] Taking lodg-

112. Smith, *Wealth of Nations* (1880), II, 345–6; *DNB*, II, 1074–9, XX, 355–6; Rouse Ball, *Hist.* (1889), 104. Cf. Wordsworth, *Scholae* (1877), 255n.

113. Gr. Br., *Sess. Pap.*, 35 (1837), 51–64, 130 (tabulated by Morrell, *Isis*, 62 [1971], 165); 37 (1837), 247–8. Cf. Dalzel, *Hist.* (1862), II, 324; Grant, *Story* (1884), II, 298.

114. Wolff to Blumentrost, 24 April 1723, in Wolff, *Briefe* (1860), 14. Other results of well-played calls: G. E. Hamberger's extra professorships and the dignity of Hofrath for refusing invitations to Altdorf, Göttingen and Halle (Börner, *Nachrichten*, I [1749], 63–5); Musschenbroek's instruments and raises (to 3000#) for remaining at Utrecht (Kernkamp, *Utrecht. U.* [1936], I, 132–3).

115. Wolff, *Briefe* (1860), 23 (7 May 1724); Hermelink and Kähler, *Phil.-U. Marb.* (1927), 347n; Gottsched, *Hist. Lob.* (1755), Beylage, 34. The fringes: 115 bushels of corn, 90 of barley, 55 of oats, 5 of peas; 10 sheep, 2 pigs, 167 pounds of fish; 164 gallons of wine; free housing in the observatory.

116. Wolff, *Briefe* (1860), 14, 25, 36, 103, 114; cf. Wuttke, *Wolff* (1841), 68.

ers was very common. Professors had large houses because their private courses, and very often their public ones as well, had to be taught in their homes. It was frequently necessary to realize some income from the extra rooms in their mansions.[117]

Wolff was rewarded for more than his physics and mathematics. A more representative entrepreneur is Kästner, who began teaching mathematics at Leipzig for 200 RT, 'which would perhaps be enough for me if I were as abstemious in pleasures of the mind as I am in those of the body.' To satisfy his lust he wrote reviews for learned journals, which often paid him in books; attendance at his courses, about twenty students a year, all beginners, brought little in fees. Vigorous effort raised his total income to between 400 and 500 RT. In 1755 he was called to Göttingen to replace Segner; Göttingen gave him 'more than I could ever hope to have had at Leipzig,' and opportunities for offering advanced courses at unusually high fees.

Salaries were not particularly high at Göttingen, and did not inflate so quickly as at other universities; Lichtenberg remarked in 1784 that Saxony (meaning Leipzig), Weimar (Jena), and Mainz paid more, while Gedike was surprised to find in 1789 that many Göttingen professors had between 300 and 400 RT.[118] (Gedike's poorest universities, Altdorf, Erlangen, and Giessen, paid their professors of philosophy 50 to 150 RT; his best, Jena, Leipzig and Wittenberg, 400 to 600, or more.[119]) But Göttingen excelled in the size of its fees. By teaching between four and five hours a day, which probably did not much exceed the average, Kästner brought his income to over 1000 RT a year.[120]

Fees charged for instruction—as opposed to premiums for degrees—became more and more important in the finances of the German professor at precisely the time that experimental physics was entering the university curriculum. In principle one gave public lectures in courses necessary for degrees in return for one's salary, and offered private instruction, for a fee, in specialized or advanced subjects. During the eighteenth century the public lectures rapidly declined, and the private courses, especially in the leading universities, came to fill most of the curriculum, even in the philosophical faculty; by the beginning

117. Paulsen, *Gesch.* (1896–7), I, 536, II, 13.

118. Lichtenberg, *Briefe* (1901–4), II, 137–8, 141; Fester, *Gedike* (1905), 17. The Catholic University of Mainz does not belong in the group because its high salaries—Lichtenberg pointed to an offer of 1800 RT—compensated for its professors' inability to charge fees (*ibid.*, 47).

119. Cf. W. Hesse, *Beiträge* (1875), 66, giving salaries of two philosophers at Duisburg in 1775; the physicist Schilling, aged 72, 270 RT after 47 years' service; J. A. Melchior, aged 54, a 20-year veteran, 128 RT. Pfaff had only 300 RT when he began at Kiel in 1793 (*ADB*, XXV, 582–3). When Duisburg professors complained about their poverty, they were turned away with the agreeable information that other Prussian universities, Königsberg and Frankfurt/Oder, suffered equally. Ring, *Gesch.* (1949²), 179–80.

120. Kästner, *Briefe* (1912), 10, 30, 36–7, 59, 65, 123, 213–14; Müller, *Abh. Gesch. math. Wiss.*, 18 (1904), 103n. Paulsen, *Gesch.* (1896–7), II, 142, estimates the average teaching at 20 to 24 hours a week, and observes that Kant once offered 34.

of the nineteenth century, fees were charged in the main courses leading to degrees. Among the causes of this remarkable evolution were the economic pressures of the secular inflation of the eighteenth century and the advance of knowledge, the creation of new subjects and new approaches that could be construed as material beyond the purview—and hence beyond the responsibility—of the public lecture.[121] Fees for a course *privatim* in the philosophy faculty ranged from 1 to 6 RT, depending on the university and the reputation of the professor; according to Gedike's numbers, Göttingen's 4 to 6 RT was about twice the average charge at German Protestant universities in 1789. A course *privatissima* came still dearer, at between 15 and 20 RT, or perhaps even twice that, at Göttingen.[122]

The right to charge fees meant little unless students enrolled in sufficient numbers. At the larger universities one could take as much or more in fees as in salary. The case of Kästner has been mentioned. Similarly Lichtenberg took 80 louis d'or (about 450 RT) from 112 auditors in 1784, a considerable improvement over the 40 students with which he started in 1777 / 8. Wolff averaged over 100 at Marburg. Gilbert had 40 in physics and 12 in mathematics at Halle in 1801–2; Kratzenstein had between 30 and 40 at Copenhagen.[123] At the smaller schools little could be got. Andreas Nunn, for example, had no takers for a course *privatim* in experimental physics at Erfurt in 1755. At Duisburg, where matriculations averaged less than 50 a year, not much could be hoped for. And even in the bigger schools difficult or specialized subjects might not pay. C. A. Hausen, professor of mathematics at Leipzig, lectured publicly, 'for no one would give money to hear about conic sections,' and even then he had few auditors.[124]

To complete the picture of professorial income we must add in premiums for degrees (which cost from 43 RT for an MA to 132 RT for a doctorate in theology at Göttingen in 1768), fees for preparing theses (30 RT at Halle), royalties, payment for collaborating in learned journals or reviews, gifts from dedicatees, and so on.[125] It is very difficult to estimate the income from these sources, which could be considerable for a well-placed man. On the whole, an able and energetic professor of physics could do better at a leading Protestant German university than at any other at the end of the eighteenth century.

121. Paulsen, *Pr. Jahrb.*, 87 (1897), 138–41; *Gesch.* (1896–7), II, 128–9.

122. Fester, *Gedike* (1905), *passim;* Müller, *Abh. Gesch. math. Wiss.*, 18 (1904), 82; Pütter, *Versuch,* I (1765), 319, reporting the higher figure—indeed 30 to 100 RT—for 1765. Other indications of fees: 2–6 RT for 5 hours' *privatim* at Halle (Schrader, *Gesch.* [1894], I, 108–9); 3.5 RT for two months', probably eight hours', *privatim* at Basle in 1760 (Spiess, *Basel* [1936], 113); 4 RT for a semester's *privatim,* 50–100 *privatissima,* at Marburg (Hermelink and Kähler, *Phil.-U. Marb.* [1927], 385).

123. Lichtenberg, *Briefe* (1901–4), II, 127, 228, 335; Hermelink and Kähler, *Phil.-U. Marb.* (1927), 391n; Schrader, *Gesch.* (1894), I, 635; Snorrason, *Kratzenstein* (1967), 53–4.

124. Stieda, *Erf. U.* (1934), 19, 27; W. Hesse, *Beiträge* (1875), 48; Kästner, *Selbstbiographie* [1909], 6–7.

125. Pütter, *Versuch,* I (1765), 320; Schrader, *Gesch.* (1894), I, 108–9.

It remains to compare the price of the standard instruments to professorial incomes. A benchmark is the gift to Harvard of a good apparatus in the style of Hauksbee, bought new in Britain in 1727 for about 3000#, or the estimate of 200 gns (5000#) for a full outfit of books and instruments made by the University of Aberdeen in 1726, or the value of Jallabert's excellent collection (4500#) assembled in the 1740s.[126] (The so-called 'complete' apparatus in the style of Musschenbroek, purchased in London by the University of Uppsala in 1740 for about 2000#, must have lacked something.) From these data it appears that the cost of a full set of instruments in the 1730s was about equal to the annual income of a well-paid professor of physics.

Fifty years later, Volta drew up a list of instruments 'needed' to bring his cabinet up to the mark. It ran to 9300# for purchases in France and England, plus an unestimated charge for items to be made locally. And that did not include apparatus for electricity, 'for which at the moment there is nothing much good in the gabinetto.' Such an expenditure was beyond Volta's means, and well beyond what was absolutely required; he wished an apparatus that would not only instruct his students but impress his many foreign visitors, who 'will view the physics cabinet with the same satisfaction and surprise—and will talk about it everywhere—as they already see, praise, and admire the botanical garden, the chemical laboratory, and the museum of natural history.'[127] We may take it that a meagre but serviceable demonstration apparatus could be bought at less than 5000# at the mid-century, and that a full and fancy one cost upward of 10,000# in the 1780s.[128]

The items in the cabinet of greatest interest to us—air pumps and, above all, electrical machines—were among the most expensive. The average cost of Volta's desiderata was 150#. Jan van Musschenbroek's double-barreled air pump sold for *f.* 300 (650#) in 1736; Martin's best pump cost 35 gns (900#) in 1765; Nairne's standard pump cost as much in the 1780s, and considerably more with accessories.[129] Nairne's standard electrical machine, a serviceable but not elaborate model with a six-inch globe, cost 170# in 1765, when Martin's best large machine brought 490#. Nairne's big cylinder machine could be bought for 480# in 1779; Cuthbertson wanted over 2000# for his completely furnished three-foot plate machine in 1782.[130] Up-to-date electrical machines and air pumps went beyond the reach of most professors in the 1780s.

126. Cohen, *Tools* (1950), 133; Rait, *Universities* (1895), 295–6; Borgeaud, *Hist.*, I (1900), 571. The Scots did not succeed in raising the money.

127. Volta to Firmian, 13 March 1780, *VE,* III, 455–67.

128. This estimate agrees with the cost of the second complete Harvard apparatus, acquired in 1765/6 for about £400 (9800#), for it contained duplicates of several expensive items. Millburn, *Martin* (1976), 131–5, 142–3.

129. Crommelin, *Desc. Cat.* (1951), 33; Lichtenberg, *Briefe* (1901), II, 6; *VE,* II, 146; cf. Henderson, *Life* (1867), 216. The Harvard air pump, bought in 1727, also cost about 650#; Cohen, *Tools* (1950), 141.

130. *BFP,* XII, 259; Hackmann, *Cuthbertson* (1973), 52; Lichtenberg, *Briefe,* I (1901), 277, 338, and *GGA* (1786:2), 2012–13; Millburn, *Martin* (1976), 131, 219. An early plate machine, of

4. INDEPENDENT LECTURERS

Electricity figured prominently in the repertoire of independent lecturers on experimental physics. They attracted many to its study, and occasionally made advances in it themselves. Their chief goal was popularization and entertainment, the reduction of the latest discoveries to the level of 'the meanest capacities' able to afford the service. One offered to explain everything 'in such a plain, easy and familiar Manner, as may be understood by those who have neither seen or read anything of the like Nature before.'[1]

The mean capacities could choose from a wide range of purveyors. At the top were the public lecturers associated with learned societies. We already know the permanent lectureships at Stockholm and Munich. Other academies, such as Bordeaux, colleges of the University of Paris, and secondary schools in France and Britain also occasionally engaged a 'physicist' to instruct and amuse them. A second class of lecturer consisted of members of learned societies who set up independently of their institutions. A notch lower, perhaps, came the unaffiliated entrepreneurs, who taught in rented rooms, and the itinerant lecturers, who performed in public houses. At the bottom of the heap were the hawkers of curiosities, the street entertainers, and the jugglers who held forth at the fairs of Saint Laurent and Saint Germain.[2]

LECTURERS OF THE BETTER CLASS

In France, public lectures of quality on the mechanical philosophy go back to the middle of the seventeenth century, when Jacques Rohault began his 'Wednesdays' dedicated to experimental illustrations of the physics of Descartes. Educated by the Jesuits in Amiens, Rohault took readily to mathematics, mechanics, and Descartes, set up in Paris as a tutor in geometry, visited artisans 'for the pleasure of seeing them work,' and won himself a fortune and a wife above his station, a lady sacrificed by her Cartesian father 'for the sake of the philosophy of Descartes.' Although he became chief of the Cartesian physicists, Rohault by no means slavishly followed Descartes; like all the successful public lecturers, he had to care more for the phenomena than for the system; not metaphysics but clarity, eloquence and manipulative skill brought in paying auditors 'of all ages, sexes and professions.'[3] Even the physicists regarded him favorably.[4]

---

five-foot diameter, was made for the Duc de Chaulnes in 1777 for 800#; Ingenhousz, *PT*, 69 (1779), 670.

1. Ferguson (1764), quoted in Harding, *Hist. Ed.*, 1 (1972), 149.

2. Cf. A. Franklin, *Dictionnaire* (1906), 570; Kästner, 'Verbindung,' in *Verm. Schr.* (1783[3]), II, 364; and Pujoulx, *Paris* (1801), 33: 'Hé comment les sciences ne feraient-elles pas des progrès rapides! Les savans courent les rues, et nos boulevards sont devenus des écoles de physique.'

3. Savérien, *Hist.*, VI (1768), 5–20; Mouy, *Développement* (1934), 108ff.; Pacaut, Ac. sci., Amiens, *Mém.*, 8 (1881), 5, 9. Mouy, *Développement* (1934), 112, is doubtless correct in rejecting the canard (Savérien, *Hist.*, VI [1768], 22) that the pedant Pancrace in Molière's *Marriage forcé* (1664) is based on Rohault.

4. Mouy, *Développement* (1934), 187n.

Among Rohault's emulators Pierre-Sylvain Régis and the physician Pierre Polinière were most conspicuous. Régis, a student of the Jesuits at Cahors, went to Paris for theology but gave it up on hearing Rohault. Admitted to discipleship, he was sent in 1665 to the provinces to lecture publicly on Cartesian physics. He returned to Paris after Rohault's death and lectured to great applause until the Archbishop of Paris shut him down, 'in deference to the old philosophy.'[5] Régis thereupon offered 'private' courses to the mighty, among them the archbishop, who is said to have become the most enthusiastic of his auditors. Régis entered the Paris Academy at its reorganization in 1699; although he was too old and ill for academic work, 'his name [as Fontenelle gracefully put it] served to ornament a list on which the public would have been surprised not to find it.'[6]

Polinière also developed against a Cartesian background. He was educated by the Jesuits at Caen and at the University of Paris, where in 1695 he initiated a course of experimental physics under the auspices of the professor of philosophy at the Collège d'Harcourt. He perhaps took as his inspiration and model his professor of mathematics, Varignon, who occasionally used experiments to illustrate his lectures at the Collège Mazarin. Polinière's course was clear, intelligent, 'a mortal blow [we are told] to the physics of Aristotle.'[7] He worked hard at improving standard demonstrations, which he interpreted in undogmatic Cartesian terms. His enterprise was sometimes rewarded by the discovery of new phenomena, like electroluminescence, which might be shown to advantage. He succeeded before general audiences, in guest lectureships at Parisian colleges, and before the Regent, young Louis XV, and Fontenelle. Yet, like Rohault, Polinière remained outside the Academy of Sciences, which first admitted such an entrepreneur in the person of Nollet, whose lecturing began just after Polinière's death.[8]

With the help of academicians whose assistant he had become, Nollet went to London to seek the advice of Desaguliers, and to Leyden to inspect the instruments of 'sGravesande. He found the apparatus so expensive that he could finance it only by building and selling duplicates: before mounting the podium he had first to enter the workshop. 'I wielded the file and scissors myself [he wrote of that time]; I trained and hired workmen; I aroused the curiosity of several gentlemen who placed my products in their studies; I levelled a kind of voluntary tribute; in a word (I will not hide it) I have often made two or three instruments of the same kind in order to keep one for myself.'[9] By 1738, when

5. *Ibid.*, 146, 166–7.

6. 'Eloge de Régis,' *Oeuvres,* V, 92.

7. Savérien, *Hist.*, VI (1768), 167–73; Hanna in Gay, ed., *Eight. Cent. St.*, 16–18. Fontenelle, *Oeuvres,* VI, 261, traces college lectureships in experimental physics to the example of the semi-private lectures arranged by the apothecary M.F. Geoffroy for the benefit of his son, Etienne-François, c. 1690.

8. Polinière, *Expériences* (1718^2), Préf.; Corson, *Isis,* 59 (1968), 402–13; Brunet, *Physiciens* (1926), 101–2; Savérien, *Hist.*, VI (1768), 185–7.

9. Nollet, *Programme* (1738), xviii–xix; Grandjean de Fouchy, *HAS* (1770), 121–37; Lecot, *Nollet* (1856), 1–13; Torlais, *Physicien* (1954), 1–40.

Nollet provided his course with a formal syllabus, his business could handle an order from Voltaire for instruments costing over 10,000#.[10]

Nollet's *Cours de physique,* which incorporated phenomena he had discovered, was perhaps the most popular exhibition of its kind ever given. In 1760 he drew 500 paying customers.[11] He aimed to be useful and agreeable, to entertain his auditors as he disabused them of their 'vulgar errors, extravagant fears, and faith in the marvellous.'[12] People of all conditions flocked to hear him, including duchesses, whose carriages piled up before his doors, and princes of the blood, who 'honored the master with their close attention, and brought away the kind of knowledge that is always an ornament to the mind, and confers luster on the most distinguished birth.' In 1739 Nollet entered the Academy as adjunct mechanic and in 1757 he became a pensionary.[13]

Several others managed to follow Nollet's example. His protégé Mathurin-Jacques Brisson marched at his heels: assistant to Réamur, public lecturer, academician, professor.[14] Similarly Sigaud de la Fond, member of the academies of Montpellier and Angers and, in 1796, also that of Paris, amused the Parisian *grand monde* with experimental physics—'Nollet improved,' he said—from about 1767.[15] The best of these later public lecturers was J. A. C. Charles, who turned to experimental physics at the age of 35, when an economizing ministry abolished his petty bureaucratic post. 'He was left with what happily suffices for those who are to excel in the arts, the free disposition of his time and talents.'[16] In 1781, after eighteen months of study, he began to lecture. His skill, plus the advertisement of a journey in his hydrogen-filled balloon, brought him a large audience. With their fees he built up what in 1795 was judged to be the most complete collection of demonstration apparatus in Northern Europe, all fashioned in the style of Nollet, Brisson, and Sigaud de la Fond.[17]

The first public physics course in London was inaugurated in 1704 by an important electrician, Francis Hauksbee, who began as an instrument maker and gave the public material like Polinière's, but interpreted on Newtonian

10. Voltaire, *Corresp.*, VI, 191, VII, 156, 176, 261: 'C'est un philosophe [Nollet], c'est un homme d'un vray mérite qui seul peut me fournir mon cabinet de physique, et il est beaucoup plus aisé de trouver de l'argent qu'un homme comme luy.'

11. Ferrner, *Resa* (1956), xliii; Tolnai, ed., *Cour* (1943), 64.

12. The same sentiment appears in Polinière, *Expériences* (1718²), Préf.; cf. Savérien, *Hist.*, VI (1768), 190.

13. Nollet, *Programme* (1738), xxxv–xxxvi; Marquis du Châtelet to Francesco Algarotti, 20 April 1736, in Du Châtelet-Lomont, *Lettres* (1958), I, 112. Cf. *ibid.*, 93; Lecot, *Nollet* (1856), 19; Torlais, *Physicien* (1954), 41–63, 203–4.

14. Torlais, *Physicien* (1954), 234–6.

15. *Ibid.*, 232–3; Torlais in Taton, *Enseignement* (1964), 630–1.

16. Fourier, *MAS,* 8 (1829), lxxiv.

17. *Ibid.*, lxxvi; Bugge, *Science* (1969), 154, 166–8. For the situation in the provinces, where demand picked up briskly in the '70s and '80s, see Mornet, *Origines* (1947²), 316; Torlais in Taton, *Enseignement* (1964), 634; G. Martin, *RHS,* 11 (1958), 214, reporting difficulty in obtaining subscribers in 1747.

principles.[18] The success of his lectures may be inferred from the eagerness of two separate parties to continue them after his death in 1713, namely his nephew, Francis Hauksbee the Younger, and Desaguliers, who perhaps left Oxford for the purpose. Desaguliers' efforts proved the more attractive to the public, who consumed, on the average, some six cycles of his lectures every year.[19]

Desaguliers also succeeded to another post of Hauksbee's, that of occasional curator of experiments to the Royal Society of London. The position, which had fallen into desuetude by the end of the seventeenth century, was apparently revived for Hauksbee when Newton became president of the Society in 1703. As we know, its incumbent had to prepare and exhibit experiments to the Fellows at their weekly meetings, an onerous task if, as in the case of Hauksbee and Desaguliers, at least some of the demonstrations rested upon original work, and reimbursement for out-of-pocket expenses was not always prompt. (When expenses were slight, payment might be immediate, as when Desaguliers 'made a present of a Worm vomit'd by a Cat, for which he had thanks.'[20]) Desaguliers' burden appears from a memorandum addressed to the Society to justify an expenditure of about £10 for the construction of four new machines. 'Before I bring anything to the Society I spend many Days about it at Home to try the Experiments before Hand; and adjust the Machines; so that the time expended and accidental Charge that Way, is often more than double the Cost of the Machines, especially because it often happens that the whole Instrument is thrown by, when I find it is not worth the Society's Notice.'[21] Doubtless apparatus tested at the Society's expense found its way into his lecture room.

Hauksbee and Desaguliers had close ties with leading applied mathematicians and physicists. Hauksbee began his lectures with James Hodgson, an assistant of the Astronomer Royal, John Flamsteed,[22] and in his curatorial capacity Hauksbee often worked with Newton. The younger Hauksbee teamed up with William Whiston, whom we have met as a Cambridge professor of mathematics.[23] Desaguliers had worked with the belligerent Newtonian, John Keill, one-time professor of astronomy at Oxford.[24] As Hauksbee's successor, Desaguliers became the most active of Newton's agents in the Royal Society.

Desaguliers had no successor of equivalent stature: unlike Paris, London proved unable to generate or to support such men in the second half of the

18. Hauksbee, *Phys. Mech. Exp.* (1709).

19. Guerlac in *Aventure* (1964), I, 228–53; Desaguliers, *Course* (1763[3]), I, ix, says that he completed his 121st lecture cycle in 1734.

20. JB, XIV, 279 (RS).

21. *DNB*, V, 850–1; Torlais, *Rochelais* (1937), 1–14; Desaguliers to RS, 29 Oct. 1733, in Misc. Corresp. D:2, ff. 71–5 (RS).

22. Rowbottom, Cong. int. hist. sci., XIe (1965), *Actes*, IV (1968), 198–9; Hug. Soc. Lond., *Proc.*, 21 (1968), 191–206.

23. Hans, *Trends* (1951), 49–50, 137, 142–3; Whiston, *Memoirs* (1749), 235–6; *DNB*, VI, 175–6 (Hauksbee Jr.)

24. *Supra*, ii.3.

eighteenth century. From time to time amateurs offered to show the public the latest scientific discoveries; such as one Rackstrow, who '[took] impressions from life, [and made] them up in plaster,' and also demonstrated electricity, about which he wrote a pamphlet at the urging of his friends, 'not being proof against flattery.'[25] There is evidence that in the late 1750s London could not maintain a single distinguished independent public lecturer. S. C. T. Demainbray, a disciple of Desaguliers', a successful lecturer in both France and England in the 1740s and early 1750s, the owner of what Franklin judged to be the best demonstration apparatus in the world, could in 1758 'hardly make up an audience in this great City [London] to attend one course a winter.'[26] In the same year James Ferguson, a painter who had been trying for a decade to support himself by his excellent courses on astronomy and experimental physics, contemplated leaving the capital: 'There are at present more than double the number [of hopeful demonstrators] which might serve the place, people's taste lying but very little that way; so that unless something unforseen happens, I believe my wisest course will be to leave London soon.'[27]

Ferguson managed to make a living, and a good one, by going on tour: Bristol and especially Bath, where, like Martin before him, he was always warmly received by the fashionable and unoccupied water-takers. Reading, Gloucester, Salisbury, Liverpool, Newcastle, and the growing industrial towns of the Midlands, which lacked facilities for adult education and recreation, supported many itinerant lecturers; at Manchester the public subscribed to at least one course a year from 1760 to 1800.[28] Naturally these courses varied in quality, from the authoritative lectures and original demonstrations of a Ferguson to the entertainments of Gustavus Katterfelto, a German who worked the Midlands with 'electricity, and a few other tricks of physics, and a little of the art of conjuring.'[29] The best of the itinerant lecturers—Adam Walker, Henry Moyes, and others—included London in their circuit, but few besides Ferguson maintained their headquarters there. Some numbers will illustrate the extent to which British public lecturers in natural philosophy shifted their attention to the provinces after 1740. Taking as a sample the fifty or so individuals about whom something is known, one finds seven public lecturers active in London in each decade from 1710 to 1740, but only four in the forties and fewer thereafter; in the case of itinerant lecturers, two were active per decade before 1740 and ten on the average from then until the end of the century.[30]

25. Rackstrow, *Misc. Obs.* (1748), i–ii. For reasons set out *infra*, iii.2, public demonstrations of electricity reached a peak in the late 1740s. As Henry Baker sneered in 1747, many then earned a 'great deal of money shewing a course of Electrical Experiments at a Shilling for each Person'; G. Turner, RS, *Not. Rec.*, 29 (1974), 64.

26. Franklin to Kinnersley, 28 July 1759, *BFP*, VIII, 416; *DNB*, V, 780–1. For colonial itinerant lecturers, see Stearns, *Science* (1970), 510–11.

27. Ferguson to A. Irvine, 17 Jan. 1758, in Henderson, *Ferguson* (1867), 225.

28. Musson and Robinson, *Science* (1969), 102n; Gibbs, *Ambix*, 8 (1961), 111; Millburn, *Martin* (1976), 38, 49–51.

29. Musson and Robinson, *Science* (1969), 101–2n.

30. Compiled from *ibid.*; Gibbs, *Ambix*, 8 (1961), 111–17; Mumford, *Manchester* (1919);

The Netherlands also supported several distinguished independent lecturers. Fahrenheit was perhaps the first; he supplemented the revenues from his thermometers by giving public instruction in experimental physics in Amsterdam from 1718 to 1729. Shortly after he retired, Desaguliers made a triumphant tour in Holland, billed by Musschenbroek as 'one of the most famous philosophers of the age.'[31] Several lesser intellects, impressed by the spectacle and the profits, set up in Amsterdam and elsewhere. After 1750 lecturers became attached to the newly-founded scientific societies, despite the prohibition against the sponsorship of public courses forced on the academies of Haarlem and Rotterdam by the University of Leyden. The most important of these men, who perhaps numbered a dozen in all, was the ubiquitous van Marum, who lectured first without sponsorship and then moved under the wing of the Teyler Genootschap.[32] A similar development occurred in Sweden, where the public lectures of Triewald, conceived in the style of Desaguliers and 'sGravesande, were later replaced by those of the Stockholm academician Wilcke.[33]

FINANCES

The London lecturers in the 1720s asked two to three guineas for a course; the itinerants of the 1760s one guinea for subscribers, or a half crown for a single lecture. Since the latter customarily gave twelve lectures and the former twice that, the average cost per session remained between one and two shillings throughout the century. Subscriptions were payable in advance, and arranged before the lecturer came to town; usually he required a guaranteed minimum, twenty or thirty paid-up clients locally, or forty or more if he had to travel, before he would agree to perform. A good lecturer could make something in this way. In four months in 1763 Ferguson grossed £139 in Bristol and Bath, and the same area yielded over twice as much in 1774. Adam Walker made 600 guineas in Manchester and Liverpool in 1792; James Bradley and Desaguliers averaged 420 and perhaps 300 guineas, respectively, in Oxford and London, in the 1720s and 1730s.[34]

Special events might bring special emoluments, such as the £120 that Whiston had from a 'numerous and noble audience' for a lecture on an upcoming solar eclipse. This was good money, and above average expectation, as we

---

Turner in Forbes, *Marum,* IV, 1–38; Fawcett, *Hist. Today,* 22 (1972), 590–5. The total for the 1730s agrees well with the figure (eleven or twelve) given by Desaguliers, *Course,* I (1734), Pref.

31. Torlais, *Rochelais* (1937), 22; Rooseboom, *Bijdrage* (1950), 21; Cohen and Cohen-de Meester, *Chem. Week.,* 33 (1936), 379–83.

32. Muntendam in Forbes, *Marum,* I, 16–17; Brunet, *Physiciens* (1926), 98; Dekker, *Geloof weten.,* 53 (1955), 173–6; Hackmann, *Cuthbertson* (1973), 15.

33. Beckman, *Lychnos* (1967–8), 187–93; Tandberg in Lund., U., *Årssk.,* avd. 2, 16:9 (1920), 4–5; *supra,* ii.2.

34. Hans, *Trends* (1951), 47–8, 139, 142–3, 147–8; McKie, *Endeavor,* 10 (1951), 48–9; E. Robinson, *Ann. Sci.,* 19 (1963), 31; Musson and Robinson, *Science* (1969), 104, 145, 164–5; Henderson, *Ferguson* (1867), 272, 340–1, 348, 376, 408; Fawcett, *Hist. Today,* 22 (1972), 590–5; Millburn, *Martin* (1976), 61–2.

learn from a letter from Smeaton to Benjamin Wilson, then (1746) contemplating an itinerant lectureship: 'I don't take y^{t} shewing ye wonders of Electricity for money is much more considerable than ye shewing any other strange . . . sight for ye same end, however if £200 could be got by a worthy employment in y^{t} way I don't see where is ye harm as there is no fraud or Dishonesty in it.'[35] One needed a capital of good will, a little information, and an apparatus costing about £300.[36]

The price of subscription shows that English lecturers did not aim at the common man. John Roebuck, acting as advance man for Henry Moyes, 'procured him the Countenance and favor'—that is the subscriptions—'of some principal gentlemen' in the neighborhood.[37] Erasmus Darwin recommended that young ladies improve themselves 'by attending the lectures in experimental philosophy, which are occasionally exhibited by itinerant philosophers.'[38] Benjamin Martin, writing of the lecture circuit of the 1740s and 1750s, permits us no doubt about his clientele. 'There are many places I have been so barbarously ignorant, that they have taken me for a *Magician*; yea, some have threaten'd my life, for raising Storms and Hurricanes: Nor would I show my face in some Towns, but in company with the Clergy or the Gentry, who were of the Course.'[39] And when gentlemen lost interest in natural philosophy, most of the audience of the independent lecturer disappeared. One cause of the depressed market for physics in London around 1760 was a diversion of interest to current events, local intrigue and the Seven Years' War. 'Some Notice may be taken abroad, of what is new and ingenious in Matters of Natural Philosophy; but here we think of nothing but Politicks, Money and Pleasure.'[40]

The Parisian purveyor of natural philosophy also suffered from changes in fashion, for his prosperity depended in large measure on pleasing the ladies. In the 1740s they flocked to Nollet, in such numbers as to drive away the gentlemen: 'It seems that among the fashionable only women are still [1749] able to meddle publicly with physics.'[41] Interest appears to have declined in the 1760s, only to rise to a new pitch in the 1780s. Paris then supported many independent lecturers. The 'lycée' of Pilatre de Rozier, 'la vogue de Paris,' had 700 sub-

35. 24 Sept. 1746, Wilson Papers, f. 22 (RS); Whiston, *Memoirs* (1749), 204–5.

36. For a good collection such as Ferguson's (Henderson, *Ferguson* [1867], 453) or John Whiteside's (Turner in Forbes, *Marum,* IV, 18); one could get by with £100, as did Caleb Rotheram (Musson and Robinson, *Science* [1969], 90). Demainbray's expenditure, estimated by Franklin at £2000 (*BFP,* VIII, 416), was probably largely for fine furniture; cf. the cost of university cabinets (*supra,* ii.3).

37. Roebuck to Watt, 14 Aug. 1777, Musson and Robinson, *Science* (1969), 145n. Mathew Bolton performed the same service for John Warltire in 1776 and 1779; McKie, *Endeavor,* 10 (1951), 48–9.

38. *Plan for the Conduct of Female Education in Boarding Schools* (1797), quoted by E. Robinson, *Ann. Sci.,* 19 (1963), 30n; cf. Millburn, *Martin* (1976), 73: 'It is now [1755] growing into a fashion for the ladies to study philosophy.'

39. B. Martin, *Supplement* (1746), 28–9n.

40. Symmer to Mitchell, 30 Jan. 1761, Add. Ms., 6839, f. 309 (BL).

41. J. B. Le Roy to the comte de Tressan, 26 Aug. 1749, in Tressan, *Souvenirs* (1897), 6.

scribers in 1785, mostly women, and Swiss guards at its doors. That year its professor of physics, Antoine de Parcieux, successor of Nollet at the Collège Navarre, offered two complete courses on natural philosophy.[42] He faced very strong competition: to mention only those of the highest quality, Brisson advertised two courses, Charles four, and Sigaud's successor, Rouland, no fewer than ten.

Descending a level, we find Jacques Bianchi, rue St. Honoré, puffing several series in electricity and offering to provide, at a cost of 55 louis (1320#) and within six months of order, a complete outfit for demonstrating the truths of experimental physics. He did not exhaust the opportunities of either buyer or seller. A competitor of Bianchi's, one Bienvenue, prepared an apparatus for his auditors to take on summer holidays, to forestall ennui; while several characters at the Palais Royal showed electricity, automatons, funny mirrors, and 'amusing experiments.'[43] These last gentlemen, from the Bianchis down, contributed nothing conceptually or instrumentally to our subject: '[Their] cabinets all contain the same items, sold by the same shops . . . A hundred such collections would not furnish the apparatus for a coherent course of instruction.'[44]

The cost of these lectures appears to have been independent of their quality. Charles and Brisson got one louis (24#) per month for courses of two or three months' duration, meeting probably three hours a week (a cost of about two livres an hour); Rouland's standard offering was 12 lessons for 24 livres; and the others asked between 1.2 (Bienvenue) and 3 (Bianchi) livres per 'séance.'[45] Public lectures in France were therefore about as expensive as those in England.[46] Who patronized them besides the *grand monde* and the 'foreigners, women and savants' known to have frequented Charles'?[47] Students, perhaps, but of what? Or maybe the highest class of artisan, of whom we occasionally find traces in the lecture halls, or rather public houses, used by itinerant lecturers in England?[48]

Most of the public lecturers in France and England—and *a fortiori* in other countries where demand was less—supplemented their incomes with other related work, particularly designing, improving, or making apparatus, teaching in

42. Torlais in Taton, *Enseignement* (1964), 634; R. Hahn, 'Sci. Lect.'; Mornet, *Origines* (1947²), 284–6.

43. R. Hahn, 'Sci. Lect.,' compiled from advertisements in the *Journal de Paris;* Daumas, *Instruments* (1953), 195–6.

44. Charles, 1794, as quoted in Daumas, *ibid.*, 196.

45. Compiled from R. Hahn, 'Sci. Lect.,' and, for Charles, from Bugge, *Science* (1969), 167. Fahrenheit had asked a little more, about 40# (*f.* 18.7) for 16 lessons; Rooseboom, *Bijdrage* (1950), 21.

46. A louis d'or had a value slightly less than a guinea; Lichtenberg, *Briefe,* II, 71.

47. France, *Elvire* (1893), 13; Fourier, *MAS,* 8 (1829), lxxvi.

48. Musson and Robinson, *Science* (1969), 108–9, 113–15, 132. If we are to credit Benjamin Donne's advertisement in the *Bath Chronicle* for 29 Dec. 1774 (Robinson, *Ann. Sci.*, 19 [1963], 28), 'few schools or academies' then offered 'courses of Lectures in Experimental Philosophy upon a proper apparatus, at only one guinea per annum additional expense.' Cf. *ibid.*, 32.

secondary schools, tutoring,[49] surveying,[50] and writing books. It is difficult to estimate what these activities might bring. One knows that the instrument business of Nollet, and, for a time, of Martin, were profitable, and a successful text, based upon tested lectures, might be rewarding as well as influential. Ferguson's books earned something: he sold the copyright to his *Astronomy* for £300, which helped to set him up as a lecturer.[51] Several English itinerant lecturers taught in secondary or vocational schools, and a few London lecturers were closely associated with the Little Tower Street and Soho Academies.[52] In the 1720s James Stirling, who owned an interest in Little Tower Street, had an annual income from his public and private lecturing of some £200, which he found to exceed his needs.[53]

We shall meet many of these frugal and able men again.

49. E.g., McKie, *Endeavor,* 10 (1951), 49; Henderson, *Ferguson* (1867), 251.
50. Robinson, *Ann. Sci.*, 19 (1963), 31, 35; *Ann. Sci.*, 18 (1962), 197, 205.
51. Henderson, *Ferguson* (1867), 52–3.
52. Musson and Robinson, *Science* (1969), 41, 119; Hans, *Trends* (1951), 82–93.
53. James Stirling to his brother, 22 July 1729, in Tweedie, *Stirling* (1922), 14.

CHAPTER III

The Case of Electricity

The early study of electricity divides into four periods, each characterized by a set of values of three variables: content, method, and support. During the first period, which comprises the seventeenth century, natural philosophers distinguished electrical from magnetic attraction and discovered that several bodies besides amber possess electricity. Most of the discoverers were either Jesuit polymaths or modern philosophers intent on making attractions compatible with their mechanical universe. Both groups resorted desultorily to experiment, and each from its own principles deduced that the agent of electricity is a subtle material emanation coaxed from susceptible bodies by the friction required to energize them. The Jesuits received their support—room, board, facilities, and leisure—through the Church; the modern philosophers usually supported themselves.

During the second period, 1700–40, electricity rose from an undistinguished variety to a subspecies of the new genus, 'experimental physics.' Seventeenth-century writers had treated it under 'attraction,' along with magnetism and the glance of the basilisk, or under whatever rubric they put amber. The discovery of electroluminescence early in the eighteenth century suggested a connection between light and fire and made electricity a fit subject for demonstration. Discovery about 1730 of the basic regularities of electrostatics gave it a prominent place in the memoirs of the leading scientific academies. Most of the work that brought this promotion was done by men associated with academies, and sometimes supported by them. The Jesuit polymath withdrew. The careful experimenter, responsible at once to Nature and a Royal Society, labored harder than his predecessors to demonstrate, if not to save, the phenomena.

During the third period, 1740–60, qualitative information increased rapidly. Electricity commanded its own monographs and independent sections of textbooks. Pleasant theories, likely Cartesian or Newtonian stories with little or no mathematics, were devised that almost fit the facts. Demonstrations, displays, and audience-participation games spread interest throughout the polite world. The interest was not merely frivolous. Electricity appeared to cure paralysis, cause earthquakes, and fashion thunderbolts. The base of support widened. Academicians, university professors, physicians, the fashionable and the common man all contributed 'experiments and observations' on the grand phenomenon.

In the fourth period, 1760–90, the qualitative theories and explorations gave way to phenomenological or instrumentalist descriptions, to quantitative formulations, and to accurate measurements. Mathematical physicists subjected part of electrostatics to the dominion of Newtonian forces. The rate of invention and degree of precision of instruments increased. Electricity had textbooks of its own, distinct branches, book-length bibliographies. The common man retired, as did the casual academician. A few professors and salaried academicians held the field.

1. THE SEVENTEENTH CENTURY

William Gilbert, an Elizabethan physician trained in the natural philosophy of Aristotle, established the subject of electricity by distinguishing overrigidly between the attractions of amber and lodestone.[1] His purpose was to remove amber's weak and trivial capacity to draw chaff so far from the hearty noble properties of the magnet that 'perverse little folk' could not find in electrical action any argument against his new magnetic philosophy. Gilbert was a zealot and even a crank about that 'wonderful director in sea-voyages,' that 'finger of God,' the compass needle, the key to the protection and prosperity of the realm. Combining his learning with his enthusiasm, he 'made a philosophy out of the observations of a lodestone.'[2]

On the authority of Averroes, Gilbert ruled that true attraction implies violence and coupled motions, as when a horse pulls a cart. Neither electricity nor magnetism qualifies. Electrical action is an 'incitation,' violent to be sure but also uncoupled or one-sided; in Gilbert's opinion, the drawing electric does not move. Magnetism is both mutual and non-violent; it makes an independent class of activity, a coming together or 'coition.'[3] This fundamental distinction expressed itself in eight differentiae, five of which Gilbert plagiarized from Girolamo Cardano, whom otherwise he criticized.

Cardano had observed that (1) amber draws many kinds of body, the lodestone only iron (2) amber draws without moving, lodestone is pulled as it pulls (3) the magnet does, amber does not, act across screens (4) the magnet pulls towards its poles, amber everywhere (5) amber draws more effectively after warming, which does not affect a lodestone. To these observations, which are acute and, except for the second, correct, Gilbert added that (6) the magnet pulls heavier weights than amber can (7) surface or atmospheric moisture inhibits electrical but not magnetic action (8) amber's power of incitation, unlike the magnet's of coition, belongs to a wide variety of substances.[4]

This last observation is Gilbert's great contribution to the study of electricity.

1. For ancient and medieval knowledge and opinion about electricity, see Benjamin, *Hist.* (1898), chaps. i–iv; Daujat, *Origines*, I (1945); Urbanitzky, *Elektricität* (1887), 67–110; T. H. Martin, Pont. acc. sci., Rome, *Atti*, 18 (1865), 97–123.

2. Bacon, *Advancement of Learning* (1605), in *Works*, I, 461, and III, 292; *DM* (Th), *ij–iiij, 9; *DM* (Mo), xlviii–xlix, 17, 181, 223, 253–4, 297–300.

3. *DM* (Mo), 74, 97–8; *De mundo* (1651), 104–5.

4. Cardano, *De subt.* (1550), 222–3; *DM* (Mo), 80, 86, 97.

Indeed, it provided the occasion for the word: from the Greek for amber, 'electron,' Gilbert coined 'electric' for a substance that draws like amber, whence 'electricity,' their common property.[5] He may have come to his grand discovery of the catholicity of electricity by following up the fact, recorded by Girolamo Fracastoro, that diamond draws when rubbed; he searched for other electrical gems using a device described by Fracastoro, a small pivoted needle or 'versorium' more sensitive than chaff to weak electrical forces. Gilbert added two dozen items to the inventory of electrics and a persuasive new count to Cardano's list.[6]

From distinctions (1), (3), (5), and (8) Gilbert deduced that the agent of electricity cannot be an innate sympathy of the kind responsible for magnetism. It is not innate because electrics must be rubbed before they will draw; and it is not a sympathy because the diversity of electrics and the objects drawn preclude a common occult quality.[7] Sympathy is not promiscuity. Electricity must therefore arise from direct action of matter. Gilbert inferred the existence of a subtle vapor, released from the electric during friction and bridging the distance to attracted objects. Of a moist constitution, it effects electrical attraction in the same way that water unites sticks that float together. The effluvia, or particles of the electrical vapor, 'lay hold of the bodies with which they unite, enfold them, as it were, in their arms, and bring them into union with the electrics.'[8]

These sticky effluvia not only bring chaff to the electric but also paste it there. Here electrical action again differs from magnetic; like magnetic poles repel, and electricity, by definition and by the nature of electrical effluvia, has no room for repulsion. Gilbert explicitly rejected the possibility that electrics might repel.[9] The second of Cardano's distinctions, which Gilbert accepted, also rules out the mutuality of electrical action. Gilbert's discoveries and theories about electricity simultaneously opened a new science and prejudiced its investigators against recognizing two fundamental properties of their subject of study.

The first to advance beyond Gilbert was the Jesuit Niccolò Cabeo, teacher, engineer, and philosopher, 'humble in all his ways, modest in dress, and blameless in character.'[10] Like Gilbert, he discussed electricity in a book, *Philosophia magnetica* (1629), devoted to magnetism; he announced new electrics and deduced that electricity is too promiscuous to be a sympathy; and he repeated the electromagnetic differentiae, to which he added two unhappy items of his own: (9) magnets can transmit their power to iron, amber cannot so endow chaff (10) magnets can repel one another, electrics flee nothing.[11] The first denies in principle the possibility of electrical conduction; although the second appears no less

5. Cf. Heathcote, *Ann. Sci.*, 23 (1967), 261–75.

6. Benjamin, *Hist.* (1898), 295–7; Fracastoro, *De symp.* (1550), 69; *DM* (Mo), 82; Gliozzi, *Per. di mat.*, 13 (1933), 1–14; Thompson, *Notes* (1901), 38–41.

7. *DM* (Mo), 86. 8. *DM* (Mo), 92–5. 9. *DM* (Mo), 88–9, 94–5, 176.

10. Libanori, *Ferrara* (1665–74), I, 145–6, and II, 213–14; *DBI*, XV, 686–8.

11. Cabeo, *Phil. mag.* (1629), 180, 182.

decisively to rule out repulsion, Cabeo has been put forward as its discoverer.[12] This incongruity arose because Cabeo replaced Gilbert's theory of watery effluvia—'words introduced for eloquence, not for explaining the cause and method of attraction'—with a theory of impact. Effluvia thin the air near the electric; remoter air moves in lest nature permit a vacuum; the breeze drives the chaff to the electric. Sometimes the breeze is so forcible that 'attracted' bodies rebound. Cabeo understood their 'resilience,' as he called it, not as electrical repulsion but as an occasional side effect important only as confirmation of his mechanism of electrical attraction.[13]

Cabeo and his work awakened interest in electricity at the center of Jesuit natural philosophy, the museum of the 'Oedipus of his age,' the professor of mathematics, physics, and Oriental languages at the Collegio Romano, the all-round polyhistor Athanasius Kircher. Kircher built his career around natural magic, which served the seventeenth century as displays of experimental physics did the eighteenth. In the museum, set up in the College in 1652, he exhibited rarities from wherever the zeal and curiosity of the Jesuits penetrated, turned water into wine, and demonstrated the curious to the discriminating.[14] Electricity had a place in his magic. He showed that the smoke from a candle dances under bombardment from the effluvia of amber; that large heavy objects, properly suspended, can be drawn by a diamond; and that bright flowers curtsey, 'to the great surprize of all,' to a bit of black coal.[15]

These illustrations and their explanation interested several of Kircher's circle. Tommaso Cornelio, a physician educated by the Jesuits, improved Cabeo's theory by attributing the breeze to the gravity of the remoter air rather than to the horror vacui, an explanation disproved, in fine scholastic style, by Kircher's friend, the Minim Emanuel Maignan. Kircher's old student Gaspar Schott prepared a compromise: gravity would do if effluvia resembled Gilbert's sticky emanation, for then they could both mobilize the air and paste the blown chaff to the electric.[16] This compromise, developed by Kircher's and Schott's student Francesco Lana, S.J., a most accomplished natural magician, became the standard account on the Continent, especially in Germany, for half a century. In Lana's version the effluvia owe their stickiness to an 'igneo-sulphureous principle.'[17]

12. By, e.g., Magrini, *Arch. stor. sci.*, 8 (1927), 37; Dibner, *Early Elect.* (1957), 14; *DBI*, XV, 687.

13. Cabeo, *Phil. mag.* (1629), 192–4. All chaff does not 'rebound' (as elementary electrostatic theory requires) because little transfer of charge takes place between it and dielectrics like amber.

14. Brischar, *Kircher* (1877); Villoslada, *Storia* (1954), 183–4; Thorndike, *Hist.*, VII, 596–621.

15. Kircher, *Mund. subt.* (1665), II, 76–7; Buonanni, *Rerum natur. hist.* (1773), 109–11; Schott, *Thaum. phys.* (1659), 377.

16. Cornelio, *Epist.* (1648), 8, 28–30, 55–60; Maignan, *Cursus* (1673²), 399–400; Schott, *Thaum. phys.* (1659), 376–83.

17. Lana, *Magisterium* (1684–92), III, 287–312; *Prodromo* (1670); *Acta erud.*, 12 (1693), 145–50; Villoslada, *Storia* (1954), 335.

Although Lana retailed the stock argument against considering electricity a sympathy, his association of electrics with a particular chemical species in effect endowed them with a common occult quality. The association, though neither universally accepted nor long held, played an important part in bringing electricity into prominence in the early eighteenth century.

MODELS, MUTUALITY, AND REPULSION

Of the several phenomena tacitly excluded by theories of effluvia—repulsion, conduction, the universality of electricity, and the reciprocity of electrical interactions—only the last forced itself on the attention of seventeenth-century physicists. Its recognition has an interest beyond the detection of a new phenomenon. It was a consequence of taking a mechanical model literally. The physicists of Gilbert's time had recourse to mechanism infrequently, and its effective explanations touched only a few disconnected phenomena. The virtuosity, inventiveness, and optimism of Descartes, however, and the counter-example of latter-day hermetists like Robert Fludd, persuaded many that mechanical models offered the only hope for a precise and comprehensible physics. Expectations rose. Physicists demanded more from models, perhaps even a complete fit with phenomena, with little or no negative analogy.[18]

Gilbert's countrymen Kennelm Digby, diplomat and philosopher, and Thomas Browne, physician and literateur, freed his watery humor from the objections of Cabeo by concocting it into an unctuous, elastic vapor.[19] Such a vapor could allow Cabeo's rebounds, occasion the reattractions of ricocheting chaff that Digby noticed, and—in its elastic contractions—draw the electric as well as the chaff. This last inference was first made about 1660, by the unconventional Cartesian fellow traveller Honoré Fabri, S.J., 'a veritable giant in science' and a liberal and candid physicist whenever his Society's obligation to combat Copernicans did not interfere.[20] The mutuality of electrical interaction, although contrary to received theory, did not menace cosmology. Fabri made public its discovery through the modernizing Accademia del Cimento, of which he was a correspondent, and in some dialogues published in 1665. There Fabri's spokesman uses a pivoted piece of amber to complete the proof that electricity cannot be a sympathy. Not only does amber draw all kinds of bodies, it also moves towards them, indiscriminately, pulled by the unctuous emanation. 'If it reacted to only one kind, like magnets to iron and vice versa, there might be some analogy; but because it approaches all, it cannot do so through a native, intrinsic force, which is silly even to think, much less to say.'[21]

18. Cf. Hesse, *Forces* (1961), 24, 100.

19. Digby, *Two Treatises* (1644), chap. xix; Browne, *Pseud. epid.* (1646), 79–82.

20. Descartes, *Corr.*, VII, 171, 210, 282–3; Leibniz, *Phil. Schr.*, IV, 241–5; Fabroni, *Lettere* (1773–5), II, 101–4.

21. Fabri, *Dial. phys.* (1665), 176–82; Magalotti, *Saggi* (1841^3), 143–7, and *Essayes* (1684), 128–32; Heilbron, CIHS, 1968, *Actes*, III:B (1971), 45–9.

Robert Boyle, head propagandizer for the corpuscularian faith in England, discovered mutuality independently of Fabri. The model of elastic threads suggested that a suspended electric might follow the body used to excite it. The experiment had its difficulties, which Boyle generously described; at last a bit of amber consented to follow a pin cushion, thus demonstrating its 'power of approaching . . . [bodies] by virtue of the operation of its own steams.' Similarly 'false locks of hair,' electrified by combing, pull themselves to the cheeks of the beauties they adorn.[22]

Two other important sets of electrical experiments based upon inferences from mechanical models were invented in the seventeenth century. One inference was not hard to find: according to Cabeo's model, electrical action should cease in vacuo. The experiment proved harder. The first to try, the Accademia del Cimento, which devoted much time and money to the study of electricity, failed. Their vacuum was the Torricelli space above the mercury in a barometer tube, into which they introduced amber, a rubber, and some bits of paper; but all in vain, as it happened, 'for whether the vessel were full or empty of air, the Amber attracted not.'[23] They did not try the air pump, which Boyle and Hooke perfected in the early 1660s; their cooperative work was effectively over by 1662, when their secretary began to draft the *Saggi* (*Essays*) that presented their results. Contrary to an opinion held by historians who can see the seventeenth century only as Galileans, the Academy's short life was not ended by its protector and benefactor, Prince Leopold de' Medici, to please the Church, but because some of the academicians left Florence and the remainder found it impossible to work together.[24]

Boyle also tried to determine whether 'the motions excited by the air had a considerable Interest' in electrical attraction. A piece of amber was excited and suspended in a glass vessel over chaff. While a sturdy lad worked the pump, Boyle lowered the electricity to its prey. '[We] perceived, as we expected, that in some Trials . . . the amber would raise it without touching it, that is, would attract it.'[25] Boyle's *experimentum crucis* did not annihilate the followers of Cabeo, who explained that he had not removed enough air to destroy the mechanism of electricity. The question was to confound electricians for well over a century, ultimately in the form whether vacuum insulates or conducts. Here the instruments of early modern physics proved inadequate. Even the best air pumps of the eighteenth century, which gave an exhaustion of about 1/600 atmosphere as against Boyle's 1/300, could not reach the region where dielectric strength

22. Boyle, *Works*, IV, 345–54; evidence for Boyle's independence may be collected from *Works*, I, 365, 451–2, 789–99, III, 279–83, 681–2, and IV, 235.

23. Magalotti, *Essayes* (1684), 43–6; *Saggi* (1841³), 51–4; Middleton, *Experimenters* (1971), 143–5; Targioni-Tozzetti, *Notizie* (1780), II, 606, 612–13.

24. Antinori in Magalotti, *Saggi* (1841³), 70, 106–9; Middleton, *Experimenters* (1971), 28–9, 34, 310–16; Favaro, Inst. ven., *Atti*, 71:2 (1912), 1173–8; Ornstein, *Rôle* (1928), 78.

25. Boyle, *Works*, IV, 34.

begins rapidly to increase and the insulating property of 'vacuum' becomes manifest.[26]

Tenacity in pursuit of mechanical models brought to light not only mutuality but also, and at last, repulsion. Christiaan Huygens, bred by his father on the physics of Descartes, gradually freed himself from its metaphysics and bric-a-brac but never surrendered its fundamental proposition: 'it is only the motion and shape of the corpuscles of which everything is made that produces all the admirable effects that we see in nature.'[27] As the leading member of the Paris Academy in the 1660s, Huygens considered the most pressing problem of physics to be the reworking of Descartes' explanations of magnetism, gravity, and elasticity. By 1672 electricity also figured in his reductionist program. That year, astounded by the results and repelled by the reasoning of Otto von Guericke, he tried some electrical experiments.[28]

Guericke had studied natural philosophy at Leyden in the 1620s. There emptiness, the great tracts of space supposed by Copernicus, filled his imagination. What could be the purpose and nature of all that nothing? After service as an engineer in the Thirty Years' War, Guericke returned to this question as a practical man accustomed to large machines.[29] He determined to make some interplanetary space and invented the vacuum pump for the purpose. The 'vacuous space' it produced had precisely the properties of Copernican emptiness: extension, and the capacities to propagate light, stifle sounds, and pass gross bodies without resistance. Guericke turned to designing a non-mechanical system that could run a Copernican world embedded in a timeless, resistanceless void. In the style of Gilbert, he assigned most of the work to planetary souls or powers. Following his own bent, he found a way to produce 'mundane virtues' in the laboratory.

These virtues include the conservative, whereby bodies gravitate and the earth retains whatever is necessary to its well-being; the expulsive, which expels harmful material like fire and keeps the moon at a distance; the directive, which fixes the earth's axis during the annual revolution; the impulsive, or inertia; and *virtutes lucens*, *soni*, and *calefaciens*, the causes of light, sound, and heat. To exhibit the virtues, the resourceful Guericke made a little earth, a collection of minerals in a ball of sulphur about a foot and a half in diameter. Since he, following Gilbert, associated gravity, and hence the conservative virtue, with electricity, he knew how to energize one power of his *globus mineralis*. After placing it in a frame (fig. 3.1a) in which he could rub and turn it conveniently, he showed that it attracted light objects like any other electric and retained them when ro-

26. Cf. Smeaton, *PT*, 47 (1751–2), 451–2; Loeb, *Fund. Proc.* (1939), 411; *supra*, i.l. Cabeo's theory appears confirmed in 'vacua' that conduct.

27. Huygens, *Oeuvres*, XXII, 710 (text of 1679).

28. *Ibid.*, XIX, 553, 611; cf. *ibid.*, 632, IX, 496, X, 22, XVI, 327, and XXII, 512, 641–2.

29. Guericke, *Exp. nova* (1672), *1–*2, 53–4; Kauffeldt, *Guericke* (1968), 14–19, 92, 141, and CIHS, X, 1965, *Actes*, III (1968), 364–8; Schimank, *Organon*, 4 (1967), 27–37.

a b

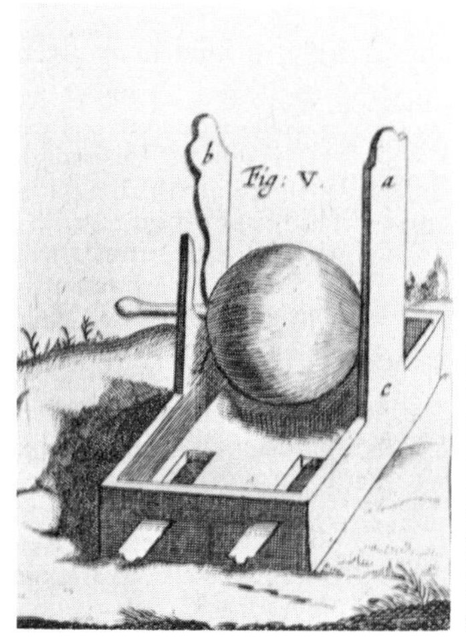

3.1 Guericke's demonstrations of the mundane virtues: (a) his so-called electrical machine (b) parading a feather supported by the expulsive virtue of the sulphur globe. From Guericke, Exp. nova *(1672).*

tated. Imagine his delight and excitement on finding that he had awakened his terrella's expulsive virtue too: a feather could be made to hover above the ball and to stay there as a moon while the terrella, in imitation of the earth's annual motion, was paraded about (fig. 3.1b). Furthermore—wonder on wonder!—the rubbed globe showed the *virtutes calefaciens* and *soni* (it felt warm and crackled) and, when seen in the dark, *lucens* (it glowed like powdered sugar). Dropped on the toe, its *virtus impulsiva* was only too evident. Had it contained a lodestone, it would have displayed the directive power as well.[30]

Guericke would have classified only one of his discoveries about the virtues as electrical. In order to secure the conservative virtue as an independent power, he devised an experiment against Cabeo. A thread attached to the globe at one end drew chaff at the other. 'We cannot concede,' he wrote, 'that the attraction occurs by the intervention of the air, because we can see by experiment that the sulphur globe, when excited by friction, can also exercise its *virtus* [*conservativa*] through a linen thread one ell or more in length.'[31]

Although several physicists, including Leibniz, Hooke, and Boyle, were intrigued by Guericke's experiments, none developed them further or recognized

30. Guericke, *Exp. nova* (1672), 125–51; cf. Schimank's notes in Guericke, *Neue Vers.* (1968), (271)–(283); Monconys, *Voyages* (1695²), III, 75–80, 481; Guericke to Leibniz, 6/16 June 1671, in Leibniz, *Phil. Briefw.* (1926), 119–20.

31. Cf. Rosenberger, *Abh. Ges. Math.*, no. 8 (1890), 89–112; Heathcote, *Ann. Sci.*, 6 (1948–50), 293–305. Among those who credit Guericke with discovery of repulsion and the electric light: Benjamin, *Hist.* (1898), 403; Wolf, *Hist.* (1952³), I, 304–5; Hoppe, *Gesch.* (1884), 4–5; Kauffeldt, *Guericke* (1968), 77, 147.

in them either electrostatic repulsion or the possibility of communicating electricity from one body to another. The sulphur ball proved as hard for the modernizing philosophers to work with as the old-fashioned concepts with which Guericke sought to explain its action.[32] Only Huygens, after completing the Cartesian theory of gravity, had the inclination and ability to search successfully for the vortical mechanism that had to underlie whatever of Guericke's magic might be reproducible.

Huygens replaced the competent but weak sulphur globe with a small amber sphere; as detector he used flocks of wool. They behaved capriciously. Many trials yielded an extraordinary regularity: wool let fall from dry, cold fingers adhered to the sphere; moist objects were driven away after attraction. 'Whence this hydrophobia? Guericke did not notice it.' The answer? Evaporation, which endows moist surfaces with atmospheres able to abet the *communication of electrical vortices*.[33] According to Huygens, no special emanation leaves an excited electric. Friction vibrates its surface, which, in turn, sets up a vortex within the surrounding subtle matter. In Guericke's demonstration of the expulsive virtue, the globe's vortex stimulates one about the feather: the primitive suppositious non-mechanical 'expulsion' is a separation caused by collisions between the vortices. Consequently, as Huygens showed, two wool flocks, each having acquired a vortex from the sphere, will repel one another without previous attraction.[34]

This acute application of the ideas of Descartes to the discoveries of Guericke is the high point of seventeenth-century studies of electricity. Guided by the vortex, Huygens recognized electrostatic repulsion, which had eluded many shrewd investigators armed, or rather disarmed, with the ordinary theory of ejected effluvia. In his picture, repulsion is coordinate with attraction; he grasped the relation among attraction, electrification by communication, and repulsion. He knew that moisture promotes communication and that two objects electrified by a third repel one another. The rapid increase in knowledge about electricity in the 1730s and 1740s started where Huygens stopped. Since he published not a work about his discoveries, they, like Guericke's that had inspired them, had first to be made again.

2. THE GREAT DISCOVERIES AND THE LEARNED SOCIETIES

Born as a side issue to the magnetic philosophy and preserved in the swaddling clothes of Jesuit compendia, the study of electricity passed a precarious child-

32. Leibniz, *Phil. Briefw.* (1926), 145–6, 158–9, 168, 221–2; Guericke, *Neue Vers.* (1968), (96)–(112); Hooke, *Diary* (1935), 10, 12; Birch, *Hist.* (1756–7), III, 59, 61, 63.

33. Huygens, *Oeuvres*, XIX, 612–16 (notes dated 1692–3); cf. *ibid.*, IX, 539, 572, and XXII, 408–10, 649, 653–4, 756.

34. An electrical vortex consisting of matter ejected by the electric had appeared in Cartesian texts in the 1680s as a development of Descartes' theory of electrical effluvia: Descartes, *Principia* (1644), in *Oeuvres* (1964^2), IX:2, 305–8, rehashed in Rohault, *Traité* (1692^6), II, 214–15; Senguerdius, *Phil. nat.* (1685^2), 361; Régis, *Système* (1691), III, 510–13; F. Bayle, *Inst. phys.* (1700), II, 314–17; Ziggelaar, *Pardies* (1971), 74.

hood among the new experimentalists of the later seventeenth century. Indications of the knowledge of the subject in 1675 may be gathered from the attempt of the fellows of the Royal Society of London to reproduce a new electrical effect carefully described to them by Newton. They were to place a telescope lens a fraction of an inch above bits of paper on a table, rub the top of the glass, and enjoy the skipping of the bits underneath it. The Fellows rubbed briskly, and often, but uselessly until, having received further instructions, they tried a brush of hog's bristles.[1] The affair exhausted them and their interest in electricity, although, as Newton pointed out, the effect might be important as well as curious: the bits come to the lower surface of the lens although only the upper is rubbed, showing that electrical effluvia can freely penetrate sensible thicknesses of glass.[2] There was no more electricity at the Society until 1705, when, literally at a stroke, its study was raised from an episodic to a continual preoccupation of natural philosophers.

ELECTRICITY AT THE ROYAL SOCIETY OF LONDON

The agent of this transformation was Francis Hauksbee, an instrument maker and specialist in vacuums, who entered the service of the Society in 1703, the year that Newton became its president. The new administration wished to revive the healthy old custom, then disused for a generation, of showing experiments at the weekly meetings. The custom had been neglected not from disinterest or poverty but because it demanded a continual inventiveness that ultimately emptied even the ablest. To arouse the interest and stimulate the thinking of the heterogeneous fellowship required demonstrations luciferous in philosophy, useful in art, ingenious in contrivance, and surprising and amusing in execution.[3] The trick was to devise them regularly. Hauksbee here had an advantage his predecessors lacked. The new president, breaking precedent, attended the Society's meetings and continually suggested illustrations, extensions, and defenses of the Newtonian systems of the world.[4]

Because of its weakness and capriciousness, electricity did not suggest itself for demonstration until Hauksbee stumbled upon it while embellishing flashy experiments encouraged by the Society. When he took office, the 'chemical phospor,' the element phosphorus, had lately been added to the natural magician's repertory. The Fellows desired to know whether it worked in vacuo. It did.[5] Hauksbee turned to the famous and problematic 'mercurial phosphorus,' which sometimes shines in the Torricelli space when a barometer is shaken. Hauksbee

1. Newton to RS, 7, 14 Dec. 1675, 10 Jan. 1675/6, *Corresp.*, I, 364–5, 393, 407; Birch, *Hist.* (1756–7), III, 260–1, 271.

2. Newton, *Corresp.*, I, 364; Westfall, *Force* (1971), 332, 364–5.

3. Cf. Wren to Brouncker, 1661, in Wren, *Parentalia* (1750), 225.

4. Cf. Guerlac, *Arch. int. hist. sci.*, 16 (1963), 113–28; *Essays* (1977), 107–19; *Mélanges* (1964), I, 228–53; Bennett, RS, *Not. Rec.*, 35 (1980), 33–4.

5. Hauksbee, *PT*, 24 (1705), 1865–6; *Phys.-Mech. Exp.* (1719[2]), 93–7; Grew, *Musaeum* (1681), 353–7.

showed that the light appears where globules of mercury tear at the glass, a display he varied several ways before identifying the friction and not the mercury as the efficient cause.[6] He incorporated this insight in an entertainment worthy a Royal Society: a globe evacuated of air and turned against his hand by a machine resembling a cutler's wheel glowed within so strongly as to make legible a book in an otherwise dark room. Hauksbee inferred that he had forced from the glass and into the yielding void the particles of light that, in Newton's opinion, enter into the composition of most or all solid bodies.[7]

Newton bethought himself of his ancient electrical experiment. Perhaps the effluvium abraded from his lens was identical with the light, or luminiferous vapor, from Hauksbee's globe?[8] (A similar inference was made about the same time by a French demonstrator of natural philosophy, Pierre Polinière, who independently discovered the electric light through experiments on the mercurial phosphorus; and later by a London physician, Samuel Wall, who began by attempting to improve the disagreeable manufacture of the chemical phosphor.[9]) It was to study this possibility, which had the remote promise of providing a bridge between the two great divisions of Newtonian science, the optical and the gravitational, that Hauksbee took up the subject of electricity. The switch, which occurred at the end of 1706, had a further recommendation. The phosphor line had played out, and, despite all diligence and application, Hauksbee had not found an adequate replacement.[10]

First he had to improve the apparatus. He used glass, in the form of a sturdy tube thirty inches long and an inch in diameter, in place of the bits of amber previously employed; and he enhanced effects by taking leaf brass or lampblack instead of straw or paper as detector.[11] The new apparatus, which quickly became standard, brought electrostatic repulsion to light. Hauksbee interpreted it in the old way, as an effect of rapidly projected effluvia, and so missed the connection between the repulsion of the detectors and their electrification.[12]

The connection between electricity and gravity proved easier to find. Threads exposed to the influence of an electrified globe—which, like Guericke's sulphur ball, represented the earth—aligned as in figure 3.2; those attached to a disk inside extended radially inward. It was 'a plain Instance of a *Repulsive* and *Attractive* force,' indeed of centripetal and centrifugal force, 'so that in these small

6. Hauksbee, *PT*, 24 (1705), 2129–35; *Phys.-Mech. Exp.* (1719²), 1–42; JB, X, 102, 105–7; Harvey, *Hist.* (1957), 271–7; Hackmann, *Electricity* (1978), 30–8.

7. Hauksbee, *PT*, 25 (1706), 2281; cf. *PT*, 24 (1705), 2131; *Phys.-Mech. Exp.* (1719²), 30, 34, 39; Newton, *Opticks*, ed. Cohen, 270–6.

8. JB, 30 Oct. and 6 Nov. 1706; *Opticks*, ed. Cohen, 341; Guerlac, *Essays* (1977), 112; Hauksbee, *Phys.-Mech. Exp.* (1719²), 'Pref.'

9. Freudenthal, *Hist. Stud. Phys. Sci.*, 11 (1981), 203–29; Polinière, *Expériences* (1718²); Wall, *PT*, 26 (1708), 67–96.

10. Hauksbee to RS, 9 Mar. 1706/7, RS, Sci. Pap., XVIII:1, 106.

11. Hauksbee appeared with the tube for the first time on 13 Nov. 1706 (JB, X, 149).

12. Hauksbee, *PT*, 25 (1706–7), 2327–31; *Phys.-Mech. Exp.* (1719²), 241–7; Home, *Arch. Hist. Exact Sci.*, 4 (1967–8), 203–17, and *Effl. Theory* (1967), 30, 37–8.

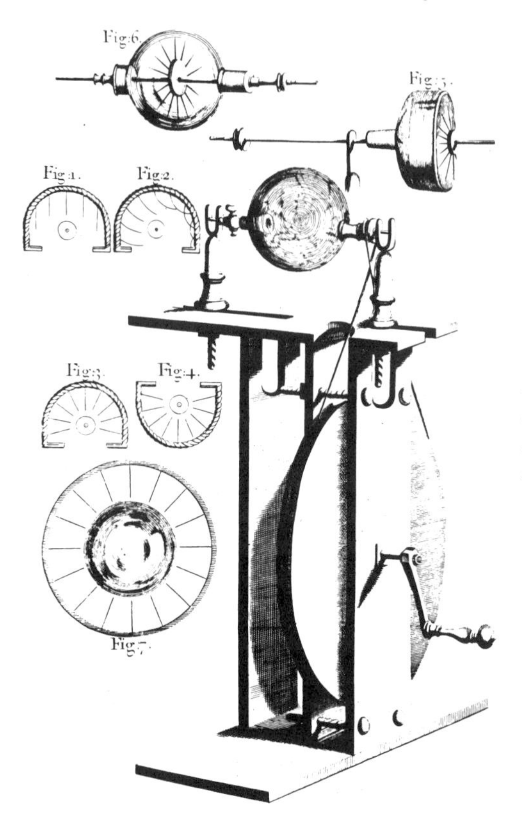

3.2 Hauksbee's prototypical electrical machine together with several deployments of 'Hauksbee's threads.' From Hauksbee, Phys.-Mech. Exp. *(1709).*

Orbs of Matter, we have some little resemblances of the Grand Phaenomena of the Universe.'[13] But only a little. The pointing threads shrank from Hauksbee's touch. Rejecting recourse to an ad-hoc repulsion between his finger and the threads, he imagined that they were supported by stiff chains of glass effluvia capable of piercing the globe and of staying intact when pushed aside together with their client threads. ''Tis very amasing,' he said; 'there are not many [phenomena] in nature more surprising.'[14] His elucidation, no less amazing, inspired one of those crude satires from which the early Society suffered. It runs as follows. A judge of a paternity suit against a eunuch, doubting the possibility of the crime 'from the various parts that were wanting,' sought advice from an Academy of Science. The academicians found for the lady on the strength of their experience with 'all sorts of Effluvia,' particularly the recently discovered sort, stiff and stout, begot by rubbing, highly penetrating, and as likely to belong to eunuchs as to glass.[15]

13. Hauksbee, *PT*, 25 (1706), 2332–5; *PT*, 25 (1707), 2374; *Phys.-Mech. Exp.* (1719^2), 54–5, 67, 74–5, 143, 154–5.

14. Hauksbee, *PT*, 25 (1707), 2373–7, 2[4]13–15; *Phys.-Mech. Exp.* (1719^2), 185.

15. King, *Useful Transactions* (1709).

Two ways of developing the deeper analogy uncovered by Hauksbee were open to Newton: either he could adapt the effluvium of the usual account of electricity or he could recast electrical theory in terms of attractive and repulsive forces. He chose the first approach, with which he had flirted before the creation of the *Principia*. This old attachment, Cartesian objections to his occult qualities, the weight of consensus about the nature of electricity, and the palpability of electrical effluvia all helped him to take this step. Not long after he had published in the second edition of the *Opticks* (the *Optice* of 1706) additional queries implying that nature effected all physical phenomena, small and great, through forces acting at a distance, he told the Society that 'most of the Phaenomena appearing in Fermentations, Dissolutions, Precipitations, and other Actions of the small Particles of Bodies one upon another were caused by Electrical Attraction.'[16] And that attraction, as he hinted in the General Scholium to the second *Principia* (1713), is caused by a 'certain most subtle spirit.'[17] In drafts for the Scholium and for the third *Opticks* (1717/8), the spirit stands revealed as the electrical effluvium: 'Do not all bodies therefore abound with a very subtle, but active, potent, electric spirit, by which light is emitted, refracted, & reflected, electric attractions & fugations are performed, & the small particles of bodies cohere when contiguous, agitate one another at small distances, & regulate almost all their motions amongst themselves?'[18]

Hauksbee died in 1713. His place went to J. T. Desaguliers, a man higher on the social scale, an Oxford graduate, a rising minister, a prominent Mason, and a successful public lecturer. These last employments were to give Desaguliers greater independence of the Society than Hauksbee had enjoyed. In his early years, however, he continued and repeated his predecessor's work and tried to strengthen Newton's analogy between electricity and gravity by cracking the old chestnut, whether electrical attraction occurs in vacuo. In his last work, Hauksbee had plumped for Cabeo's theory in order to explain the orientation of the threads. Were he right, the critical analogy would fail since Newton's gravity reaches across spaces devoid of matter.[19] Desaguliers found for Newton, who nonetheless decided not to electrify the *Opticks*. The suddenness of his decision may be inferred from the oddity that the volume as published ends with 'Book I, Part I' and that Desaguliers refers to the electrical universe in a lecture syllabus published in 1717.[20] Without Newton's protection, the Society's interest in electricity declined; and not until two years after his death in 1727 did the subject again seriously engage the attention of the Fellows.

The agent of revival was Stephen Gray, who in 1729 had been a camp follower

16. JB, 19 Apr. 1710.

17. *Princ.*, ed. Cajori, 547; Hall and Hall, *Unpubl. Sci. Pap.* (1962), 336–7, 350–4.

18. McGuire, *Ambix*, 15 (1968), 175–6. Other pertinent drafts are in Newton, *Corresp.*, V, 366–7, and in Westfall, *Force* (1971), 393–4, 416–18.

19. Hauksbee, *Phys.-Mech. Exp.*, 'Appendix;' JB, 9, 16, 23 May 1717.

20. Desaguliers, *Phys. Mech. Lectures* (1717), esp. 2–3, 40, 76; cf. Guerlac, *Essays* (1977), 120–4.

of the Society for thirty years: a contributor to its *Transactions*, a correspondent of its secretary, an occasional visitor to its meetings, but not until 1730 a member of its fellowship. A dyer by trade, Gray did not belong to the society the Society preferred; and his shyness prevented him from claiming the honor that his accurate observations in physics and astronomy deserved.[21] His merit will appear from the results of his first study of electricity. Using Hauksbee's new apparatus as generator and a feather as detector, he found again the phenomena that Guericke had attributed to the expulsive virtue. He took the assimilation of electrical effluvia with the particles of light literally and inferred that attraction occurs through effluvia reflected from neighboring objects and that common electrics, like wax and amber, will give light on rubbing. He soon modified the first proposition to require all bodies near excited electrics themselves to emit electrical matter, 'the attraction [or apparent repulsion being] made according to the current of these Effluvia.'[22] The letter of January 3, 1708, in which Gray described these acute observations and conjectures, was passed by the Society's secretary to Hauksbee, who put forward one or two as his own and suppressed the rest. He could not afford to encourage rival experimentalists: paid only for what he could invent, he had to balance science and self-preservation.[23]

Twenty years later the conception of stimulated emission led Gray to an important discovery. He was then a gentleman pensioner at the London Charterhouse, a place he had acquired in 1719 through the intervention of influential Fellows. Desultory play had revealed the existence of a new class of semi-rigid electrics, such as feathers and ox guts.[24] Perhaps even metals could be made electric? After frustrating trials by heat, friction, and percussion, he thought to try the effluvia of a glass tube. The tube had corks at both ends to keep out dust; and Gray was surprised to find that feathers went to the corks rather than to the glass. 'I . . . concluded that there was certainly an attractive Virtue communicated to the cork by the excited Tube.' Chance favors the prepared mind. Gray stuck a pole in the cork, and a string on the pole, and so reached fifty-two feet, at which a tea kettle or pint pot on the end of the string could still attract leaf brass. He could go no further. No greater free drops were available at the Charterhouse, and horizontal propagation appeared impossible, for Gray could not transmit through a line suspended from the ceiling by short lengths of string.[25]

At this point, June 30, 1729, Gray visited a country Fellow, Granville Wheler, who proposed to try silk supports, the narrowness of which might prevent the loss of virtue. It worked. But brass wire, although no less skinny, did not. Their endeavor to increase the mere distance or quantity of transmission thus netted a

21. Taylor to Keill, 3 July 1713, in RS, Corresp., LXXXII, 5.

22. Full text in Chipman, *Isis*, 45 (1954), 33–40.

23. Hauksbee, *PT*, 26 (1708–9), 82–6; JB, X, 178; RS, Sci. Pap., XVIII:1, 113, under 4 Feb. 1708/9.

24. Gray, *PT*, 31 (1720–1), 104–7.

25. Gray, *PT*, 37 (1731–2), 18–44.

3.3 Gray's charity boy as pictured in Doppelmayr's Neu-ent. Phaen. *(1744).*

qualitative distinction of the utmost importance: 'the Success we had before depended upon the Lines that supported the Line of communication being Silk, and not upon their being small.' With the help of another country Fellow, John Godfrey, they sought substances that might serve as supports, or that, like ivory, metals, and vegetables, were good 'receivers' of virtue. They found glass, hair, and resin in the first category, and water, an umbrella, and a charity boy, always plentiful around the Charterhouse, in the second.[26]

Gray's group always understood that effluvia from the tube flow into the string and along the charity boy, whence they force themselves across the air to agitate the brass (fig. 3.3). This conception had a fundamental and fatal ambiguity, the conflation of the mechanisms of conduction and attraction. Gray's effluvia run easily through threads, wires, and flesh, with difficulty through air, and not at all through silk, glass, or resin whereas the usual or attractive effluvia easily penetrate air and glass but cannot pass screens of metal or cloth. Trouble might have been glimpsed as early as 1731, when Gray, after recommending glass bricks as supports because they prevent escape of effluvia, mentioned that Wheler had drawn leaf brass across five superposed glass panes.[27]

The Society recognized the importance of Gray's discoveries. It admitted him a Fellow and awarded him the first prize it ever gave for meretorious work in science. It also left the study of electricity to him, for (if we credit Desaguliers) he had become crotchety in his old age, 'of a Temper to give it entirely over, if he imagined that anything was done in Opposition to him.'[28] In the four years that he enjoyed his monopoly, he examined and classified the sparks obtainable from electrified objects of diverse sorts and shapes. While following up Hauksbee's

26. Gray, *PT*, 37 (1731–2), 399–404.
27. Gray, *PT*, 37 (1731–2), 399, 405–6.
28. Desaguliers, *PT*, 41:1 (1739–40), 187.

experiments, he had shown that a conical glow often proceeds from the finger as apex to the tube and had identified the luminous with the electrical effluvia. In 1734 he returned to the subject on reading a report of experiments by the Parisian academician C. F. Dufay, who, having strung himself up like a Gray boy, had heard a snap, felt a shock, and seen a spark when he reached out to touch a colleague standing on the floor.[29] When a metallic object replaced the colleague, similar effects occurred; when an electric did so, nothing happened.

Gray confirmed the prickles and glimmers, and added that shape as well as substance played a part: a metallic object in place of the boy gave the same single snap and spark if blunt, but conical glows if pointed; the snap occurred only between other non-electrics, the glow sometimes also from a single body, at each approach of the tube. Many came to the Charterhouse to see the hoary sage 'rouse the pours that actuate Nature's frame / the momantaneous shock, th'electrick flame;' until, in 1736, his earthly light extinguished, he sped 'where Bacon waits with Newton and with Boyle / to hail thy genius, and applaud thy toil.'[30]

Desaguliers then came forward to repeat and clarify the distinction between 'conductors' and 'insulators,' terms he introduced to describe substances that can, and those that cannot, communicate the virtue. Since he held a theory akin to Gray's and, as a scholar, tried to make it explicit, he ran into contradictions in describing the motion of effluvia along conductors and through the air.[31] Wheler too delivered an instructive paper, a report of experiments made in 1732, which, had it been published in its time, would have given him priority in announcing that that oxymoron, electrical repulsion, occurs only between objects both of which are electrified. Hauksbee's several statements about repulsion had given rise to the idea that friction causes the tube to vibrate and its effluvia to move in and out periodically, causing chaff to be alternately attracted and repelled without changing its electrical state. Wheler showed that a thread may be attracted and repelled, repelled only, or attracted only, according as it is insulated, electrified by communication, or grounded. He understood that this regularity, which will be denoted 'ACR'—first attraction, next communication, finally repulsion—required a new definition of electricity. It is neither an attractive power, as the seventeenth century held, nor an attractive and repulsive virtue, as Hauksbee and his expositors believed; but a 'Virtue attractive of those bodies that are not attractive themselves, and repulsive of those that are.'[32]

METHOD AT THE ACADÉMIE DES SCIENCES, PARIS

The procrastination that cost Wheler credit for announcing ACR was typical of the gentlemen of the Royal Society of London. They had nothing to gain from

29. Dufay, *PT*, 38 (1733–4), 258–66.

30. Gray, *PT*, 39 (1735–6), 16–24; Williams, *Miscellanies* (1766), 42–3.

31. Desaguliers, *PT*, 41:1 (1739–40), 186–93; cf. Desaguliers, *Course* (1763²), II, 331; *FN*, 377.

32. Wheler, *PT*, 41:1 (1739–40), 98–117; cf. 'sGravesande, *Math. Elem.* (1721), II, 2–13.

publishing their results except reputation, which frequent exposure might not enhance. The Parisian academicians were in this respect not gentlemen. They took, or aspired to take, money for their science, and their statutes obliged them to present new papers at a steady clip. Constantly exposed to one another's criticism, they cultivated thoroughness and method as well as productivity. Charles François de Cisternay Dufay was an adept in the method and a gentleman in everything save pace of work. He proved the ideal successor to the curators and virtuosi of the London society. He ordered their scattered insights and discoveries, disentangled what he called 'simple rules' (the dominant electrostatic regularities), and demonstrated that electricity was a common property of sublunary matter.

When he came across Gray's experiments, Dufay had been a member of the Academy for ten years. Two of the many investigations that he had by then completed will illustrate his technique. The earlier concerned the mercurial phosphorus, which had retained its air of mystery by not always appearing when expected. Dufay disliked mysteries. After thorough study of both the literature and the phenomena, he was able to give effective rules for purging the mercury and insuring the glow. In the same way he examined the Bologna stone (BaS) and the 'hermetic phosphor' (CaS), phosphorescent substances then prized for their supposed rarity and hedged around by trade secrets. Against this mystery Dufay brought his guiding principle, that a given physical property, however bizarre, must be assumed characteristic of a large class of bodies and not of isolated species. He contrived to make almost everything but metals phosphorescent; he depressed the phosphor market by publishing his recipes; and he became sensitive to the endless small variations in the physical properties of bodies. 'How different things behave that seemed so similar, and how many varieties there are in effects that seemed identical!' [33]

Dufay's method demanded that he first establish the extent of the classes of electrics, of bodies electrifiable by communication, of insulators, and of conductors. His specification of the first two classes put an end to the tedious listing of individual new electrics. Every body properly handled can be electrified by friction ('par lui-même,' as he put it) except metals and fluids; and all substances whatsoever, except flame, can acquire the virtue through contact or near approach of the tube. (In this universal dictum he conflated conduction with induction followed by spark-over, a mixture that a generation of electricians was required to separate.) In order that electrification by communication succeed, the test body must rest upon a body electric 'par lui-même.' A consequential application of this 'Rule of Dufay' was a prescription for electrifying water. Dufay directed that the fluid be held in a glass jar on an *insulating stand*; 'one would try in vain using a platform of wood or metal.' [34] Here again we confront the circumstance emphasized in connection with Gilbert's theory: a strong organizing prin-

33. Dufay, *MAS* (1723), 295–306; *MAS* (1730), 524–35.
34. Dufay, *MAS* (1733), 73–84.

ciple may rule out the possibility of effects critical to the field it delimits. No one who accepted the Rule of Dufay could intentionally have invented the Leyden jar.

In subsequent experiments Dufay had the help of J. A. Nollet, who was to succeed him as the 'chief of the electrifying physicists of Europe'[35] but who then struggled to survive on the fringes of the academic establishment. They confirmed Gray's results and emphasized that moistening the line promoted communication. Then, following up another observation of Gray's, Dufay found that the tube's virtue acted weakly or not at all through wet silk curtains, which, when dry, it easily penetrated. These facts may appear paradoxical on the received theory. In the one case, moisture facilitates the flow of effluvia down strings; in the other, it inhibits their crossing of curtains. No more than the similar puzzle about glass, which obtruded in Gray's experiments, could this paradox be resolved by any plausible reworking of the theory.

The next business was to dispel the fog in which Hauksbee and his expositors had left electrical repulsion. First, Dufay doubted the existence of the phenomenon; then he considered that since bad or non-electrics, like metals, respond the most, and good electrics, like amber, respond the least to the tube, perhaps a negative attraction, or repulsion, obtains between two electrified objects. This analogy, a product, perhaps, of Dufay's presuppositions about unity and variation in physical properties, led directly to his discovery of ACR: a gold leaf dropped on the tube fled it when electrified; two leaves each independently electrified by it repelled one another.[36] This recovery of the unpublished results of Huygens and Wheler immediately brought a discovery still more splendid. Dufay knew to look for diversity in similarity as well as for uniformity underlying inhomogeneity. Would a leaf electrified by the tube flee electrics other than glass? Yes and no. It fled crystal but came to wax, resin, and gum copal. Dufay reached the bold hypothesis of two electricities: an object can be electrified either like glass, 'vitreously,' or like wax, 'resinously;' communicated electricity is of the type of the communicator; objects with dissimilar electricities attract, those with like electrifications repel one another.[37] He believed that each electric has a fixed electrical type; in a rare lapse, he had satisfied himself by experiments on too few pairs of substances that the electrification expressed does not depend upon the nature of the rubber.

Defay died in 1739, at forty-one, having further enriched the study of electricity with the discoveries about sparks and shocks that intrigued Gray. Both emphasized a point later recognized as critically important: in passing a spark the electrified body gives some or all of its electricity to its partner.[38] Defay, who followed Hauksbee rather than Gray in distinguishing between luminous and

35. *Ibid.*, 233–54; Paulian, *L'électr.* (1768), xvii.
36. Dufay, *MAS* (1733), 458.
37. *Ibid.*, 467–9.
38. Gray, *PT*, 37 (1731–2), 227–30; *PT*, 39 (1735–6), 166–70; Dufay, *MAS* (1737), 95–7.

electrical effluvia,[39] explained that a 'matter of electric light' accompanying electrical effluvia becomes manifest in crossing the 'atmospheres' of third matter, which, in Dufay's moderate Cartesian physics, surround all ponderable bodies. In the atmospheres of living and metallic bodies, this matter meets something that renders it sensible fire. Whence the suggestion that was to inspire the best trick of the showmen of electricity: 'one might contrive therewith to ignite dry cumbustible materials wrapped about a living body.'[40]

A few months before his death Dufay had the satisfaction of seeing his protégé Nollet an adjunct academician. Nollet had acquired a following and an income as a public lecturer and instrument maker, and a knack for demonstrating the new Anglo-French discoveries in electricity.[41] Like Dufay, he was attentive to small differences; but, unlike his master and unfortunately for his reputation, he too often allowed detail to obscure regularity. He also took more literally than Dufay the Cartesian vow to refer all physical phenomena to inert matter in motion. The 'système Nollet,' which guided study of electricity in Francophone Europe for over a decade, is an artifact of a mechanician, academician, demonstrator, and instrument maker.

The occasion for the invention of the system was a report of electrical games invented by German professors and of their application of Hauksbee's machine to the experiments of Gray and Dufay. The report interested Nollet as demonstrator (the Germans were said to kill flies with sparks from their fingers) and as instrument maker. 'I did not sleep until I had a great wheel built.'[42] His labor-intensive arrangement appears in figure 3.4, which also suggests the furniture of his patron Louis XV; the electricity, excited by the hands of his research assistant (fig. 3.5), jumps to the metal bar supported by insulating cords, where it is available for experiment. Now entered the systematizing Cartesian academician. In the dark, sparks spring from points on the bar, especially from its corners, in divergent conical jets; we evidently witness an outgoing or 'effluent' stream of electrical effluvia with their associated luminous or fiery matter. Whoever doubts the inference may be reassured by his feeling of meeting a spider's web as he presents his forehead to the bar, or, better, by the smart smack the effluent gives it when it comes too near. But electrics attract as well as repel. An 'affluent' stream evidently sets in toward the bar from neighboring bodies including the air. Since the affluent arises everywhere, its flow toward a cylindrical gun barrel (fig. 3.5) is homogeneous and radial. A small unelectrified object away from the barrel, such as E, is likely to be more strongly pressed by the homogeneous affluent than by the divergent effluent and so appears attracted. On contact or

39. Hauksbee finally rejected assimilation of the two sorts of effluvia on the ground that light can be generated by friction when electricity does not appear, and vice versa; *Phys.-Mech. Exp.*, 'Appendix'; *FN*, 35.

40. *MAS* (1734), 520, 522.

41. Nollet, *Programme* (1738), 99–104.

42. Nollet to Dutour, a corresponding member of the AS, 27 Apr. 1745 (Burndy).

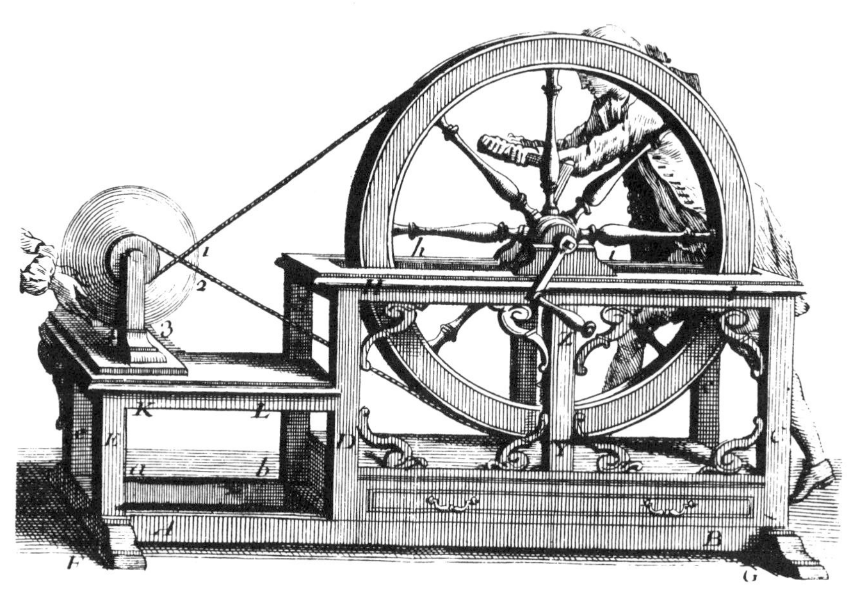

3.4 The classic electrical machine, as pictured in Nollet, Essai *(1750[2]).*

3.5 The système Nollet, showing the divergent effluents and isotropic affluent; below is the first depiction of the Leyden experiment. From Nollet, Essai *(1750[2]).*

close approach it is stimulated to produce its own effluent, which, according to Nollet, increases its effective volume and brings it under the influence of the barrel's jets. And that, Nollet told the Academy, is the whole secret of the sequence ACR.[43]

Nollet did not assume, or want, 'two electricities.' His currents differ in direction, not in nature; he had to distinguish vitreous from resinous electrification by quantity not quality. On the assumption that flow from wax is weaker than flow from glass, he accounted roughly for Dufay's rules of electrostatics. He also brought Hauksbee's pointing threads, 'the most celebrated experiments of the last forty years,' within the system: they align with the affluent currents toward the concave and convex surfaces of the globe and move when the effluent from the finger reaches them. When applied to the inner threads, the explanation requires the assumption, then universally held, that electrical effluvia penetrate glass at least up to thicknesses of bottle walls.[44]

Voilà le système Nollet. 'At first sight nothing is simpler, nothing more ingenious than this hypothesis;' 'a more probable and natural explanation can scarcely be expected;' 'not only does it suffice to explain the facts on which it was founded, but all others since discovered.'[45] Savants liked its boldness and clarity, its merciless reduction of all electrical phenomena, not sparing the two electricities, to quantitative differences in the direction of flow of one universal, fiery, electrical matter. Unfortunately, its close modelling on ACR made it virtually inapplicable to novelties like the Leyden experiment.

PROFESSORIAL GAMES

In 1740 electricity occupied but a small part of the repertory of up-to-date public lecturers like Nollet and Desaguliers and a much smaller part of the minds of most academicians and professors. In 1745 polite learning knew nothing more fashionable. Aristocrats travelled to see the electricity of famous professors; the Académie des Sciences in Berlin chose electrical experiments to entertain its king; a Mr. Smith at Bath offered 'all lovers and judges of experimental philosophy' the sight of his 'electrical phenomenon' from ten in the morning to eight at night.[46] The agents of this transformation were as unlikely as the event itself: playful German professors.

German expositors of natural philosophy had been content with a form of Lana's theory until the 1730s, when J. J. Schilling, professor of mathematics and philosophy at the University of Duisburg and a fellow traveller of the neighboring Dutch Newtonians, took up the subject. To confirm his suspicion that 're-

43. Nollet, *MAS* (1745), 110–12, 124–42; *Essai* (1750²), 65–93, 148–52, 157–64.

44. Nollet, *MAS* (1745), 145–7; *Essai* (1750²), 93–115, 164–5.

45. Respectively, Sigaud de Lafond, *Précis* (1781), 112; Réaumur, letter of 25 May 1747, in Réaumur, *Lett. inéd.* (1886), 60; Grandjean de Fouchy, *HAS* (1745), 4.

46. Gralath, Nat. Ges., Danzig, *Vers. Abh.*, 1 (1747), 278–9, and 2 (1754), 399; *HAS*/Ber (1745), 11–12; [Haller], *GM*, 15 (1745), 194; Millburn, *Martin* (1976), 5.

pulsive force' played the main part in electrical attractions, he showed that a hollow glass ball would swim around in water in obedience to the waving of the tube.[47] The tube's repulsion drives away 'atmosphere' on the near side of the ball, which then moves in obedience to the greater air pressure on its far side. Frightful physics for a Newtonian and a professor but a pretty demonstration, which 'almost drove mad with delight' a poet-physicist at the University of Leipzig, young Georg Matthias Bose.

A frantic search among the glass makers of Leipzig failed to turn up a tube. In his extremity, 'with the vigor of youth,' he ripped the beak from an alembic just removed from the fire. It did for Schilling's experiment, while the rest of the vessel, turned on a spit, allowed the performance of Hauksbee's. A few years later, in 1737, Bose encountered Dufay's memoirs, which he inhaled 'in one breath, at one sitting.' Repeating the experiments required more energy. The panting professor recalled his cannibalized alembic: why not use it rather than the tedious tube to generate electricity? 'Tout ce que tu [Dufay] fis, fut par des tubes caves, / Qui sont bons, mais qui font l'essay tardif et grave.'[48]

With his plentiful generator and an insulated storage bar or 'prime conductor' (PC), Bose could mount impressive practical jokes. Here are a few. Insulate a dinner table and one chair, which you occupy; run a hidden wire from a concealed PC to within reach; have an accomplice run the machine; grasp the wire, touch the table, and watch your guests watch the sparks fly from their forks. Try the Venus electrificata, the kiss of an insulated electrified lady: 'La peine vint de près. Les levres me tremblerent, / La bouche se tourna, presque les dents brisent!' Or electrify an insulated saint wearing a pointed metal cap (fig. 3.6), a pretty beatification of which Bose long retained a monopoly as he neglected to mention the headgear that supplied the halo.[49] Or the 'wonder of wonders': electrify a full water glass in accordance with the Rule of Dufay, and extract sparks from it at the point of a sword.[50] Fire from water! A nice paradox, and also a profitable one, as it was to inspire that pivotal event in our history, the invention of the Leyden jar.

Bose's machine was adopted by C. A. Hausen, mathematics professor at Leipzig, whose experiments captivated his younger colleague, J. H. Winkler, professor of Oriental languages. Hausen died in 1743; Bose had left for Wittenberg in 1738; Winkler became the chief of the Leipzig electricians and exchanged his chair in Hebrew·for one in physics. His electricity attracted two princes of Saxony, a cabinet minister, a Russian ambassador, the major general

47. Schilling, *Misc. ber.*, 4 (1734), 334–5. Cf. van Sanden, *Diss.* (1714), 5, 13, 17, 24, 29–33; Sturm, *Kurz. Beg.* (1713), 560–3, and *Phys. el.*, II (1722), 1096, 1106–7; Hamberger, *El. phys.* (1735²), 473–8.

48. Bose, *Tent.*, I:i (1738), and I:ii (1743), 53ff.; *Electricité* (1754), 1–2.

49. Bose, *Tent.*, I:ii (1743), 58, 61–2, 65–9, I:iii (1744), 79–80, and II:i (1745), 16–17; *Electricité* (1754), 54.

50. Bose, *Electricité* (1754), 47–8.

3.6 The beatification as practiced and revealed by Rackstrow, Misc. Obs. *(1748). Electricity runs from a prime conductor to the plate above the crown, which is evacuated to enhance the effect.*

of the Polish army, and an Austrian arch-duchess.[51] Meanwhile electricity was communicated to Berlin. The show stopper at the public inauguration of Frederick's new academy on January 23, 1744, was the ignition of warmed alcohol by a spark from a PC. Thus was accomplished the inflammations imagined by Dufay and Bose; when electricians substituted themselves for the PC, the curious flocked to see them throw thunderbolts from their fingers.[52]

But it is not all games for professors and academicians. They must have theories. Bose, who twitted attractionists for worshipping 'either God and Newton, or no God at all,' and for disliking his poetry, plumped for the system of effluence and affluence, of which he claimed to be an independent discoverer.[53] Hausen, who hankered after English physics, supposed the effluvia to run around in a vortex held to the tube by an attractive force.[54] Winkler thought similarly, although his 'electrical atmospheres'—which surround all bodies electrified or not—pulsate rather than rotate when excited.[55] These theories were but hors d'oeuvres. The Berlin Academy gave a fuller treat, a prize competition about the 'causes of electricity.' Four responses were deemed print-worthy.[56] The winner, J. S. von Waitz, engineer and courtier, considered that rubbing deprives an electric of its normal content of electrical fire. Neighboring mobile objects, making good its loss, squirt themselves toward it by jet propulsion; on arrival they surrender their remaining fire and withdraw under the pull of the fire beyond. Thus ACR. The three runners-up offer ingenious variations. Electrical matter is elastic owing to Newtonian repulsion or to mini-vortices;[57] attraction occurs via vortical motions, density gradients, or centrifugal forces in the effluvial envelopes; contact confers or stimulates an atmosphere; repulsion ensues via the action of the air or from the elasticity of the electrical matter.

These models, the work of three professors, an engineer, a small-town mayor, and an unknown, are typical of qualitative baroque physics. Post factum, with less predictive value than the phenomenological rules they inexactly re-express, they confronted no strong constraints, no standards of precision that might help to decide between one set of mechanical pictures and another. That the self-conscious new academicians of Berlin rewarded Waitz' and three similar essays says much about the character of the non-mathematical physical sciences in the middle of the eighteenth century.

51. Hausen, *Novi prof.* (1743), 'Praef.,' xi–xii, 1; Winkler, *Gedanken* (1744), 'Vorr.'

52. *HAS*/Ber (1745), 3–13; Bose, *Tent.*, I:ii (1743), 26–7, and I:iii (1744), 76–7.

53. *Tent.*, I:i (1738), 8–11; Bose to Formey, 2 Nov. 1756 (Formey Papers, Deutsche Staatsbibl., Berlin); Bose, *Recherches* (1745), xviii–xl.

54. Hausen, *Novi prof.* (1743), 2, 7, 10–12, 27–9, 34–45.

55. Winkler, *Gedanken* (1744), in *Recueil* (1748), 69–78, 88–114, 131–5, 145.

56. Waitz et al., *Abhandl.* (1745).

57. Malebranche's vortices had already been adapted to electricity by Molières, *Leçons* (1734–9), III, 429–37, and by Mazières, *Traité* (1727).

3. THE AGE OF FRANKLIN

The little lightning of the German professors won electricity new investigators. Most were casual and transient like John Wesley, who wondered momentarily how 'flame issues out of my finger, real flame, such as sets fire to spirit of wine;' but some were determined and resourceful, like Daniel Gralath, the first good historian of our subject, and William Watson, F.R.S., an apothecary who started his career as electrician by igniting in the German manner all the combustibles he had in stock.[1] Three of these new recruits—a German cleric, a Dutch lawyer, and an American printer—precipitated a revolution in electrical theory. None at the time of his discoveries was formally affiliated with a university or national academy, but each had access to professors or academic correspondents who encouraged and publicized his work. The phenomenon deserves notice: the major progress in the study of electricity after its codification by Dufay and its popularization by sparks occurred on the boundary between the unaffiliated middle class and organized institutions of learning.

THE LEYDEN JAR

The cleric, E. J. von Kleist, dean of the cathedral chapter in Kammin (now Kamien, Poland) in Pomerania, tried to augment the power of the flare by increasing the amount of electrified matter from which it sprang: the bigger the PC the bigger the *Schlag*. He ran a wire from the PC into a big glass full of water as a convenient method of effecting the increase. No sensible improvement resulted since, in accordance with the Rule of Dufay, he insulated the vessel while filling it with effluvia. He next brought a nail stuck in a little bottle filled with alcohol up to the PC; hoping, apparently, to create a portable flare, he held the bottle in his hand. On touching the nail with the other hand he got a big surprise. He shared it with some children, who were knocked off their feet by the shock. 'With such sparks Herr Bose would give up kissing his charming Venus.'[2] Kleist described his new power to at least five persons, of whom one was a veteran professor of experimental philosophy (J. G. Krüger) and one an able member of the Berlin Academy (J. N. Lieberkuhn). None was able to reproduce his results. Not until March 1746, three months after Kleist had announced his striking news, did anyone working from his instructions succeed.[3]

He had forgotten to emphasize the counter-intuitive step that made a condenser from a nail in a bottle: he did not say that the experimenter must grasp the

1. Wesley, *Journal* (1906–16), III, 320–1; Watson, *PT*, 43 (1744–5), 481–9.

2. Kleist to Krüger, 19 Dec. 1745, in Krüger, *Gesch.* (1746), 177–81; Gralath, Nat. Ges., Danzig, *Vers. Abh.*, 2 (1754), 402, 406–11; Winkler, *Eigensch.* (1745), 31–57, 77–83.

3. Gralath (the man who succeeded), Nat. Ges., Danzig, *Vers. Abh.*, 1 (1747), 512–13, and 2 (1754), 407–11; Kleist to Krüger, n.d., in Krüger, *Gesch.* (1746), 181–4. Cf. Hoppe, *Gesch.* (1884), 18–19; Priestley, *Hist.* (1775^3), I, 103–4.

outside of the bottle and stand on the floor during electrification; he did not say that the bottle's exterior must be *grounded*. Without this prescription, the knowledgeable investigator, observing the Rule of Dufay and recognizing the semitransparency of glass, would use a thick bottle and insulate it and/or themselves while trying to fill it from the PC.

While Kleist was knocking down children, Pieter van Musschenbroek, who had succeeded 'sGravesande as professor of physics at the University of Leyden, was trying to draw fire from water in the manner of the lyrical Bose. We know how Musschenbroek would have proceeded from Andeas Gordon, professor of physics at the Benedictine convent in Erfurt, a good experimenter who made important improvements in the electrical machine. Gordon directed that the water-filled vessel rest on an insulating stand while the effluvia run into it via a wire to the PC; otherwise they would leak out the bottom to ground.[4] Here our lawyer, Andeas Cunaeus, intervened. After visiting Musschenbroek's laboratory, he tried Bose's experiment on his own. Alone and ignorant of the Rule of Dufay, he electrified the vessel in the manner most natural, holding it in his hand (fig. 3.7a); when he drew a spark from the PC with the other, he let the genii out of the bottle.[5]

Cunaeus showed Musschenbroek and his assistant, J. N. S. Allamand, how they too could blast themselves with electricity. 'I thought I was done for,' the professor wrote Réaumur, his correspondent at the Paris Academy, adding precise directions for realizing the 'terrible experiment' and advice not to try it.[6] The courageous Nollet, informed by Réaumur, bent himself double and knocked out his wind. Others who tried reported nose bleedings, temporary paralysis, concussions, convulsions, and dizziness. The gallant Winkler warned that his wife was unable to walk after he used her to short a Leyden jar. In a characteristic conceit, Bose wished he might die of the electric spark so as to do something worth noting in the Paris Academy's annual *Histoire*.[7]

These exaggerations suggest how flagrantly the action of the condenser violated received ideas about electricity. Theories could not predict the outcome of events; overstatement was a natural consequence of uncertainty, even of fear, before a manifestation of electrical force far stronger than any yet experienced. If, by a slight simplification of Bose's little game, one could create a blow of unprecedented power, might not a trivial alteration of the Leyden experiment

4. Gordon, *Versuch* (1745), 2; Musschenbroek, *Elements* (1744), I, 190; Hackmann, *Electricity* (1978), 82–4.

5. Allamand, *Bibl. brit.*, 24 (1746), 432; Allamand to Nollet, n.d., *MAS* (1746), 3–4; Musschenbroek to Bose, 20 Apr. 1746, in Bose, *Tent.*, II:ii (1747), 36–7; F. A. Meyer, *Duisb. Forsch.*, 5 (1961), 31–4.

6. Musschenbroek to Réaumur, 20 Jan. 1746, in AS, Proc.-verb., LXV (1746), 5; *HAS* (1746), 10; Heilbron, *Isis*, 57 (1966), 264–7, and *RHS*, 19 (1966), 133–42.

7. Nollet, *MAS* (1746), 2–3; Gralath, Nat. Ges., Danzig, *Vers. Abh.*, 1 (1747), 513–14, and 3 (1757), 544; Sigaud de Lafond, *Précis* (1781), 260–4; Priestley, *Hist.* (1775^3), I, 85–6; Winkler, *PT*, 44:1 (1746), 211–12; Bose, *Tent.*, II:ii (1747), 39.

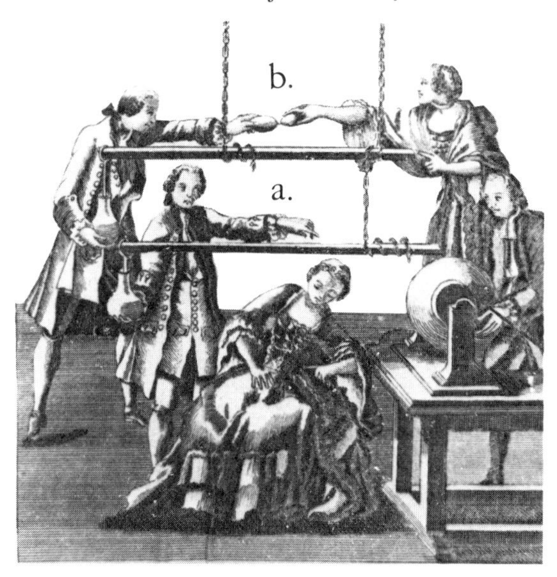

3.7 (a) A one-person and (b) a two-person discharge train. From Nollet, Leçons, *VI (1748).*

transport an unlucky electrician into the next world? 'Would it not be a fatal surprise to the first experimenter who found a way to intensify electricity to an artificial lightning, and fell a martyr to his curiosity?'[8]

Frank admissions that the jar shattered accepted theory appeared on every hand. Musschenbroek, hitherto an authority, now 'understood nothing and could explain nothing' about electricity. Gordon, Gralath, and Winkler were nonplussed at the violation of the Rule of Dufay. It appeared that the Leyden experiment was 'different in kind' from the classical repertory.[9] Perhaps the powers of frictional and communicated electricity differ, or—as one resourceful commentator supposed, in anticipation of the Copenhagen interpretation of quantum mechanics—electricity expressed itself in contradictory ways according to the experiment tried.[10]

No one was killed or even injured by the Leyden jar. Reports of experiments became soberer. By the end of 1746, consensus had been reached on several facts and errors. Nollet found that the shape of the vessel did not figure but that its substance had to be glass or porcelain. The latter proposition, a deduction from experiments with flawed wax and sulphur specimens, placed an unproductive emphasis on the supposed uniqueness of glass. The former proposition

8. [Squario], *Dell'elett*. (1746), 379; cf. *Nollet* (1746), 23.

9. Gordon, *Versuch* (1746²), 111; Gralath, Nat. Ges., Danzig, *Vers. Abh.*, 2 (1754), 447; Winkler, *Stärke* (1746), 'Vorr.,' 72; cf. Le Monnier, *PT*, 44:1 (1746–7), 292–3.

10. Respectively, Södergren, *De rec. quib*. (1746), 12; Needham, *PT*, 44:1 (1746), 259.

eventuated in the 'Franklin square,' a glass pane armed on both sides with metal foil, after Watson and his friends discovered that a bottle filled with lead shot and coated externally with lead sheet electrified more strongly than Cunaeus' device. Physicists found three ways besides external coating to enlarge their shocks: Gralath connected jars in parallel into what he called a battery; Nollet increased the size of the PC; Gralath, Watson, and Benjamin Wilson, an ambitious painter then trying to recommend himself to the Royal Society by his fearless electricity and Newtonian piety, found the bottle the fiercer the thinner the glass.[11] A battery of thin-walled jars shorted through the body hits like a bludgeon. 'The first time I experienced it, it seemed to me, used as I am to such trials, as though my arm were struck off at my shoulder, elbow and wrist, and both my legs, at the knees, and behind the ankles.'[12]

Science is a social enterprise. Let a gentleman hold the jar and a lady the PC (fig. 3.7b); both feel the shock when they touch. How many others can be inserted in the train? Academician L. G. Le Monnier tried 140 courtiers, before the king; Nollet shocked 180 gendarmes in the same presence, and over 200 Cistercians in their monastery in Paris (fig. 3.8). 'It is singular to see the multitude of different gestures, and to hear the instantaneous exclamations of those surprised by the shock.'[13] Only persons in the train felt the commotion; those in side chains branching from the main line felt nothing. Thus electricians discovered that the discharge—to use the word they introduced for the climax of the Leyden experiment—goes preferentially along the best conducting circuit between the inside and outside of the jar. Using the same sensitive detectors, they learned that the best is not always the shortest.[14] In one demonstration only those at the extremes of the chain felt the shock, which appeared to avoid one of the company suspected 'of not possessing everything that constitutes the distinctive character of a man.' Some wits deduced that eunuchs cannot be electrified. Three of the king's musicians, 'whose state was not equivocal,' held hands, and jumped as other men. The shy shock was found to occur only when the train stood on moist ground; apparently the discharge went through the arms and legs of the extreme members and completed its course in the soil.[15]

The charged jar was also intriguing when innocently insulated. It unaccountably preserved its punch for hours or days; even when grounded externally, it remained potent provided its top wire was not touched. With its bottom or tail

11. Nollet, *MAS* (1746), 7–10, 20–2, and *MAS* (1747), 161; Watson, *Sequel* (1746), §§13, 26, 28; Gralath, Nat. Ges., Danzig, *Vers. Abh.*, 1 (1747), 506–41, esp. 513–14; Wilson, *Essay* (1746), 89. Cf. Sigaud de Lafond, *Précis* (1781), 226–7, 279; Hoppe, *Gesch.* (1884), 19–20.

12. Watson, *Sequel* (1746), §§29–30.

13. Nollet, *MAS* (1746), 18; Priestley, *Hist.* (1775³), I, 125–6; Sigaud de Lafond, *Précis* (1781), 293; *HAS* (1746), 8.

14. Gralath, Nat. Ges., Danzig, *Vers. Abh.*, 1 (1747), 518; Watson, *Sequel* (1746), §§25, 31, 34–41; Le Monnier, *MAS* (1746), 452; Priestley, *Hist.* (1775³), I, 109–14, 130.

15. Sigaud de Lafond, *Précis* (1781), 284–91; cf. *VO*, III, 222–5, and Lichtenberg, *Briefe*, II, 144–5.

3.8 A Japanese version of Nollet's many-person discharge train. From a drawing of 1813 reproduced by Prinz, Schweiz. Elektr. Ver., Bull., *61 (1970), 8.*

insulated, no explosion, but only a small spark, could be elicited by touching the head wire connected with its internal coating. It could then be revivified, or made capable of the Leyden shock, merely by holding it in the hand.[16] This behavior, as well as the apparent violation of the Rule of Dufay, defied elucidation in terms of Nollet's double fluxes, Watson's coursing aether (a scheme he thought similar to Nollet's), Winkler's pulsating atmospheres, or, indeed, any articulation of the old approach that presupposed the semi-permeability of glass and the qualitative similarity of the electrifications on either side of the bottle.[17]

BENJAMIN FRANKLIN

Franklin took up electricity in the winter of 1745–6, in his fortieth year, when his business no longer needed his full attention and yielded an income that could support learned leisure. Printing was not an inappropriate preparation for an Enlightenment experimentalist; it taught pertinent skills, the coordination of head and hand, familiarity with wood and metal, exactness, neatness, dispatch. All

16. Nollet, *MAS* (1746), 12, 454–5; Watson, *Sequel* (1746), §§23, 54; Bose, *Tent.*, II:ii (1747), 36–7; cf. Priestley, *Hist.* (1775^{3}), I, 110, 122–3.

17. Attempts at elucidation: Nollet, *MAS* (1746), 11–18; *Essai* (1750^{2}), 194–206; *MAS* (1747), 195–6; Watson, *Sequel* (1746), §64; *PT*, 45 (1748), 102; Winkler, *Stärke* (1746), 73–9, 103ff.; *PT*, 41:1 (1746), 211.

the productive English electricians of the 1740s came from the higher trades: Watson the apothecary, Wilson the painter, and Wilson's collaborator John Ellicott, clockmaker, whose theory of attraction and repulsion resembled Franklin's.[18] Printing inculcated not only manual skills but also practice in straight and accurate thinking. In editing or composing, the successful printer had to be clear, economical, and pertinent; everything was set by hand, and paper cost as much as labor. These experiences helped to frame Franklin's style. The same power of extracting the heart from the matter and expressing it plainly that delights us in the sayings of Poor Richard was ready to serve Franklin when he examined the phenomena of electricity. He also drew upon his experience of men and institutions. His success in building up his business and shaping his community, in mastering people and machines, supported his characteristic optimism, the expectation that he could control or cajole his environment.

A report of the German professors' fireworks reached Pennsylvania in the fall of 1745, together with a glass tube suitable for producing them. The coincidence was the work of Peter Collinson, F.R.S., the London agent of the Library Company of Philadelphia; having seen an account of the new experiments in the *Gentleman's Magazine* for April, 1745, he thoughtfully shipped the apparatus with the journal. The anonymous account provided not only the occasion but also the vocabulary, orientation, and experimental procedures of Franklin's first investigations of electricity. It was not mere journalism. The *Gentleman's* stole it from the *Bibliothèque raisonnée*, a French review conducted by Dutch academicians, who had had it from the Swiss biologist Albrecht von Haller, then a professor at Göttingen.[19] Despite his distance from the centers of European learning, Franklin was initiated into the study of electricity by the Continental professoriate.

Haller's account includes a suggestive experiment in which the usual human condensor, rotated 90 degrees, stands upon an insulator of pitch (fig. 3.9). Should anyone approach the boy, a spark will jump between them, 'accompanied with a crackling noise, and a sudden pain of which both parties are but too sensible.' This suggestive demonstration went back to Gray and Dufay, neither of whom, however, saw its significance. Franklin made it the basis of a new system of electricity. Let two persons stand upon wax. Let one, A, rub the tube, while the second, B, 'draws the electrical fire' by extending his finger towards it. Both will appear electrified to C standing on the floor; that is, C will perceive a spark on approaching either of them with his knuckle. If A and B touch during the rubbing, neither will appear electrified; if they first touch afterwards, they will experience a spark stronger than that exchanged by either with C, and in the process lose all their electricity. Gentleman A, says Franklin in explanation,

18. Randolph, *Wilson* (1862), I, 8; Wilson, *Essay* (1746), viii–ix; Ellicott, *PT*, 45 (1748), 195–224.

19. Heilbron, *Isis*, 68 (1977), 539–45; [Haller], *Bibl. rais.*, 34 (1745), 3–20, and *GM*, 15 (1745), 193–7.

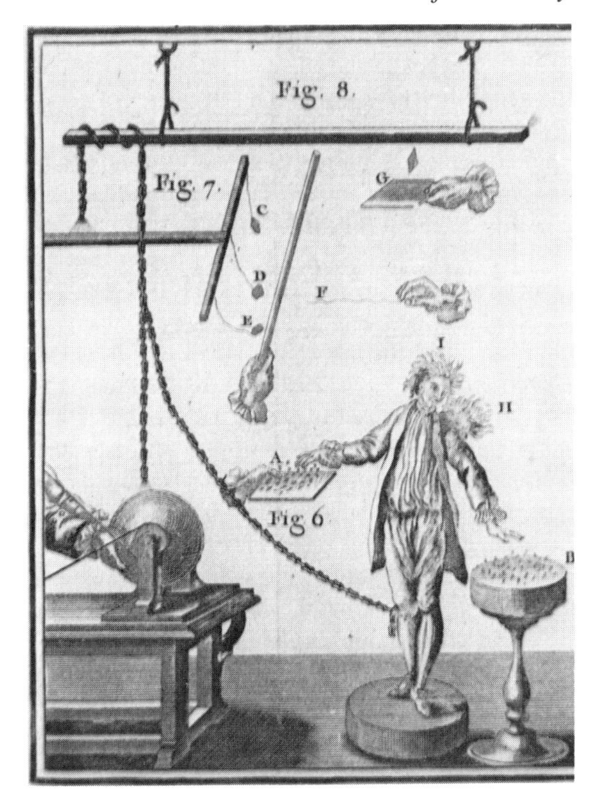

3.9 Gray's boy rotated, as pictured by Nollet, Essai *(1746).*

the one who collects the fire from himself into the tube, suffers a deficit in his usual stock of fire, or electrifies minus; B, who draws the fire from the tube, receives a superabundance, and electrifies plus; while C, who stands on the ground, retains his just and proper share. Any two, brought into contact, will experience a shock in proportion to their disparity of fire, that democratic element striving to attach itself to each equally.[20] Bose, Allamand, Watson, and Wilson had remarked with surprise that on rubbing the tube or globe while insulated, they electrified it less strongly than they did while grounded. They inferred that the rubbing depleted or rarified their stock of electrical matter; but none hit on the notion of qualitatively different, mutually annihilating, positive and negative electrifications.[21]

The form of Franklin's analysis was characteristic. In his first published work, *A Dissertation on liberty and necessity, pleasure and pain* (1725), he considered

20. EO, 171–7; cf. Gliozzi, *Elettrol.* (1937), I, 188–9.

21. Bose, *Recherches* (1745), xli; Allamand, *Bibl. brit.*, 24 (1746), 406–37; Watson, *PT*, 42:2 (1747), 704–49; Wilson, *Essay* (1746), 'Pref.'

the freedom of the will using terms he later found convenient for classifying sparks.[22] Since God is omniscient, omnipotent and all good, our world, His creation, must be arranged for the very best: there is no room for liberty of action. The only cause for motion in the universe is desire to avoid pain: 'The fulfilling or satisfaction of this desire, produces the sensation of pleasure, great or small in exact proportion to the desire.' Consider A, an animate creature, and B, a rock. Let A have 10 degrees of pain. Ten degrees of pleasure must therefore be credited to his account; 'pleasure and pain are in their nature inseparable.' Let him then have his pleasure; he thereby returns to the indifference that B has enjoyed throughout. One cannot miss the analogy between the animating pain, the inseparable, equal, compensating pleasure, and the inert rock, on the one hand, and negative electricity, positive electricity, and the neutral state, on the other.

The chief result of this analysis, as it pertained to electricity, was the discovery of contrary electrical states. The originality of the discovery is perhaps best gauged by the reluctance of Europeans to accept it. They did so because Franklin's theory could stand up to the Leyden jar, which had nullified theirs. According to him, the jar's internal coating can acquire a large charge of electrical fire only with the outer surface grounded because only then can the answering negative or deficit establish itself. Franklin believed that the charging continues until the outer surface of the bottle empties: 'no more can be thrown into the upper part when no more can be driven out of the lower.' He demonstrated the equality by making a cork play between wires attached to the coatings (fig. 3.10a). The cork pendulum swings to and fro, carrying fire from the top to the bottom, until the original state has been restored.[23]

How does the accumulation produce a deficit, the plus a minus? Franklin supposed that the bottle's glass is absolutely impermeable to electrical matter; that particles of electric matter repel one another; that the repulsion operates over distances as great as the thickness of the jar; and that this macroscopic force, arising from the accumulation within the bottle, drives out the electrical matter naturally resident in its exterior surface. Most of these suppositions were peculiar to Franklin, particularly the odd notion that the potent bottle contains no more electricity when charged than when not (its pleasure and pain separate but equal) and the revolutionary concept of the impenetrability of glass. Recall that earlier electricians had allowed glass limited transparency in order to account by projected effluvia for both attraction across glass screens and insulation by glass bricks; and they were consequently perplexed that a Leyden jar of thin glass, grounded on one side, could preserve a large charge. Characteristically, Franklin cut through the paradox by firmly choosing one alternative and ignoring or downgrading the phenomena that supported the other.

Along with the impenetrability of glass, Franklin perforce admitted action

22. *BFP*, I, 55–71.
23. *EO*, 180.

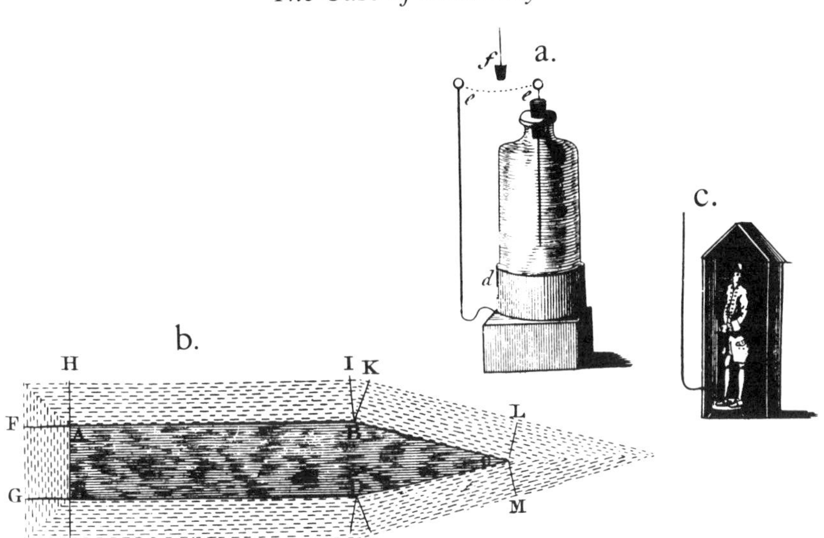

3.10 (a) Franklin's demonstration of the opposite electrifications on the coatings of a Leyden jar (b) his depiction of the statics of electrical atmospheres (c) the sentry box for collecting lightning. From EO[5].

over macroscopic distances, a proposition that, despite the success of the gravitational theory, was still a bugaboo even among Newtonian physicists. But he did not try to relate what he took to be the macroscopic results of the charging to the primitive repulsive forces that he understood to generate them. For example, his proposition that the positive charge on the inner coating equals the negative on the outer conflicts with his account of charging. The condition for cessation must be the vanishing of the pressure driving electrical matter into the grounding wire; if the primitive force decreases with distance, as Franklin supposed it to do, the pressure or macroscopic force can only be annulled when the farther accumulation exceeds the nearer deficit. The discovery by later Franklinists of the inequality of the charges on the two surfaces of the condenser was to mark an important advance over the theories of the founder. He left the second, and sometimes also the first, approximation to others.

The implausible principle that the jar does not accumulate electricity implied that its entire charge resides in the glass, the coatings and wires serving only to fetch and dispose of the fire. Franklin charged a jar, removed its head and tail wires and their connections, and decanted the water; on refilling and rearming it, he could cause the usual explosion without further application to the machine. The same trick worked for a plate of glass electrified like the jar.[24] The search for the charge on the condenser, begun to confirm an inference from the theory of

24. *EO*, 187.

contrary electricities, ended in the invention of the electrophore and the recognition of dielectric polarization and residual charge.

Franklin's first harvest of electricity would not have been complete without elaborate new diversions. He and his collaborators—Philip Syng, silversmith, Thomas Hopkinson, lawyer, Ebenezer Kinnersley, preacher and lecturer—planned a fanciful picnic on the banks of the Schuylkill. After firing whatever spirits they had not drunk, they were to dine on an electrocuted turkey, roasted on an electrical jack before a fire kindled by the electrical bottle. To conclude, 'the healths of all the famous electricians in *England*, *Holland*, *France* and *Germany* are to be drunk in electrified bumpers, under the discharge of guns from the electrified battery.'[25] The bumper, a small tumbler nearly filled with wine and electrified as the bottle, gives an exquisite electrical kiss.

Mechanics of the electrical fire In disregarding the hypothetical dynamics of the electrical matter, Franklin differed from those famous electricians whose health he drank and whose practice he did not know. His cicerone Haller had omitted their systems as imperfect and premature. He went his way until he met with Watson's account of a matter that causes attraction, repulsion, shocks, sparks, etc., by its streaming, and Wilson's attempt, in violation of the ordinary laws of motion, to relate the same phenomena to pressures set up in Newton's stagnant aether.[26] The encounter resulted in Franklin's inconsistent system of conformal atmospheres.

A positively charged spike (fig. 3.10b) retains its redundancy owing to the attraction between electrical and common matter. Franklin reasoned that the portions HAF, IKB, and LCM, because they rest on smaller surfaces and experience less attraction than do HABI or KCLB, are more easily drawn away.[27] Discharge does occur more readily from corners than from unbroken surfaces, but Franklin's mechanism does not explain why. He relied on two inconsistent sets of forces, one to establish the conformal atmosphere, the other to preserve it; if the forces that maintained it determined its shape, it would be shallow opposite points and deep opposite planes. Again, the primitive forces proposed—attraction between the elements of common and those of electrical matter, repulsion between particles of the electrical—conflict with the fact that neutral bodies do not interact electrically. Let the quantities of electrical and common matters in the first body be E and M, those in the second e and m. Then E will be attracted by m and repelled by e, but M will be drawn by e without compensating repulsion. Similarly m experiences unbalanced attraction. Franklin's electrical mechanics require unelectrified bodies to run together.

Another set of inconsistencies afflicted these atmospheres. First, air must readily allow them passage since they move freely through it when the bodies

25. *EO*, 200.
26. Watson, *PT*, 44:2 (1747), 704–49, or *Sequel* (1746); Wilson, *Essay* (1746), 6–26, 63–77.
27. *EO*, 213–38.

they surround are transported. *Sed contra*, pure air must resist atmospheres since they cannot be maintained in vacuo (as Franklin's pump, like Boyle's, misleadingly disclosed) or in moist or smoky, that is, impure air. Second, metal draws electrical matter more strongly than air does because a charged shot holds its atmosphere. *Sed contra*, air draws more powerfully than metal because, as Franklin affirmed, insulators differ from conductors only in their greater power to 'attract and retain' electrical matter.[28] No solution to these conundrums appears to exist within the system.

The obscure theory of atmospheres enabled Franklin to gesture toward an explanation of the less obscure phenomena of attraction and repulsion. Somehow an atmosphere causes its host and neutral bodies to run together, and confers atmospheres on them; which atmospheres, without mixing, then force their positively charged carriers apart (ACR).[29] Whether the atmospheres rotate, gyrate, pulsate, or vegetate does not appear. Compared with earlier models, they seem lifeless, even vestigial. Nonetheless, they remained essential to Franklin's thought and made the explanation of the mutual repulsion of negatively charged bodies, a fact he did not know until 1749, a contradiction in terms, an interaction between objects that, by definition, lacked the mechanism of interaction.[30]

And what of that most compromising phenomenon, attraction across a glass screen, the standard demonstration of penetrability? Franklin suspended a feather in a bottle and approached a tube to its exterior. Just as the surplus thrust by the machine into the Leyden jar expels electrical matter from the tail wire, so the tube's atmosphere forms another from material it drives from the internal surface of the bottle. It is this internal atmosphere that draws the feather via mechanics Franklin did not specify.

Many aspects of Franklin's system did not depend upon the models he devised to illustrate them. The doctrines of impenetrability of glass and of electricity plus and minus, and their embodiment, the theory of the Leyden jar, had a logic and appeal of their own. Their advantages would eventually have brought the new system the favorable consideration of European electricians. As it happened, Franklin's ideas rapidly gained a hearing owing to the success of an experiment having no logical connection with the most important and characteristic of his theories.

Lightning That lightning and electricity agreed in many properties was a commonplace when Franklin took up the subject. Haller, for example, emphasized the parallel between the transmission of electricity along an insulated string and the direct path of lightning along wires used to ring church bells. In 1748 the Bordeaux Academy of Sciences offered a prize for an essay on the relation between lightning and electricity. It was won by a physician who took as his by-

28. *EO*, 214, 248; cf. Home, *Br. J. Hist. Sci.*, 6 (1972), 131–8.
29. Cf. Polvani, *Volta* (1942), 46–7; Finn, *Isis*, 60 (1969), 363–4.
30. *EO*, 199, 365.

word an old conceit of Nollet's: 'l'électricité est entre nos mains ce que le tonnerre est entre les mains de la nature.' [31]

One of Franklin's collaborators discovered that a grounded metallic point could quietly discharge an insulated iron shot at a distance whereas a blunt object could extract the electricity only when very near, and then suddenly, noisily, and with a show of sparks. Whence the difference? According to Franklin's homemade physics, the point acts only upon the small surface of the shot directly facing it and so can pull away the redundant electrical matter a little at a time: 'As in plucking the hairs from a horse's tail, a degree of strength not sufficient to pull away a handful at once, could yet easily strip it hair by hair; so a blunt body presented can not draw off a number of particles [of the electrical matter] at once, but a pointed one, with no greater force, takes them away easily, particle by particle.' [32]

Franklin boldly assimilated the shot to a thunder cloud and the pointed punch or bodkin to an instrument capable of robbing the heavens of their menacing electricity. [33] He illustrated and confirmed his thought with a misleading laboratory demonstration. Take a pair of brass scales hanging by silk threads from a two-foot beam; suspend the whole from the ceiling by a twisted cord attached to the center of the beam, maintaining the pans about a foot above the floor; set a small blunt instrument like a leather punch upright on the ground. Now electrify one pan and let the cord unwind; the charged pan inclines slightly each time it passes over the punch until the relaxation of the cord brings it close enough for a spark to jump between them. If, however, you mount a pin atop the punch, the pan silently loses fire to the point at each pass and, no matter how close it approaches, never throws a spark into the punch.

Franklin's faith in the power of points pushed him beyond European electricians, who had contented themselves with suggesting analogies. Why not catch lightning for parlor tricks? The receiver would have to be insulated, but Franklin apprehended no danger; he expected that a sharply pointed rod would bring down lightning slowly enough to allow the experimenter to prevent dangerous accumulations. He accordingly proposed, in 1750, that a sentry box containing an insulating stand A (fig. 3.10c) be mounted on a tower or steeple; an iron rod, projecting twenty or thirty feet above the box, would fetch the lightning, which the sentry would draw off in sparks. [34] The sight, sound, smell, and touch of the sparks would confirm the identity of lightning and laboratory electricity.

Franklin's dismissal of his agent's peril would appear disingenuous apart from his belief in the uniformity and benignity of nature. As Haller observed, lightning liked to run down wires and ropes attached to church steeples. Often enough the other ends of these ropes were held by men ringing bells to break up thunder clouds. The destruction of bellringers by lightning had been remarked; and

31. Nollet, *Leçons*, IV (1748), 314–15; Brunet, *Lychnos*, 10 (1946–7), 117–48.
32. *EO*, 219. 33. *EO*, 210–11, 334. 34. *EO*, 222.

Franklin's sentry stood in the same relation to his box and rod as the bellringer did to his church and steeple.[35] Franklin tacitly admitted the danger of his sentry in his consequential proposal to replace bellringers by *grounded* rods as protection against lightning. Yet he assured travellers overtaken by thunderstorms in open country that they need not fear taking a lightning stroke. Their dripping garments would protect them, if analogy held; for, as Franklin had shown in unsavory experiments with a Leyden jar, it is harder to electrocute a wet rat than a dry one. 'What an amazing scene is here opened for after ages to improve upon.'[36]

THE RECEPTION OF FRANKLIN'S IDEAS IN EUROPE

Franklin reported his results to Collinson, who informed Watson, who relayed what he thought significant to the Royal Society. In 1751 Collinson and his friends had Franklin's letters published. Watson reviewed the little book favorably; he had managed to conflate his states of electrical aether at different densities with Franklin's contrary electrifications although he rejected the fundamental principle of the electrical opacity of glass. Watson's collaborators did as he did. Wilson as usual took his own path, ignored Franklin, reaffirmed his electro-optical-Newtonian aether, and gave out the Leyden jar as an unanalyzable violation of the Rule of Dufay.[37] Franklin's work was first promoted on the Continent, in a way that illustrates the variety of methods by which science advances through organized learning.

France In 1751 Nollet's second-magnitude star was still rising; his lectures were the fashion, his authority and resourcefulness respected, his bonds with his academic patron, Réaumur, firm and cordial.[38] Another rising savant, Georges-Louis Leclerc, later Comte de Buffon and already Nollet's superior in the Academy, disliked Réaumur, his protégé, and his science. Réaumur represented the worst traits of the older generation; plodding, cautious, thorough, painstaking, Cartesian, he had written among other tedious works seven quarto volumes on insects, a labor that the quick and superficial Buffon, self-appointed Newtonian panjandrum, polished popularizer and propagandist, affected to despise. 'After all,' he said, 'a fly ought not to occupy a greater place in the head of a naturalist than it does in nature.' Réaumur reciprocated this good opinion; and he helped a friend to savage the first volumes of the *Histoire naturelle* of his enemy, 'whose way of reasoning is even more revolting than his opinions.'[39] While still smarting from this attack, Buffon chanced on Franklin's book. Properly handled, it

35. J. N. Fischer, *Beweis* (1784), recorded that 386 hits to bell towers over thirty-three years killed 103 bell ringers; cf. Cohen, Frank. Inst., *J.*, 253 (1952), 393–400.

36. Wesley, *Journal*, IV, 53 (17 Feb. 1753); cf. Nollet, *Lettres* (1753), 10–11.

37. Watson, *PT*, 45 (1748), 94–100; *PT*, 47 (1751–2), 203, 210, 362–72; Wilson, *Treatise* (1752^2), 34.

38. Spiess, *Basel* (1936), 99.

39. Torlais, *Réaumur* (1933), 20–42; *Esprit* (1961^2), 237–47, 343–5, 383; *Physicien* (1954), 234–6; *Presse méd.*, 66:2 (1958), 1057–8; *RHS*, 11 (1958), 27–8.

could embarrass Nollet, annoy Réaumur, and confound others among his 'great many enemies.'[40]

Buffon had Franklin rendered into French by T. F. Dalibard, who prefaced the translation with an 'abridged history' of electricity that does not mention the leading French authority on the subject. Contemporaries did not miss the slap, and Nollet was not altogether wrong in supposing the unknown Franklin to be a creation of his enemies.[41] In February 1752 Buffon's group poached on Nollet's royal preserve and showed Franklin's experiments to the king. In May Dalibard set up a pole to catch lightning at Marly-la-ville, six leagues from Paris. On May 10 it thundered there. An old dragoon left in charge by Dalibard ran to the insulated pole, presented to it an insulated brass wire, and saw the first spark man intentionally drew from the sky (fig. 3.11). The local curé repeated the experiment and wrote Dalibard, who read the letter to the Academy.[42]

Every electrician in Europe rushed to collect thunderbolts. For a year they studied the electricity they carelessly brought from the sky. They learned that collectors need not be pointed and that captured electricity, contrary to Franklin's expectations, was more often negative than positive.[43] In 1753 a German at the Petersburg Academy of Sciences, G. W. Richmann, did the experiment Franklin had designed (fig. 3.12); instead of picking up minor electrical disturbances as the Marly group had done, he caught a thunderbolt and died.[44] Franklinists turned their attention to protecting buildings and to developing theory under continuous sallies by the discomfited Nollet.

Nollet countered in three chief ways: he multiplied demonstrations of the permeability of glass; he tried to show that the Leyden discharge occurs via simultaneous effluences from both coatings rather than by one-way flow from top to bottom; and he pointed out weaknesses, inconsistencies, and oversimplifications in Franklin's theory. A typical demonstration of permeability: a jar, sealed into the receiver of an air pump with its neck emerging, can be charged as usual if the receiver is evacuated; sparks play in the vacuum, and a person touching the outer surface of the receiver and the head wire gets the Leyden shock. The fiery electrical matter can almost be seen squeezing through the glass bottom into the void beyond. 'If, after seeing these things, anyone still maintains that the electrical fluid does not penetrate glass, his bias must be invincible.'[45] A typical demonstration of the double flux in discharge: thin cards placed in the train near the top and bottom coatings receive similar punctures when the jar explodes, proving, according to Nollet, that electrical mátter springs from the negative as well as

40. Montesquieu, letter of 11 Nov. 1747, in *Oeuvres*, VII (1879), 328–9.

41. *EO* (Dal); Torlais, *Physicien* (1954), 129–45; *RHS*, 9 (1956), 340; Nollet to Dutour, 30 Mar. 1752 (Burndy).

42. *EO* (Dal²), II, 99–125; *EO*, 257–62; *BFP*, IV, 303–10.

43. Brunet, *Lychnos*, 10 (1946–7), 117–48.

44. *PT*, 44 (1755), 61–9; Sigaud de Lafond, *Précis* (1781), 355–7; Nollet, *Lettres* (1753), 136.

45. Nollet, *MAS* (1753), 435–40; *Lettres* (1753), 61–73.

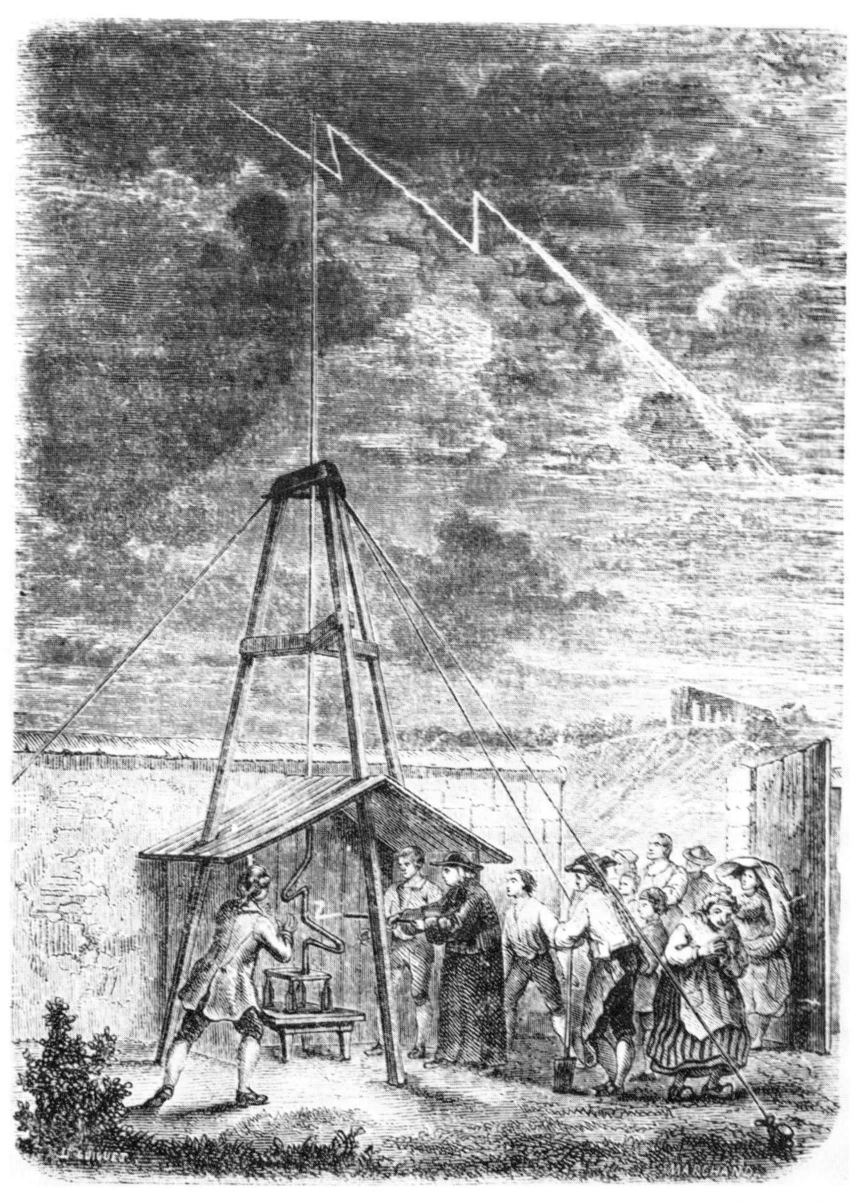

3.11 The experiment at Marly according to a 19th-century artist. From Figuier, Les merveilles de la science *(Paris, n.d.), I, 521.*

3.12 A 19th-century depiction of the sort of experiment that killed Richmann.

from the positive coating. Both demonstrations owe their plausibility to the old, literal concept of electro-luminous effluvia: to turn their force, the Franklinists had to claim that the fire in the receiver is not that supplied from the PC to the jar and that the shape and position of holes blasted by the discharge do not show the course of the electrical matter. Later work was to disconnect the appearances still further from their agent: the lights are not electrical matter aglow, and the holes are not punched by the current whatever its direction.[46]

Having thus tried the strengths of Franklin's system, Nollet turned, and told, against its weaknesses. What had it to say about the mechanism of attraction and

46. Nollet, *Lettres* (1753), 126–7. The holes result from outward explosions caused by attractions between the wires and charges they induce in the cards; *VO*, III, 222–5; Atkinson, *Elements* (1887), 81–3.

repulsion, the oldest, most dependable, and most challenging electrical phenomena? Nothing but impossibilities or admissions of ignorance. 'That,' snapped Nollet, 'is almost like saying to a man who points to a weather vane as proof of the wind: Ho! we don't know what turns the vane!'[47] What retains positive atmospheres around bodies? How can bodies without electrical matter, negative bodies, act electrically? Many of the older electricians, who thought these objections weighty, remained with Nollet. Younger men, as is their wont, ran after novelty.

Buffon's group could not meet Nollet's fire and the single competent French Franklinist, J. B. Le Roy, Nollet's junior at the Academy, could not best him. Le Roy's discovery of the difference in appearance of discharges from points—a brush from positive ones, a star from negative ones—gave strong support to Franklin's concept of one-way currents and so helped to stymie discussion at the Paris Academy until Nollet's death in 1770.[48] The Academy's place in the van of electrical studies fell to Italian professors.

Italy In 1748 the chair of physics at the University of Turin, long the preserve of Cartesian Minims, went to a Newtonian Piarist, Giambattista Beccaria. It was rumored that jingoism had won the new professor preferment over his competitor, the distinguished French Minim and Newtonian commentator François Jacquier, a charge that gained plausibility as years went by and Beccaria gained no academic glory. The electors grew concerned to justify their choice. One of them, intrigued by news of the spectacle at Marly, summoned the beleaguered professor: 'Here is a new branch of physical science for you; spare no expense, but cultivate it so as to make yourself famous.'[49]

Professors at Bologna, astonished equally by the spectacle and its inventor, had already set up poles. 'Who would have believed that electricity had learned students in North America? In Philadelphia, a city in Canada?'[50] They learned that atmospheric electricity could occur in tranquil weather. Beccaria discovered the same when his pole began to spark in the summer of 1752. He did not stop, as did the Bologna group, at the outworks of Franklin's system. He meticulously examined the power of points, the impenetrability of glass, the pumping of the machine, and the action of the condenser and refound, among much else, Le Roy's convenient distinction between appearances at positive and negative points. He rushed to prepare the first European Franklinist treatise. His enemies resolved to suffocate the book. They brought copies of Nollet's works from Paris to prove the inanity of the system from Philadelphia in Canada. They purloined

47. Nollet, *MAS* (1755), 293–317; *Lettres* (1760), 63; cf. Yamazaki, *Jap. Stud. Hist. Sci.*, 15 (1976), 45, 50, 59.

48. Le Roy, *MAS* (1753), 459–68; Home, CIHS, XIV (1974), *Actes*, II (1975), 270–2.

49. Vassalli-Eandi, *Lo spett.*, 5 (1816), 101–5, 117–22, and Acc. sci., Turin, *Mem.*, 26 (1821), xxvii.

50. Verratti, Acc. sci. inst. Bol., *Comm.*, 3 (1755), 200–4; Tabarroni, *Coelum*, 34 (1966), 127–30.

the sheets as they came from the press and published a refutation of one-half of the book before Beccaria had corrected proofs of the other. To no avail. *Dell'elettricismo artificiale e naturali* appeared in 1753, with an answer to Nollet and a blast at its critics, to 'the applause of the learned and the despair of the spiteful.'[51]

Throughout reigns the spirit of system; where the printer had been disorganized, parochial, unassertive, and open, the pedagogue ordered, developed, polished, and generalized.[52] From the appearances of the discharges at points, he deduced the direction of electrical currents; from the properties of objects electrified by contact with the PC, or with the frame of an insulated machine, he defined plus and minus electricity, respectively. Against Nollet, Beccaria effectively argued the impenetrability of glass and the contrariety of the electricities: the supposed proof of the seepage of fire across the bottom of the Leyden jar into the evacuated receiver in fact is an occular demonstration of ejection of electrical matter from the lower plate of a condenser; that a cork will only play between head and tail wires, but not between the heads of bottles charged to different degrees, shows the contrary character of the electrifications top and bottom. Buffon's clique was delighted with the response, as it saved them the trouble of making one themselves, and they issued it in French for the worldwide 'extirpation' of the followers of Nollet.[53]

Beccaria did not succeed in meeting Nollet's objections to Franklinist mechanics of attraction and repulsion. Sometimes he dismissed the demand for mechanism as impertinent and stuck to phenomenological rules; other times he put forward intricate machinery, including a revival of Cabeo's choreography.[54] His increasing attention to the mechanics of the electrical matter ran against the trend towards instrumentalism among new recruits to Franklinism, and he gradually lost influence over the discipline his treatise helped to define.

THE ATMOSPHERES ATTACKED

The plausibility and vulnerability of the old system of effluvia rested on a conflation of induction with conduction, of charge with force, and of presence with action. Franklin challenged the comfortable old approach in two ways that could not be ignored. First, the activity of lightning conductors in serene weather, when no discharge from clouds could reasonably be supposed, forced a clear distinction between conductive and inductive effects. Second, the theory of the Leyden jar, which already separated by the thicknesses of bottle bottoms the presence of electrical matter from its place of action, was found to apply in ar-

51. Vallauri, *Storia*, III (1846), 141–3; Vassalli-Eandi, *Lo spett.*, 5 (1816), 101–5, 117–22; Beccaria, *Dell'elett.* (1753), 160–7, 235.

52. Cf. Gliozzi, *Arch.*, 17 (1935), 20–1, 32; *Elettrol.* (1937), I, 209, 246.

53. Beccaria, *Dell'elett.* (1753), 9–12, 17, 23–30, 40, 48–9, 55, 149–54, 235–45.

54. Beccaria, *Dell'elett.* (1758), 36–49, 85–97, 113–15; cf. Home, *Br. J. Hist. Sci.*, 6 (1972), 146–9.

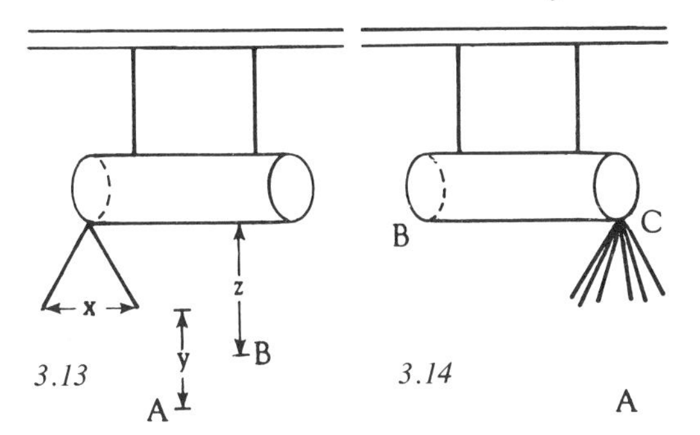

3.13 Canton's induction experiments.

3.14 Franklin's improvements of Canton's inductions.

rangements without glass, in circumstances demanding a clear choice between distance forces and atmospheres.

A schoolmaster in London, John Canton, F.R.S., brought the issue of the charging of lightning conductors into the laboratory. He hung a cylinder (the conductor) on insulating cords (fig. 3.13) and brought up a glass rod (a positive cloud). Cork balls (lightning detectors) attached to the cylinder diverged as the rod approached, collapsed as it receded. Next he charged the cylinder with a spark from the tube (a lightning stroke); the corks spread to a distance *x*, then fell together as the tube approached from A along *y*. Similar effects occurred with a wax rod (a negative cloud); but if wax were brought along *y* to a cylinder charged from glass, the balls spread further rather than collapsing.[55] Canton elucidated these effects partly via analogy to Franklin's theory of the Leyden jar and partly via appeal to electrical atmospheres. The intricate hybrid account inspired an attempt at clarification by the founder.

Franklin replaced the corks by a tassel of threads and electrified the cylinder by a spark from the glass tube.[56] The tassel spread, and spread still more when he brought the tube to B (fig. 3.14) but closed as much when he withdrew it. The threads fell further with the tube at A, but recovered when it was withdrawn. He discharged the cylinder and placed the tube at B; the tassel spread, but collapsed when, with the tube still at B, he took a spark from C. He took away the tube and the threads opened. To explain these clear and striking effects, Franklin introduced three principles that, although couched in terms of palpable atmospheres, in practice reduced them to ghosts. Contiguous atmospheres 'do not readily mix and unite into one atmosphere, but remain separate, and repel one another;' 'an electric atmosphere not only repels another electric atmosphere, but will also repel the electric matter contained in the substance of a body approaching it; and, without joining or mixing with it, force it to other parts of the body containing it'; 'bodies electrified negatively, or deprived of their natural quantity of elec-

55. Canton, *PT*, 48:1 (1753), 350–8.

56. *EO*, 302–6. Like Canton, Franklin developed these experiments chiefly as analogies to the action of atmospheric electricity; cf. *BFP*, XIII, 248, and *FN*, 516.

tricity, repel one another (or at least appear to do so, by a mutual receding) as well as those electrified positively, or which have electrified atmospheres.' In the last proposition, in which Franklin characteristically capitulated to the facts,[57] the atmospheres, already made unmechanical and even metaphorical by their unity and inviolability, have become otiose.

Franklin came very close to excising all vestiges of effluvia from his system while criticizing Wilson's attempt to express the contrary electricities by differently directed pressures in the Newtonian aether. Perhaps Wilson's absurd mechanics encouraged Franklin so to reduce his atmospheres that they no longer required mechanical properties. He now spoke freely of negative atmospheres, meaning thereby not material substances but states in the neighborhood of negatively charged objects. He could not fully transfer this insight to the positive case but compromised with the doctrine that positive atmospheres do not extend very far into the space through which their influence is felt.[58]

When he took on Wilson, Franklin was living in London as agent for the Pennsylvania Assembly. He became prominent in the Royal Society and chief of a group of electricians that included Canton, Watson, and later, Joseph Priestley and Jan Ingenhousz. The group differed from Wilson and his less distinguished followers not only in electrical theory but also in politics and social status, Franklin's friends gravitating about the Club of Honest Whigs, Wilson's levitating toward aristocratic patrons of the arts.[59] Their opposition culminated in the mock-heroic battle of the knobs and spikes, which broke out in the late 1770s when a British powder magazine, defended by sharp grounded rods as directed by Franklin, suffered minor damage from lightning. Wilson located the trouble in the points. In elaborate experiments conducted in a London dance hall (fig. 3.15), he showed, what no one doubted, that pointed rods discharge electrified bodies at greater distances than blunt conductors can. Since, he said, Franklin's long spears evidently do not draw lightning silently but are struck just like blunt poles, it is only prudent to use short lightning conductors with obtuse ends: loaded clouds that would strike to tall pointed rods might, if high enough, pass harmlessly over little blunted ones.[60]

Wilson's large-scale experiments had been made possible by George III; the American colonies, then in full revolution, were represented by Franklin. The shape of lightning conductors became a matter of politics. The king, according to a story perhaps invented by the French, instructed the president of the Royal Society that lightning rods would henceforth end in knobs. 'Sire,' he is said to have said, the 'prerogatives of the president of the Royal Society do not extend

57. Cf. *EO*, 365–9; *FN*, 531–3.

58. Hoadley and Wilson, *Observations* (1756), §§75–93, 271–4, criticized by Franklin in *BFP*, VIII, 241–63.

59. Cf. Crane, *Will. Mary Q.*, 23 (1966), 210–33.

60. Henley et al., *PT*, 68:1 (1778), 236–8; Wilson, *ibid.*, 239–42, 247.

3.15 Wilson's demonstration of the superiority of blunt lightning rods: the model of the Purfleet arsenal is at the right, the electrostatic generator in the center background. From Wilson, PT, *68:1 (1778), 245–313.*

to altering the laws of nature,' and forthwith resigned.[61] The laws of nature do not discriminate much between points and knobs on lightning conductors. Wilson was correct in asserting that elevated pointed rods do not silently discharge distant clouds, and his adversaries were right in insisting that protection increases with the height of the conductor.[62] Both parties erred in believing that the paltry discharges of the laboratory copied the grand processes of nature.

Atmospheres in Berlin and Petersburg In the mid 1750s Leonhard Euler, his son Johannes Albrecht, and some colleagues in and around the Berlin Academy took up the study of electricity. A prize competition adjudicated in St. Petersburg in 1755 stimulated, if it had not occasioned, their interest. The younger Euler won with an old-fashioned account of ACR, which he, adapting ideas of his father's, associated with the streaming of a luminiferous aether.[63] He did not

61. Pringle et al., *PT*, 68:1 (1778), 313–17; Nairne, *PT*, 68:2 (1778), 825, 835, 859–60; Cuvier, *HAS*, 5 (1821–2), 220–1.

62. Schonland, RS, *Proc.*, 235A (1956), 440–2; Viemeister, *Light. Book* (1972), 110–23, 197–8.

63. Home, *Isis*, 67 (1976), 22–3; Home in Home and Connor, *Aepinus's Essay* (1979), 68–72; L. Euler to G. F. Müller, 20/31 Dec. 1754 and 26 Sept. 1755 in Euler, *Berl. Petersb. Ak.*, I (1959), 79, 92.

mention Franklin or the two electricities and eschewed distance forces. The announcement of the prize had encouraged attention to the mechanics of the medium, to the 'progressive, rotational, and vibratory' motions of the electrical matter. These terms reflected the Cartesianism of M. V. Lomonosov and of his friend, the late electrical martyr Richmann, both of whom rejected contrary electricities and tried to make do with the 'agitation' of a 'certain electrical matter that surrounds electrified bodies to a certain distance.'[64]

In the year of the prize, F. U. T. Aepinus, an excellent applied mathematician, became the Berlin Academy's astronomer. He and his student J. C. Wilcke soon went further than the Eulers in electricity. Wilcke was then preparing a dissertation on the contrary electricities. He studied Franklin thoroughly, translated him into German, and examined Nollet's objections.[65] The arguments against the opacity of glass and certain details of the induction experiments of Canton and Franklin bothered him. He applied to his professor, who was completing a theory of the electricity of the tourmaline. Aepinus had found it useful to consider the stone a miniature Leyden jar. With this analogy in mind, he watched Wilcke repeat the English induction experiments. In one brilliant, enduring insight he saw that the insulated tin cylinder, the approaching glass tube, and the spreading tassel made an imperfect Leyden jar. The tube corresponded to the interior coating, the cylinder to the external, and—here is the breakthrough—the air gap to the glass. Had the cylinder been grounded, the analogy would have been complete. To test the proposition, he and Wilcke built a condenser of metal-covered frames separated by air instead of glass. They had the satisfaction of a shock comparable with one from a well-charged bottle.[66]

The air condenser made it impossible to maintain the theory of literal atmospheres. Although the repulsive force of the upper plate certainly reached the lower, its redundant electrical matter as certainly did not; for in that case the condenser, being shorted internally, could not have charged. The only recourse, according to Aepinus, was to banish the effluvia, admit that similarly charged bodies repel and dissimilarly charged bodies attract one another, and neglect the mechanism of the forces.[67] The true spirit of Franklinism is instrumentalism. Aepinus noticed the consequence mentioned earlier, that the proposition that ordinary matter attracts the electrical, the particles of which repel one another, implies that unelectrified bodies interact electrically. In order to compensate the pulls Em and Me (to use the earlier notation) Aepinus feigned a repulsion Mm

64. J. A. Euler et al., *Diss.* (1757), 'Programma;' Lomonosov to Shuvalov, 31 May 1753, in Lomonosov, *Ausg. Schr.* (1961), II, 190–1; *ibid.*, I, 175, 305–7, 326–7; Richmann, *NCAS*, 4 (1752–3), 305, 323–4.

65. Oseen, *Wilcke* (1939), 18–35; *EO* (Wilcke), esp. 219–21, 271–2, 280–6, 290, 308–9, 348.

66. Aepinus, *MAS*/Ber (1756), 106–7, 117–20; *Recueil* (1762), 50–1; *Tentamen* (1759), 2, 75–83; *Essay* (1979), 281–5; *EO* (Wilcke), 306–9.

67. Aepinus, *MAS*/Ber (1756), 108; *Tentamen* (1759), 5–10, 35–40, 257–9; *Essay* (1979), 239–40, 256–9, 392–3.

between constituents of ordinary matter. He conceded that physicists who took Newtonian gravity to be a primary quality might be 'horrified' at his proposition; for his part he understood forces as descriptions and was not revolted at ascribing both gravitational attraction and electrical repulsion to the same particles.[68]

Wilcke could not bring himself entirely to accept so severe an instrumentalism, and the Eulers rejected it altogether. Wilcke reached Franklin's compromise: negative atmospheres are mere spheres of activity, spaces distorted by the presence of a deficient object; positive atmospheres belong to bodies, surround bodies, are body, and by their physical presence directly cause ACR.[69] The Eulers, having learned about the contrary electricities, patched their hypothesis to suit and submitted it under the protection of heady and irrelevant mathematics for a second Petersburg prize awarded them in 1757. Neither they nor the runner-up, Paolo Frisi, could bring their calculations usefully to bear upon their aether models.[70] Their failure helps to measure the achievement of Aepinus. His mathematical reformulation of Franklinism, published in 1759 in St. Petersburg, where he had gone for academic advancement, yielded unexpected results both useful and true.

One example of Aepinus' mathematizing has already been given: the necessity to place the mutual repulsion of matter particles among the fundamental postulates of the theory. That it accounted for minus-minus repulsion was an added virtue. Three more examples will indicate the nature of the calculations, in which the dependence on distance r of the elementary attractions and repulsions is represented as $f(r)$, f being a continuous, decreasing function of its argument. The first example concerns the action of $A(a, Q+\delta)$ on $B(b, q)$, where a and b are the common matter, Q and q the normal contents of electrical matter, and δ the charge. Aepinus writes for the force F on B:[71]

$$F = [(Q+\delta)b + qa - ab - (Q+\delta)q]f(r).$$

This formidable expression is zero: $(Qb+qa-Qq-ab)f(r) = (Q-a)(b-q) \cdot f(r) = 0$, as it represents the force between unelectrified bodies; the remainder, $(b-q)f(r)$, is therefore also zero since from the preceding equation $Q = a$, $q = b$ (in Aepinus' measure the amounts of ordinary and electrical matter in a body are equal). A charged body exerts no force on a normal one. Only if the electrical matter in B can segregate, making the side toward (or away from) A negative (or positive), can attraction ensue. Aepinus observed that a perfect insulator cannot suffer electrical attraction.[72]

68. Aepinus, *Tentamen* (1759), 16–37; *Essay* (1979), 256–9.

69. *EO* (Wilcke), 221–4, 233–6, 262–3, 270–1, 307, 340–1.

70. J. A. Euler, *MAS*/Ber (1757), 126–8, 139–47; J. A. Euler et al., *Diss*. (1757).

71. Aepinus, *Tentamen* (1759), 39, 114–17; *Essay* (1979), 260–1. A factor of proportionality has been omitted from the equations and $f(r)$ is written for Aepinus' r.

72. *Tentamen* (1759), 127–8; *Essay* (1979), 304–6. The point is also made in *EO* (Wilcke), 310; cf. Wilcke, *AKSA*, 24 (1762), 220–2.

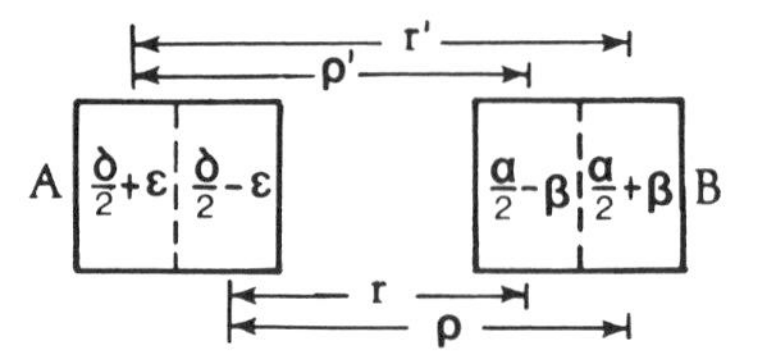

3.16 Diagram of Aepinus' theory of mutual induction.

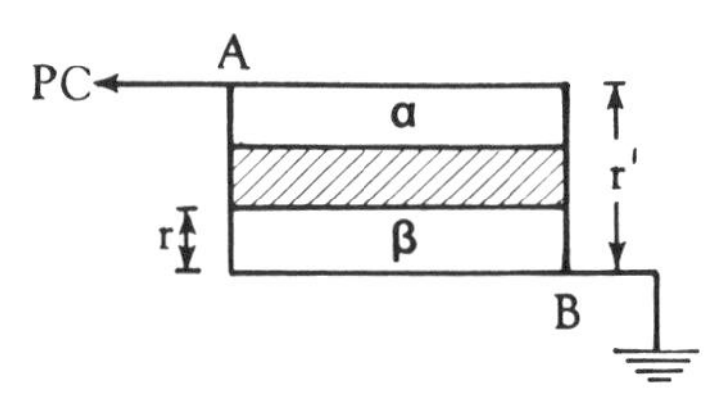

3.17 Diagram of Aepinus' theory of the charging of a condenser.

The second example concerns the ordinary case, in which the electrical matter segregates in both A and B. Let A and B have the surplus charges δ and α, and let them segregate the amounts ϵ and β, respectively (fig. 3.16). The force on B is

$$F = (\tfrac{1}{2})[(\delta - 2\epsilon)((\alpha - 2\beta)f(r) + (\alpha + 2\beta)f(\rho)) + (\delta + 2\epsilon)((\alpha - 2\beta)f(\rho') + (\alpha + 2\beta)f(r'))].$$

Since all quantities on the right of this equation are positive and since, for large distances, the induced charges ϵ and β are much less than δ and α, F too is positive. Positively charged bodies repel one another. But since β and ϵ depend on r, there may be a region where F is attractive. For simplicity, assume A to be a good insulator ($\epsilon \simeq 0$). F will vanish, and the force change sign, where

$$(2\beta - \alpha)(f(r) + f(\rho')) = (2\beta + \alpha)(f(\rho) + f(r')).$$

The same can happen with both A and B negative. After confirming theory by experiment, Aepinus declared that the rule requiring repulsion between bodies similarly electrified does not hold in general.[73]

An equally curious prediction emerged from our third example, an analysis of the condenser. Guided, as usual, by careful analysis of the forces, Aepinus saw that a Leyden jar ceases charging when the repulsion at the tail wire drops to zero, not, as Franklin affirmed, when the outer surface has lost its electrical matter. If the charge inside is α, that outside β, the repulsive force on unit charge at B (fig. 3.17) is $(\alpha f(r') - \beta f(r))$; at cessation, $\alpha = \beta f(r)/f(r')$, whence $\alpha > \beta$. Franklin erred again in thinking that the charges on the two surfaces of the condenser are equal. The experiments apparently confirming his error succeeded because, for thin Leyden jars, $f(r) \simeq f(r')$. An exploded air condenser with plate

73. Aepinus, *Tentamen* (1759), 132–6; *Essay* (1979), 267–8, 315–20.

separation of one or two inches, however, retains a sensible fraction of the charge initially put into the upper plate.[74]

Aepinus' book received less attention than it deserved. It was hard to procure and difficult to read; it had too much mathematics, Latin, and detail for most electricians of the time. Wilson may speak for them: 'The introducing of algebra in experimental philosophy is very much laid aside with us,' he wrote Aepinus, 'as few people understand it.'[75] So also was the instrumentalism too advanced. Euler, for example, praised his friend's experiments but deplored reliance on 'arbitrary attractive and repulsive forces.'[76] The acceptance of Aepinus' approach awaited the work of younger men who independently entered into the same path.

4. QUANTIFICATION

THE ATMOSPHERES DESTROYED

The quantification of electrostatics rested on a strict distinction between the location of a charge and its sphere of action. The distinction came most forcibly to hand and mind in the manipulation of pieces of Franklin squares. Here three classes of experiments may be recognized: Franklinist, Pekingese, and Symmerian. All eventuated in a similar invention, known in its definitive form as the 'elettroforo perpetuo,' which became the vehicle of an instrumentalist theory in the style of Aepinus.

The Peking experiments were invented in the early 1750s by Jesuits in China who had been encouraged to devote their leisure to electricity by gifts of equipment from Richmann and the Petersburg Academy. They placed a glass pane, rubbed side down, on the glass lid of a compass case; the needle rose to the lid, stayed, then fell to its normal position; removal of the pane caused a second ascent, replacement another fall, and so on many times, without their rerubbing the plate.[1] 'These experiments are certainly remarkable,' wrote Aepinus, who knew of them through the Jesuits' letters to St. Petersburg; 'it seems quite paradoxical that electricity, become almost extinct, after some time, as if spontaneously, without further rubbing, can be resuscitated.' He reduced the paradox to his brand of Franklinism by recognizing the experimental arrangement as an imperfect Leyden jar and introducing the postulate, derived from analogy to the action of a magnet on iron, that an electric force can polarize an insulator like glass. The lid polarizes negatively on its upper surface, positively on its lower; the needle rises under the net attraction of the charged pane and the polarized lid; electrical matter leaks from the under surface of the lid through the needle to ground until the net force becomes zero, as in the charging of a condenser,

74. *Tentamen* (1759), 82–7; *Essay* (1979), 269–70, 286–9.

75. Add. Mss. 30094, f. 91; Home, *Isis*, 63 (1972), 196, 202–3; CIHS, XIII (1971), *Actes*, VI (1974), 287–94; and in Home and Connor, *Aepinus's Essay* (1979), 189–224.

76. Euler to Müller, 30 Dec./10 Jan. 1761, in Euler, *Berl. Petersb. Ak.* (1959), 166.

1. Gaubil to Deslisle, 25 Oct. 1750, and to Kratzenstein and Richmann (deceased), 30 Apr. 1755, in Gaubil, *Corresp.* (1970), 617, 810–11.

whereupon the needle retires. On removal of the pane, it reascends through the influence of the negative charge remaining on the top of the cover.[2]

In Symmerian experiments, which were idealizations by Nollet of inventions by Robert Symmer, F.R.S., a silk ribbon is charged by friction while resting on a glass plate, or two panes are superposed, armed externally, and charged as a single Franklin square.[3] Each pair cohered strongly but attracted weakly externally; yet, on separation, every member showed itself to be strongly electrified. Symmer had used these experiments to argue for the existence of two electrical fluids, as will appear; Nollet thought they supported the double flux; and Gianfrancesco Cigna, M.D., nephew, former student, colleague and academic opponent of Beccaria's, whom Nollet tried to enlist against his uncle, declared after thorough examination that the experiments made difficulties for everybody. 'How [can] the contrary electricities destroy themselves when mixed and how, when unable to combine, [can] they draw mutually, restrain one another, and act as if their reciprocal attraction were obstructed?' Cigna observed that two panes charged as a single square continued to cohere even after the explosion, whereas Franklinist theory called for the total destruction of the equal and opposite electricities supposedly residing in the external surfaces of the glass. 'What seems singular is that, although discharged and exercising no electrical force on external objects, yet the glasses strongly cohere, and on separation display opposite electricities on their unarmed faces.'[4]

Cigna proposed to save the phenomena by introducing a new, Symmerian, state of electricity, which dissipates more slowly than the ordinary kind. One of Cigna's demonstrations of its action proved inspirational. He brought a ribbon charged with positive Symmerian electricity to an insulated lead plate, which he momentarily grounded, giving it a negative Franklinist charge. When together, ribbon and plate gave no external signs; during separation, a small spark passed between them; when apart, each showed strongly electrical. Cigna then discharged the plate, brought up the ribbon with its tenacious Symmerian electricity, and repeated his manipulations; 'and thus,' he said, 'we find an easy way of multiplying electricity without friction.'[5] He had found the principle, if not the Franklinist explanation, of the electrophore.

The Franklinist form of these manipulations, developed by Wilcke, derived from the problem of the location of the charge of the Leyden jar. With support from the Stockholm Academy of Sciences, of which he had become demonstrator, he fashioned a multiply dissectible Franklin square (fig. 3.18) consisting of the glass plate ABCD, the coatings b,B, and the leads L,C. He charged the

2. Aepinus, *NCAS*, 7 (1758–9), 277–302.

3. Symmer, *PT*, 51:1 (1759), 340–89; Nollet, *MAS* (1761), 244–58, and *Lettres* (1767), iii–v, 1–151.

4. Cigna, *Misc. taur.*, 3 (1762–5), 31, 54–9, 72; Vassalli-Eandi, Acc. sci., Turin, *Mem.*, 26 (1821), xiv–xx, xxix.

5. Cigna, *Misc. taur.*, 3 (1762–5), 43–51.

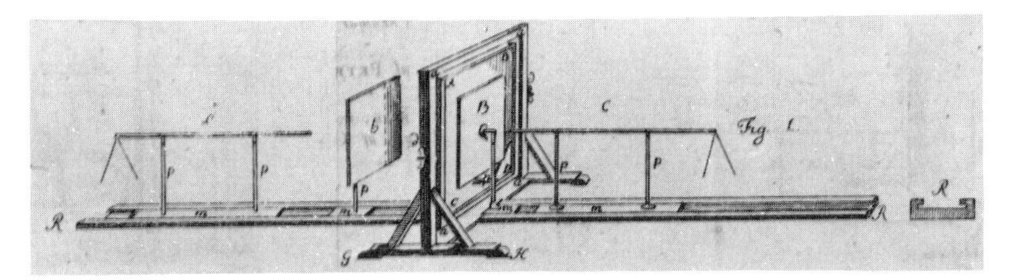

3.18 Wilcke's dissectible condenser. From Wilcke, AKSA, *24 (1762), 213–35.*

square intact, then stripped the pieces singly or in combination. Here is a typical experiment: electrify the square (C positive), explode it, remove B and C, take a spark from them, briefly join C and L, remove B and C, take another spark, and so on. 'In this way the glass can keep electrifying the coatings for many days or weeks, as often as the experiment is repeated.' Wilcke explained its behavior by rigorously distinguishing induction from sparking over (conduction). In the latter case electrical matter is literally present where it glows; in the former its atmosphere exists only metaphorically where it acts.[6]

Piqued by Cigna, Beccaria examined a dissectible condenser, the glass of which consisted of two separable panes. He reproduced and multiplied all the Pekingese, Symmerian, and Franklinist experiments, which he tried to reduce to the Philadelphia system as amended by himself. In order to secure the basic principle that contrary electricities mutually annihilate when juxtaposed, he introduced still another state of electrification, the 'vindicating.' Where the panes of the exploded glass join, or where Cigna's Symmerian ribbon rests on its Franklinist lead plate, the contrary electricities not only neutralize but disappear, to be 'vindicated' when the panes, or ribbon and plate, separate. Charge and explode a dissectible square with single glass, annihilating all electricity; remove the coating via insulating strings from the positive side; electricity vindicates, the bare glass showing positive and the coating negative. The business can be repeated without again charging the condenser. Even Aepinus had missed this notable phenomenon, which Beccaria called the 'oscillation of the electricities.'[7] It arises from the return of positive absorbed charge, which overcomes the depolarizing negative electrification at the upper surface. The top of the glass goes plus and makes the coating minus by induction. As Cigna found, the absorbed charge released can be enough to give a second Leyden shock.[8]

Volta Beccaria's beautiful experiments, clumsy theory, and argumentativeness challenged a young man who was to become the greatest of all Italian elec-

6. Wilcke, *AKSA*, 24 (1762), 213–35, 253–74.

7. Beccaria, *Experimenta* (1769), 1–3; *Treatise* (1776), 395–413; Zeleny, *Am. J. Phys.*, 12 (1944), 329–39.

8. Beccaria, *Experimenta* (1769), 6–10, 36–44; *PT*, 57 (1767), 298, 310; Cigna, *Misc. taur.*, 5 (1770–3), 98–9; Barletti, *Physica* (1772), 41.

tricians. In 1770 Alessandro Volta was twenty-five, a teacher in a secondary school in Como, and a correspondent of Nollet's and Beccaria's. His messages to these authorities scarcely pleased them; for Volta, having embraced the method of the *Principia* and the program of Boscovich, insisted that electricity arises from an attractive force acting, or appearing to act, at a distance. Nollet replied, correctly, that the approach was hopeless; Beccaria sent his works as correctives and encouraged Volta to experiment, but broke off in 1768 when the ingrate published an essay critical of vindicating electricity and the mechanistic residues in Turinese Franklinism.[9] The old, comfortable, commonsensical mechanism being unsustainable, Volta insisted on attractive forces; 'following the example of the most distinguished men, I will then seek to explain certain natural phenomena that do not arise from impulse and from known laws.'[10] Beccaria reaffirmed his odd view that the contrary electricities destroy one another in the union of a charged insulator with a momentarily grounded conductor only to reappear 'vindicated,' as Beccaria hoped also to be, in subsequent experiments.[11] Volta undertook so to prolong the duration of electrification as to destroy the theory of its alternate destructions and recuperations. In June 1775 he announced to Joseph Priestley, as official historian of electricity, the invention of an inexhaustible purveyor, which, 'electrified but once, briefly and moderately, never loses its electricity, and although repeatedly touched, obstinately preserves the strength of its signs.'[12]

The device consisted of a metal disk B (fig. 3.19a) containing a cake made of three parts turpentine, two of resin, and one of wax; a wooden shield CC covered with tin foil; and an insulating handle E. It can easily be charged, as became standard, by rubbing the cake while grounding the plate; but as that procedure would have obscured the connection with vindicating electricity, Volta told Priestley to electrify the instrument like a Leyden jar and to take sparks alternately from shield and plate until the 'oscillation of the electricities' (fig. 3.19b). Here, where Beccaria supposed vindicating electricity to set in, the shield can be removed and its negative charge given, say, to the head of a Leyden jar; then replaced, touched, and again brought to the head; and so on until the condenser is moderately charged. Any number of jars can be electrified without regenerating the original; and, if it should decline, it can be reinvigorated by lightly rubbing its cake with the coating of a jar previously charged by it (fig. 3.19c).[13]

The instrument was so successful that several physicists, particularly Beccaria, Cigna, and Wilcke, stepped forward to claim its invention, while others pressed the priority of Stephen Gray, Aepinus, and the Jesuits of Peking.[14] Many

9. *VE*, I, 4, 33–43, 64–5, 91; *VO*, III, 23; Polvani, *Volta* (1942), 58–9, 64.

10. *VO*, III, 25–9; cf. Boscovich, *Theory* (1763), §§511–12.

11. Beccaria, *Elettr*. (1772), 407ff.

14. Cigna, *Scelte*, 9 (1775), 83; Vassalli-Eandi, Acc. sci., Turin, *Mem*., 26 (1821), xxx–xxxi; *VO*, III, 135–6; *VE*, I, 81, 90; J[acquet], *JP*, 7 (1776), 501–8; Henley, *PT*, 66 (1776), 514–15; Barletti, *Dubbî* (1776), 47–50.

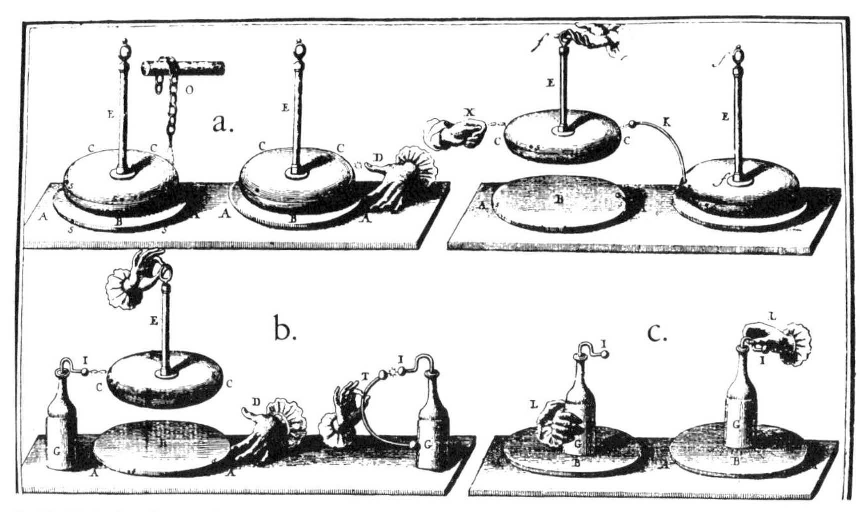

3.19 Volta's electrophore: (a) charging by 'oscillation of the electricities' (b) charging a bottle by an electrophore (c) charging an electrophore by a bottle. From VO, *III, 101.*

who had played with dissectible condensers had anticipated some of the maneuvers of the electrophore; but it was Volta alone who had made a usable instrument, had developed the dielectric, the armatures, and the play with the bottle.[15] He had not only made a new instrument: he had literally incorporated, for all to see, the difficulties and paradoxes still clinging to the concept of atmospheres.

The electrophore 'daunted the most celebrated [electricians] of Germany and Italy;' it was 'the most surprising machine hitherto invented,' 'a riddle as great as the Leyden jar,' 'a menace to the Franklinist system.'[16] In Paris it gave heart to the Nolletists. In London William Henley, a linen draper by trade and momentarily the Royal Society's most prolific Franklinist, hunted unsuccessfully for the source of the shield's endless charge: 'it is hard to say how or where the electricity is deposited,' he fretted, 'there is so much of it.'[17] It appeared that once again an instrument would overthrow electrical theory.

The electrophore did not, however, have the explosive force of the Leyden jar. The earlier instrument came into existence violating theory; the later was the outcome of anatomizing condensers under the guidance of Franklinist principles. The electrophore forced not the overthrow of received theory but the excision from it of incongruous relics of earlier approaches. As F. K. Achard, chief electrician of the Berlin Academy in the 1770s and 1780s put the point: 'The an-

15. *VO*, III, 120, 137–43; JP, 8 (1776), 23.

16. Respectively, de Herbert, *Th. phen.* (1778^{2}), 'Praef.;' *Encycl. Brit.* (1797^{3}), VI, 424; Achard, *Vorl.* (1791), III, 60; Baldi, *De igne* (1790), 59.

17. Gehler, *Phys. Wört.*, I (1787), 817; Achard, *Vorl.* (1791), III, 60.

nouncement of the electrophore was the spark that drew the favorable attention of electricians to the long-neglected approach of [Aepinus].' Achard himself was a convert. He had tried to explain Volta's purveyor in the style of Euler; he taught himself that it would not do, and accepted the explanations proposed by Wilcke and by Jan Ingenhousz, M.D., F.R.S.:[18] the cake electrifies the shield by induction at a distance; each retains its charge on, not near or around, itself; the contrary 'atmospheres' do not mix when the instrument is assembled because they have no corporeal existence; shield and cake each show strong signs when apart not because their 'atmospheres' have been revindicated but because the charge of each can manifest itself free from the overlapping influence of the other.

Among the most influential popularizers of Aepinus' approach was Volta, who met with a copy of the 'incomparably profound' *Tentamen* in the late 1770s. He started about 1780 to 'rend the veil that prevents little electricians from seeing clearly into the nature of electrical atmospheres.' The enlightened admitted *actio in distans*, whose dominion, 'already extensive in physics and chemistry, is becoming daily more evident in the phenomena of electricity.'[19] Volta propagandized via an extensive international correspondence and on trips beyond the Alps, to which he, now professor of physics at the University of Pavia, was treated by the governor of Lombardy. In Paris he taught Buffon, Franklin, Lavoisier, Le Roy, and the expatriate Genevan J. A. Deluc, who saluted him as the Newton (Franklin being the Kepler) of electricity.[20] In London he made an agent of Tiberio Cavallo, a writer of authoritative books on natural philosophy.[21] In the Germanies his chief conquest was G. C. Lichtenberg, professor of physics at the University of Göttingen.

Lichtenberg had made a reputation in electricity by discovering the 'Lichtenberg figures' described by dust falling on a charged electrophore cake.[22] After the visit from Volta, a 'raisonneur sans pareil,' a lusty electrical genius, a 'Reibzeug für die Damen,' Lichtenberg promoted the new interpretation of atmospheres in his influential editions of Erxleben's physics texts: the rule of the working of atmospheres, and the recognition that they do not exist, provide the 'key to the secrets of electricity.'[23] The same message, coupled always with explanations of the electrophore and/or the dissectible condenser, occurs throughout the respectable literature on electricity published in Germany in the late 1770s and 1780s. Joseph de Herbert, ex-Jesuit, professor of physics at the University of

18. Achard, *MAS*/Ber (1776), 130–1; *Chem. phys. Schr.* (1780), 226–33; Wilcke, *AKSA*, 24 (1762), 213–35, 253–74; Ingenhousz, *PT*, 68:2 (1778), 1037–48; *PT*, 69 (1779), 661; *JP*, 16 (1780), 117–26.

19. *VO*, III, 206, 210n, 236, 373; Gliozzi, *Physis*, 11 (1969), 231–48; Barletti to Volta, 24 Mar. 1776, *VE*, I, 121.

20. *VO*, III, 33–4, 301–5; *VE*, II, 104–5, 163–5.

21. Cavallo to Volta, 7 Feb. 1791, *VE*, III, 118; Cavallo, *Comp. Treat.* (1795³), III, 282.

22. *VE*, II, 225–73; Lichtenberg, Ak. Wiss., Göttingen, *Novi comm.*, 8 (1777), 168–80, and 1 (1778), 65–79.

23. Lichtenberg, *Briefe*, II, 150, 153–4, 203 (letters of 1784–5).

Vienna: 'electrical actions do not originate in the transition of fluid . . . but by . . . action at a distance.'[24] 'Corpora elelctrizata vires suas ad aliquod intervallum exerunt.' 'Dieser Raum nennt man den Wirkungskreis oder die elektrische Atmosphäre.' Atmosphere, Einflüsse, spherae activitatis electricae, influences électriques, Wirkungskreis, all mean the same thing, 'the space through which electrical action reaches.'[25]

In Italy Volta and Carlo Barletti, his colleague at the University of Pavia, advanced the new interpretation.[26] In England electricians gave the rules of the Wirkungskreis without reference to atmospheres.[27] The French shook off the torpor induced by Nollet: in 1785 Coulomb began to report his celebrated experiments, which rest on a theory equivalent to Aepinus'; and two years later R. J. Haüy, member of the Paris Academy, incorporated Coulomb's results in the nonmathematical epitome of Aepinus that he issued to the great applause of reviewers in France and Italy.[28]

TWO FLUIDS OR ONE?

The instrumentalism implicit in the shearing of the atmospheres was reinforced by the Augsburg peace in which the dispute over the number of electricities, reopened in 1759, concluded. In that year the eponym of Symmerian electricity reported to the surprised fellows of the Royal Society that his stockings, of which he wore one black and one white on each leg, would snap and sparkle on separation, attract chaff, swell as if they still contained a limb, and perform disembodied cancans. On reunion each pair of socks gave only the feeblest electrical signs. The perfectly symmetrical and yet 'contradictory powers' of the black and white stockings—equally strong when apart, similarly impotent when united—showed Symmer that the contrary electricities arose from two distinct positive principles, perhaps materialized as two different, counter-balancing fluids.[29]

Soon he was charging Leyden jars with his socks. 'Having thrown into a small phial filled with quicksilver, the electricity of one black stocking, I received from the explosion a smart blow upon my finger . . . ; by means of four, I kindled spirits of wine in a teaspoon . . . [and] felt the blow from my elbows to my breast.' It appeared that a real fluid sprang from each surface of a discharging jar, as Nollet claimed; but Symmer broke completely with him in supposing that the two flows differ in quality as well as in quantity and direction. To demonstrate

24. De Herbert, *Th. phen.* (1778²), 72–80.

25. Respectively, J. B. Horvath, ex-S.J., *Physica* (1782), 318–19; Achard, *Vorl.* (1791), III, 32–3; Gehler, *Phys. Wört.* (1787), I, 737, and IV, 800–1.

26. Barletti, *Phys. sp.* (1772), 30–4; Soc. ital., *Mem. mat. fis.*, 1 (1782), 28, 43.

27. E.g., Cuthbertson, *Abhandl.* (1786), 1–11, in *Pract. El.* (1807), 1–10; Nicholson, *Intro.* (1782), II, 403–6.

28. Haüy, *Exposition* (1787), of which lengthy reviews are in *JP*, 31 (1787), 401–17, and *Bibl. oltr.*, 2 (1788), 139–60, and 3 (1788), 221–38.

29. Symmer, *PT*, 51:1 (1759), 340, 354, 382.

the opposing streams, Symmer sought to puncture cards; lacking the necessary apparatus he applied to Franklin, who, 'with great civility,' assisted him in puncturing quires of paper placed in the discharge train of a Leyden jar.[30] Their results resembled Nollet's. Franklin concluded that the holes meant nothing, Symmer that they confirmed the existence of two distinct fluids.

Despite the collaboration of their chief, the English Franklinists paid little attention to Symmer.[31] He guessed rightly that he would have better luck with Nollet, who admired the anti-Franklinist experiments and discovered the effluences and affluences responsible for the behavior of dancing socks (fig. 3.20). He also simplified the demonstrations, eliminated the socks, and emphasized, as Symmer had, the analogy between a pair of united stockings and a Franklin square with two superposed panes.[32] These were the experiments that Nollet described to Cigna and that through him and Beccaria stimulated the invention of the electrophore.

Cigna concluded that Symmer's demonstrations did not prove whether electricity came in one fluid or two. By the 1780s his agnosticism had won the day; most physicists had by then stopped trying to show the exclusive truth of either view and affected to consult only their convenience in the choice of system. Before this positivistic paralysis could set in, however, Symmer's schism had to find leaders more influential than its founder. The first of these heresiarchs were Franklinist apostates like Wilcke and Barletti.

Wilcke attacked the favorite argument from the luminous appearances of points. He observed that candle smoke is blown away from a glowing point regardless of the sign of its charge and argued that electrical matter must rush outward from both the brush and the star.[33] Franklinists answered that the wind blowing the smoke consists of charged air particles driven from the point without the intervention of the electro-luminous matter.[34] The round went to the Franklinists, who, however, became uneasy about the arguments to prove the direction of circulation of their single fluid. Several called attention to the imposing sparks that jumped from machines like Wilson's big generator (fig. 3.15) and the huge electrostatic machine built at the Teyler Foundation of Haarlem in 1785. The physicist in charge of the Dutch white elephant, Martinus van Marum, had commissioned it in the belief that knowledge increases with the size of the machine used to procure it. Following this happy anticipation of a later philosophy, he obtained a fat spark, two feet long and as thick as a quill pen, not includ-

30. *Ibid.*, 358, 371–87.

31. Wilson, *PT*, 51:1 (1759), 329–30; Wilson to Bergman, 26 Oct. 1761, in *Bergman's For. Corresp.* (1965), 418; Symmer to Mitchell, 19 June 1760, and 17 Feb. and 30 Jan. 1761, Add. Mss. 6839, ff. 183, 209, 214.

32. Symmer to Mitchell, 30 Jan. and 15 May 1761, *ibid.*, ff. 209, 225; Nollet, *Lettres* (1760), esp. 56–7, 63, 155–62, 195–200; *Lettres* (1767), 159.

33. Wilcke, *AKSA*, 25 (1763), 216–25.

34. Priestley, *Hist.* (1775^3), II, 184–91; Nicholson, *Intro.* (1782), II, 82–5.

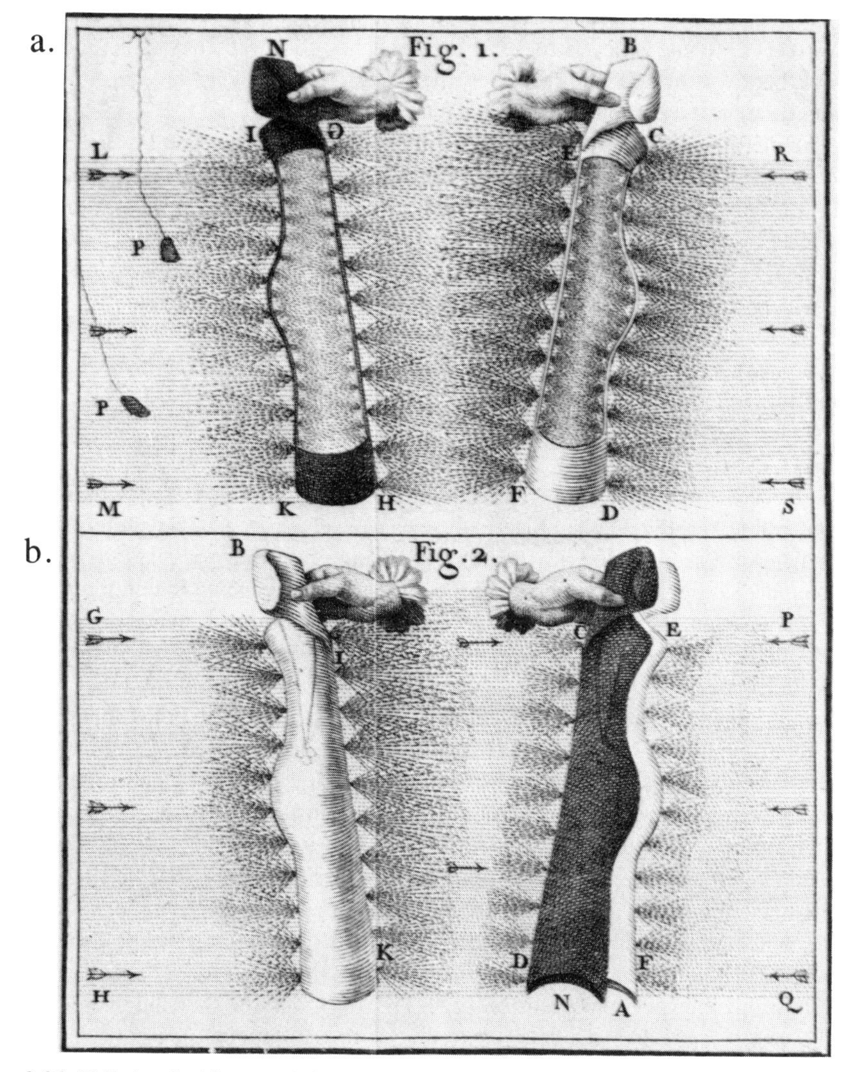

3.20 Nollet's elucidation of Symmer's socks. (a) When apart, the stockings swell owing to collisions between effluents from their internal surfaces, and 'attract' exterior bodies P by the usual affluent (b) they collapse when combined because of the interference between superposed resinous and vitreous jets. Note the short range of the jets from CD and EF. From Nollet, Lettres *(1767).*

ing prominent ramifications that always made acute angles with a line from the positive PC to the object struck.[35] The branches followed the main, one-way flow: there was no trace anywhere of the contrary dualist current. Priestley and Volta thought the demonstration decisive.[36]

The dualists defended themselves by conceding what they had never denied, that positive and negative electricity give different optical signs. They noted that van Marum had not shown a spark from a negative PC. This desideratum was supplied by William Nicholson, a London schoolteacher and scientific publicist, and neither a Franklinist nor an F.R.S. He showed that negative electrodes give characteristic sparks (fig. 3.21) and that luminous signs cannot decide the number of electrical fluids.[37]

The dualists then went on the offensive. Unitarians parried the favorite argument from perforation of cards by pointing to its passé effluvialist presuppositions: the perforations arise from attractive forces between the coatings and charges induced in the cards; they are indirect consequences, not footprints, of the electrical matter; holes reveal nothing about the direction of circulation.[38] More telling was the dualist objection to negative electricity. Many preferred the postulate of a second electrical matter consisting of particles mutually repellent to Aepinus' desperate hypothesis of the antipathy of elements of common matter. Symmerian experiments enhanced the implausibility of taking a relative void—the absence of electrical matter—as the essence of the negative state. Why does not the surplus of one of the stockings fill the void of the other when they are joined? 'When well considered,' says Father Barletti, Franklinism 'has an air of magic and sympathy that smells of Gothic philosophy.' Symmer's approach avoids all the difficulties of Franklin's 'moral theory,' which requires a vacuum to resist, 'with heroic scrupulosity,' the entry of the matter it 'ardently desires.'[39]

This argument, freed from its monkish rhetoric, comes to asserting that Franklinist negative electricity is more difficult to picture than Symmer's second fluid, and less easy and convenient to work with. Many electricians therefore recommended dualism not for its demonstrable truth but for its greater simplicity, convenience, and plausibility.[40] The only answer that singlists could make was to appeal to the same criterion, with the comical consequence that each side claimed to offer greater facility of thought than the other. Most British

35. Van Marum, Teyler's Tweede Genoot., *Verh.*, 3 (1785), x, 26–30, 92; *JP*, 27 (1785), 148–55; Dibner, *Nat. Phil.*, 2 (1963), 87–103; Hackmann, *Cuthbertson* (1973), 14–15, 24–5.

36. Priestley, *Hist.* (1775³), II, 260; Volta to van Marum, *VO*, IV, 66.

37. Nicholson, *PT*, 79 (1789), 277–83, 288.

38. Franklin to Ingenhousz, 3 May 1780 and 16 May 1783, *BFP* (Smyth), VIII, 191–2, and IX, 45; Volta to Landriani, 27 Jan. 1776, *VO*, III, 157–8.

39. Barletti, *Dubbî* (1776), x–xi, 32–4, 45–50; Soc. ital., *Mem. mat. fis.*, 7 (1794), 444–61.

40. E.g., Caullet de Veaumorel, *Descr.* (1784), iv–vi, xxxi; Gehler, *Phys. Wört.*, I (1787), 764–6, 830–1; Haüy, *Traité* (1803), I, 334; E. G. Fischer, *Physique* (1806), 257; Saxtorph, *El.-laere*, I (1802), 28–30.

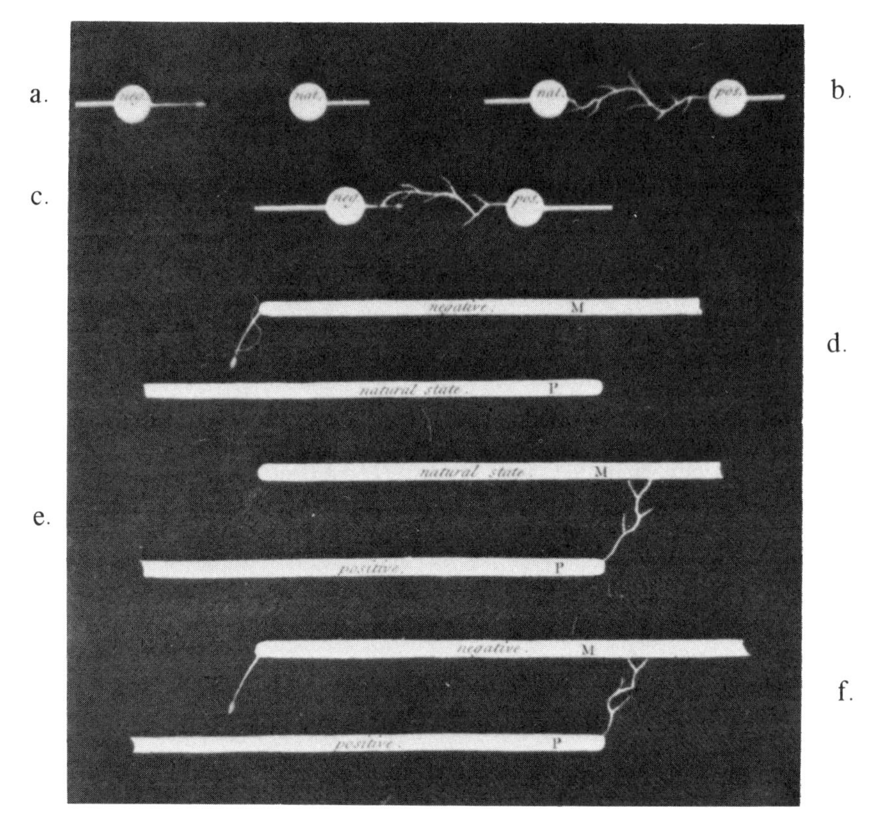

3.21 Nicholson's sparks between (a) − and 0 (b) 0 and + (c) − and + balls, and between (d) − and O (e) 0 and + (f) − and + rods. From Nicholson, PT, *79 (1789), 265–88.*

minds, strengthened or immobilized by Franklin's eminence, found greater security in one fluid while most Continentals preferred to reason about two. The cardinal point was the recognition that, as Priestley put it, 'no fact can be shown to be positively inconsistent with either [theory].'

One took one's choice or stayed agnostic. 'Oh monsieur, il faut être unitaire,' cries Volta. 'I'm neither unitarian nor dualist,' replies Lichtenberg, 'but I'm ready to become the one or the other as soon as I've seen a decisive experiment.'[41] 'In physics we have not yet reached puberty.'[42] Volta had to concede that in its present immaturity his science could not prove 'la nostra cara dottrina,' his dear singlist theory; 'I am very far from regarding the dualist hypothesis as absurd, only as very implausible, too complicated, and needing hypoth-

41. Priestley, *Hist*. (1775[3]), II, 44, 52; Lichtenberg to Wolff, 20 Dec. 1784, in Lichtenberg, *Briefe*, II, 174–7.

42. Quoted in Mauther and Miller, *Isis*, 43 (1952), 226.

esis on hypothesis,' that is, two suppositious fluids rather than one. We leave the last word on this matter to schoolmistress Margaret Bryan, who exploited the ambiguity of electrical theory to inculcate 'the love and practice of justice.' After discussing without prejudice the arguments on both sides, she allows that the question cannot be decided. 'Cull the Sweets of religion as you rove the flowery paths of Natural Philosophy,' she tells her charges, '[and don't] make hasty judgments.'[43]

QUANTIFIABLE CONCEPTS

The successful quantification of physical theory presupposes the existence of appropriate concepts expressible mathematically and amenable to test and refinement by measurement. The heart of the process is constructing the concepts. As the painful emergence of the notion of localized charge (Q) abundantly illustrates, even the easiest concepts may be very difficult to find. Effective quantization of electrostatics required, in addition to Q, the concepts of capacity (C) and potential (V), and the reduction of ponderomotive forces to microscopic *forces* (italicized to distinguish them from macroscopic ones) assumed to act between *elements* of electric charge. These ingredients were in hand by the late 1780s.

Capacity and tension The business of measurement had rapidly outpaced understanding of the quantities measured. In the 1740s Gray's 'thread of trial' gave way to a pair of strings whose angular separation α could be determined precisely. Nollet added a circular scale and an optical system (fig. 3.22); Beccaria, thinking that not α but its sine measured the force sustaining the threads against their gravity, determined the appropriate chord directly.[44] Others, like Richmann, substituted a fixed object for one of the threads and read $\alpha/2$ on a scaled quadrant. The chief inadequacies of the design—leak and insensitivity—were lessened by Canton, who terminated the threads in small, light balls made from elder pith. The English used the pith balls until 1770, when Henley invented a serviceable version of Richmann's instrument that quickly became standard (fig. 3.23).[45] With it and the later sensitive bottled electrometers of Bennet and Volta (who employed, respectively, gold leaves and flat, thin straws in place of the original threads), electricians could determine α as reliably and exactly as they wished.[46]

But what did it measure? And how should it be associated with other measures, such as those proposed to rate the output or power of electrical machines? Here a commercial element entered to darken a situation already sufficiently

43. *VO*, IV, 270, 359, 380; Volpati, *Volta* (1927), 75–7; Bryan, *Lectures* (1806), 'Pref.,' 163, 168, 190

44. Nollet, *MAS* (1747), 102–31; Beccaria, *PT*, 56 (1766), 105–18.

45. Richmann, *CAS*, 14 (1744–6), 299–324, and *NCAS*, 4 (1752–3), 301–40; Canton, *PT*, 48:2 (1754), 780–5; Priestley (re Henley), *PT*, 62 (1772), 359–64.

46. Bennet, *PT*, 76 (1786), 26–34; Volta to Lichtenberg, July and Aug. 1787, *VO*, V, 35–57. Cf. Polvani, *Volta* (1942), 136–40.

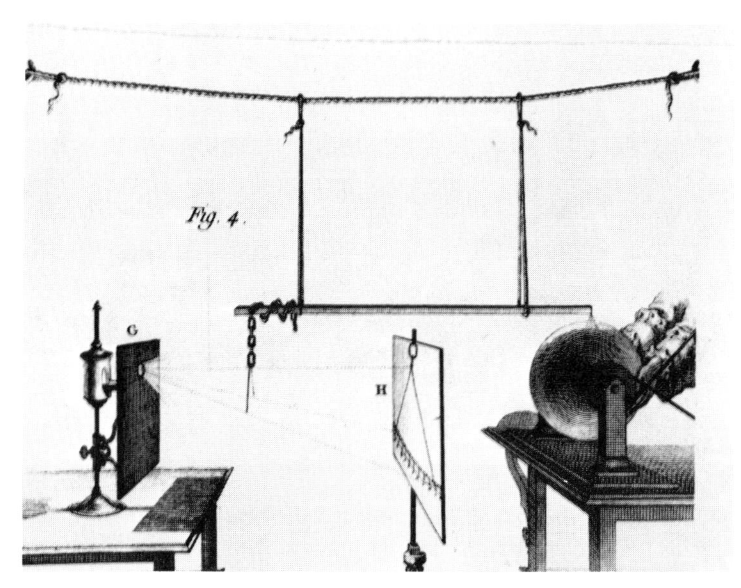

3.22 Nollet's electroscope: the lamp at G images the threads from the prime conductor on the screen H. From Nollet, MAS *(1747), 102–31.*

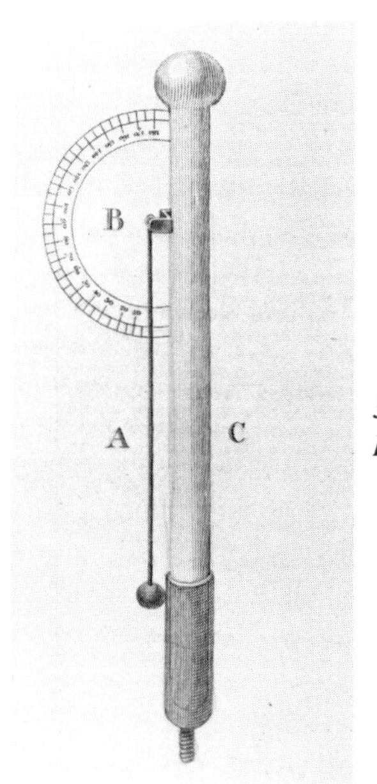

3.23 Henley's electrometer. From Priestley, PT*, 62 (1772), 359–64.*

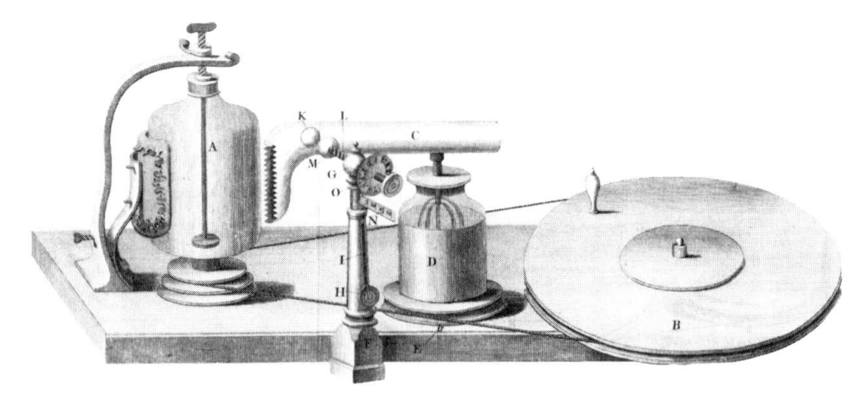

3.24 Lane's electrometer (the spark gap KM), as used with an inexpensive table electrical machine. From Lane, PT, *57 (1767), 451–60.*

obscure, for the various manufacturers used the measure that showed their instruments to best advantage: the area of glass rubbed per turn of the wheel, the length of the longest spark from the PC, the length of a standard fine wire fused by discharge of a Leyden jar, the rate of sparking across a gap between the PC and ground. None of these quantities proved as useful as the measure most obvious to the laboring man, namely n, the number of times he must crank the machine to charge objects to the same degree.[47] He knew that it was more tiresome to electrify a Leyden jar than a charity boy. Apparently α measured a property of electricity not readily visualized whereas n indicated the relative amounts of Q on objects equally electrified according to the electrometer. These concepts found clear expression in the Lane electrometer, a device invented by an electrifying apothecary to meter his patients' shocks. It consisted of the adjustable spark-gap KM (fig. 3.24), put in series with the patient and the jar(s) D. Lane found n to be proportional to the coated surface of D: the Q needed to force the gap was evidently inversely proportional to the effective size, or electrical capacity, of the bottle.[48]

The notion of capacity also emerged in experiments on simple conductors by Nollet, Richmann, Beccaria, and Franklin, whose characteristically direct demonstration survives in courses on elementary electrostatics.[49] Place three yards of brass chain in an insulated silver can fitted with an electrometer. Charge the can with a spark from a Leyden jar and note the reading α. Now draw up the chain with a silk line: as the chain rises, the threads fall; the can will take another spark, which returns its electrometer to α. In Franklin's old-fashioned terminol-

47. *VO*, III, 217; van Marum, *Ann. Phys.*, 1 (1799), 81–4.

48. Lane, *PT*, 57 (1767), 451–5; Walker, *Ann. Sci.*, 1 (1936), 72–3.

49. Nollet, *MAS* (1747), 125ff.; Richmann, *NCAS*, 4 (1752–3), 307–9; Beccaria, *PT*, 51 (1759–60), 517, and 57 (1767), 309; *EO*, 272–5.

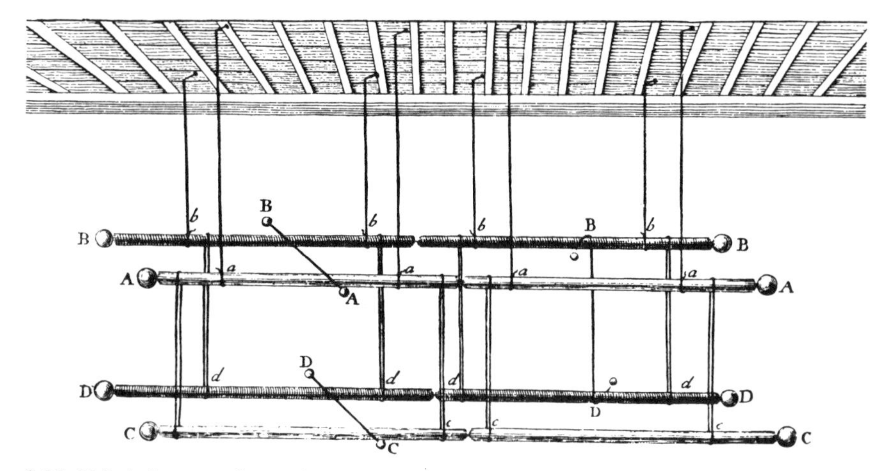

3.25 Volta's long conductor hung in eight sections in two parallel lines. From VO, *III, 207.*

ogy, the experiment shows that 'increase of surface makes a body capable of receiving a greater electric atmosphere;' in terms of the emerging quantifiable concepts, it suggests that, for the same charge, a body's 'degree of electrification' varies inversely as its capacity.

The inexhaustible electrophore helped Volta to bring the concept further. The shield has a greater capacity when lying on the negative cake than when apart from it. Volta inferred that an electrified body must *decrease* the capacities of neighboring insulated conductors for electricity homologous to its own: the presence of a plus body creates a 'tension' in others that engages part of the room usually available for accumulating excess fluid.[50] The various surface elements of a charged conductor must similarly inhibit one another; for a given surface the longest conductor will have the greatest capacity. Volta demonstrated this last inference with twelve interconnected gilded cylinders, each eight feet long and one-half inch in diameter, hung in separated rows and charged like a prime conductor (fig. 3.25). Although the total surface did not much exceed that of a standard PC, it had six or seven times the capacity, as measured by wheel turns or by its 'intolerable shock, which shatters the whole body.'[51]

While Volta was hanging up his conductors, Wilson rigged his gigantic artificial cloud, 3900 yards of wire plus a sectioned tin-foil cylinder 155 feet long when entire. He too knew that for best effect the conductors had to be placed so that the electricity of their parts did not 'interfere too much with one another.' The greater the length the greater the blast. The big cylinder alone struck like a

50. Volta, *PT*, 72:1 (1782), x–xi, xix–xxi.
51. Volta, *JP*, 12 (1779), 249–77 (*VO*, III, 201–29).

Leyden jar when charged by only six turns of Wilson's cylinder machine. There the experiment stopped, for, says the prudent investigator, 'I could not prevail upon anyone present to take a higher charge.'[52]

In analyzing the action of his long conductor, Volta made do with the macroscopic concepts Q, C, and T (tension) and established a quantitative relationship among them. He measured T by α and understood it to represent 'the force exerted by each point of an electrified body to free itself of its electricity, and communicate it to other bodies.'[53] Experiment showed that, for a given T, the Q localized on an isolated conductor increases with its capacity. He guessed that $Q = CT$. The relationship could not then be determined, or even confirmed, by comparing a measure of Q (say n) with a measure of T (say $\alpha[n]$) because the electrometers were not linear. Volta later devoted much time and ingenuity to calibrating his straw electrometer by subdivision: he would fix equal scale divisions by finding the spread for a charge Q upon a standard conductor, then for $Q/2$ (obtained by touching the conductor with an identical one), $Q/4$, and so on.[54] With such an instrument α or T is proportional to Q for a given isolated conductor; to show that the factor is capacity, Volta touched an insulated cylinder electrified to α degrees to a similar one $n-1$ times as long. If all went well, the electrometer read α/n.

Volta brought his ideas about Q, C, and T before electricians in 1782 in his usual way, incorporated in a new instrument, a 'condensatore' for rendering sensible electricity otherwise too weak for detection. It is nothing but a small electrophore with a layer of varnish as cake: the thinness of the dielectric gives the assembled condensatore a large capacity, which allows it to soak up all the charge conveyed from an atmospheric probe, while the small capacity of the separated shield enables the accumulated electricity to display itself. The device had a vast progeny, including a mechanized version by Nicholson, whose ingenious doubler (fig. 3.26) anticipated the influence machines of the nineteenth century.[55]

Force The demonstration that gravity diminishes as a simple function of distance encouraged even those who refused to accept Newtonian apologetics to look for similar rules for other apparent actions at a distance. The case of magnetism, examined earlier, illustrates both the appeal and the difficulty of the task. A few inconclusive attempts on electrical force were made in the 1740s.[56] The first to obtain regular and reproducible results was Daniel Bernoulli, professor at

52. Wilson, *PT*, 68:1 (1778), 252, 296–7.
53. Volta, *PT*, 72:1 (1782), 259.
54. *VO*, V, 37–42; Saussure, *Voyages* (1786), II, 205–9; Polvani, *Volta* (1942), 136–8.
55. Volta, *PT*, 72:1 (1782), ix–xxxiii; Nicholson, *PT*, 78 (1788), 403–6; Walker, *Ann. Sci.*, 1 (1936), 88–93.
56. E.g., Kratzenstein, *Th. el.* (1746), 34–5; Richmann, *CAS*, 14 (1744–6), 323; Gralath, Nat. Ges., Danzig, *Vers. Abh.*, 1 (1747), 525–34; Bose, *Tent.*, II:ii (1747), 42; Ellicott, *PT*, 44:1 (1746), 96.

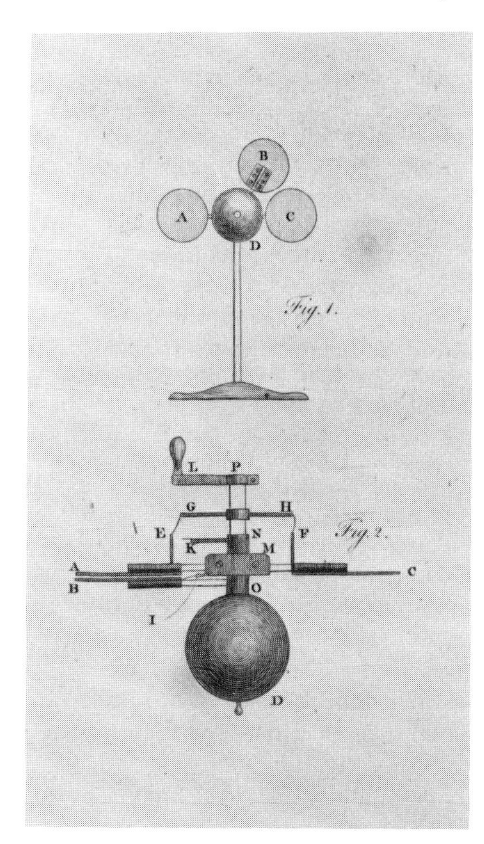

3.26 Nicholson's doubler, from Nicholson, PT, *78 (1788), 403–7. The doubler consists of two fixed metal disks A and C, a movable one B, and a metal ball D. (a) Give a small charge* Q *to A and bring B opposite; at that instant the pins E and F touch the protruding wires at G and H, connecting A and C, and B comes into contact with D via the wire at I. Owing to the great capacity of the conjugated plates A and B, the result of their confrontation is that most of* Q *remains on A and* −Q *is induced on B. (b) Bring B opposite C, breaking the first contacts and connecting C and D via the pin at K; C obtains a charge* Q *by induction. (c) When B returns to A, the connections between it and D, and between A and C are restored: A charges to almost* 2Q *at the expense of C and B charges to almost* −2Q *by induction. The charges may be doubled again at the next revolution.*

the University of Basle, who about 1755 began to study Franklin's system with the help of his former student, Abel Socin, M.D. Their results point to difficulties in applying the concept of force to which eighteenth-century physicists and their historians were, and are, exposed.

At a distance x below a disk four inches in diameter suspended from a PC, Bernoulli placed a similar disk supported on a vertically floating calibrated glass tube. He found the weight w required beneath the float to counter the attraction of the upper disk for several values of x. It appears that he kept the PC connected to an operating machine; such an arrangement, which gives the upper disk a different charge at each setting, would suggest itself to one endeavoring to determine the force of a given degree or tension of electricity. Bernoulli obtained a force rule that looks like Coulomb's: $w \sim f \sim 1/x^2$. If the lower disk was effectively earthed through the moist tube, the pair made a condenser of fixed potential V; since $V \sim sx$, s being the charge density of the plates, and since the ponderomotive force f on either goes as s^2, $f \sim 1/x^2$. Plainly Bernoulli's relation refers to macroscopic force, not to *force* between *elements* of electrical matter. Nonetheless, some historians have thought it an anticipation of the law of squares.[57]

All the early quantifiers failed because they had no theory that enabled them to move from microscopic interactions to macroscopic measures, from *force* to force, from the general to the particular. The earliest known reductionist argument was sketched by Priestley to account for Franklin's odd discovery that an insulated cork lowered into a charged metal cup does not go to its sides, nor, having touched bottom, does it show any electricity when withdrawn. 'The fact is singular,' Franklin had said. 'You require the reason; I do not know it.' Priestley did: 'the attraction of electricity is subject to the same laws with that of gravity, and is therefore according to the squares of the distances; since it is easily demonstrated that were the earth in the form of a shell, a body in the inside of it would not be attracted to one side more than another.'[58]

The argument is to the point. Yet it is not a demonstration, or even a proper enunciation, of the law of squares. Priestley had no warrant for equating the 'attraction of electricity' with the *force* of a particle of electrical matter, and he required not Newton's theorem but the unproved converse (if a mass point experiences no force within a uniform gravitating shell, the *force* of gravity is inverse-square). Again, a bucket is not a spherical shell. The force of gravity does not vanish within a uniform can; electricity's does, except near the mouth, because the mobile electrical matter arranges itself on the external surface so as to nullify its action within. Cavallo noticed this difficulty and rejected Priestley's law as an artifact of the can, in which opposing electrical forces fortuitously cancelled.[59]

To determine the law of *force* directly, one must balance an appropriate macroscopic electrical force by a measurable agent, gravity, for example, or torsion. The first alternative was exploited by John Robison, who became professor of natural philosophy at the University of Edinburgh after an adventurous career that included a stay in St. Petersburg, where he met and admired Aepinus. Robison first published his determination of *force* in the excellent account of his friend's work that he contributed to the *Encyclopedia Britannica* of 1801; he then dated, or perhaps misdated his measurement in 1769, before he had met Aepinus, whose 'reasoning,' he acknowledged, had inspired it.[60]

Robison's apparatus appears in figure 3.27, where A represents a brass ball, B and D gilt cork balls, BD a stiff waxed thread free to turn in a vertical plane in the amber fitting C, LAEF a glass frame. After A and B are charged, their angular separation may be measured for various orientations of LA and the force of repulsion calculated in terms of the angles and the moments of momentum of the

57. Socin, *Acta helv.*, 4 (1760), 224–5; Whittaker, *Hist.* (1951²), I, 53; Roller and Roller, *Harv. Case St.* (1957), II, 610–11. Cf. *VO*, V, 78–9; Stanhope, *Principles* (1779), 7–8, 34–61, 73–4.

58. Franklin to Lining, 18 Mar. 1755, *EO*, 336; Priestley, *Hist.* (1775³), II, 372–4.

59. Cavallo, *Comp. Treat.* (1782²), 199.

60. Playfair, RS, Edinb., *Trans.*, 7 (1815), 495–539.

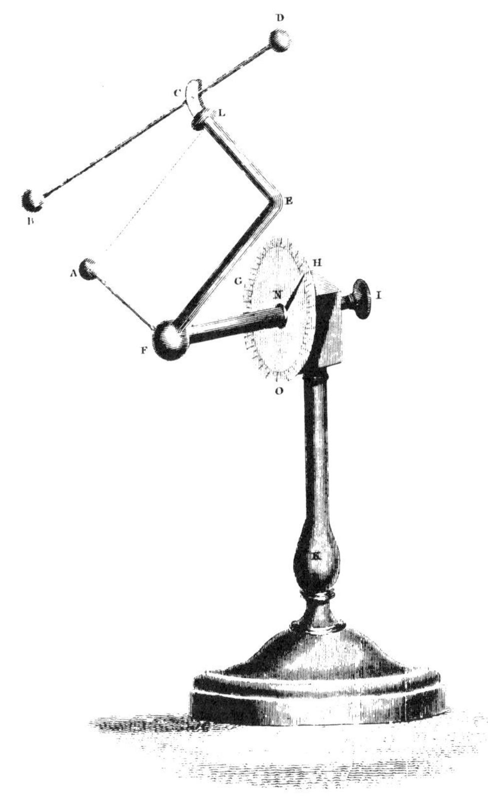

3.27 Robison's apparatus for measuring f(r) *between small charged spheres. From Robison,* System, *IV (1822).*

balls around C. The force diminished nearly as the squares of the distance between the centers of A and B. Robison deduced that electrical *force* follows the same law: since the separation of the balls is large compared with their diameters, their charges may be assumed uniformly spread, and if the *force* is inverse-square, then, by application of a fundamental theorem in the gravitational theory, the distributed charges will act as if they were collected at the centers of the balls.[61]

The definitive measure of *force* exploited torsion in place of gravity. The technique was invented by C. A. Coulomb, an experienced military engineer, during a prize competition, the winning of which assisted his entry into the Paris Academy in 1781. The competition concerned the diurnal variation of the earth's magnetism, which was customarily measured by a needle on a pointed pivot. Coulomb, who knew all about ropes, perceived that the needle would be much more responsive if hung from a fine silk thread so as to be free to oscillate in a horizontal plane. He showed that such oscillations are isochronous for thin wires

61. Robison, *System* (1822), IV, 68.

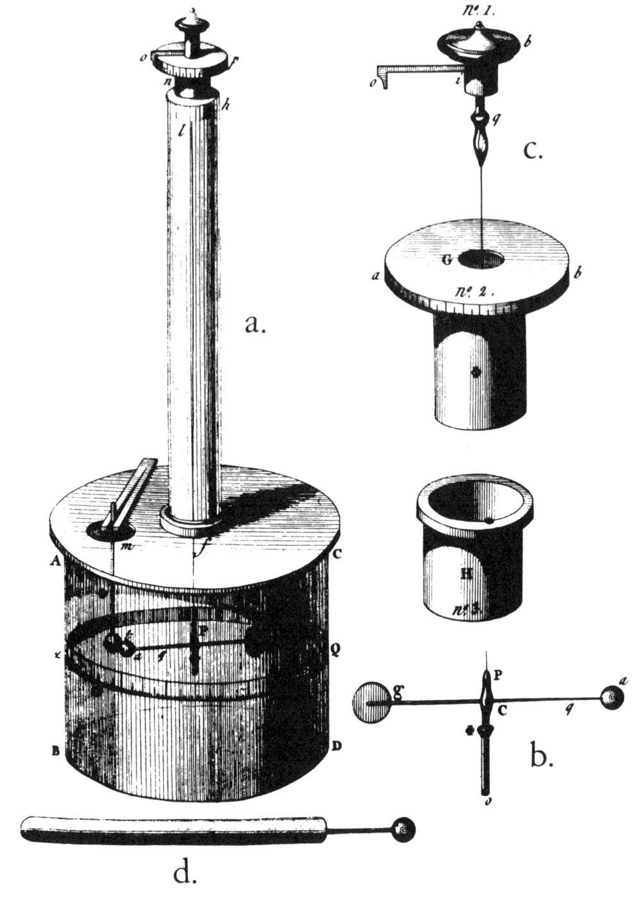

3.28 Coulomb's apparatus for measuring f(r). *(a) The torsion balance (b) the electrometer stalk (c) detail of the knob by which the wire is twisted (d) the probe. From Coulomb,* MAS *(1785), 569–77.*

as well as for threads, and he deduced from the second law of motion that the restoring force of torsion must be proportional to the angle of twist.[62] After perfecting the technique by measuring minute mechanical forces, Coulomb tried electricity.

Within the glass cylinder ABCD (fig. 3.28a) he suspended a wire carrying the waxed thread *q* (fig. 3.28b) bearing an elder ball *a* on one end and a counterbalance *g* on the other. A second ball, *t*, was fixed with its center in the plane of

62. Coulomb, AS, *Mem. par div. sav.* (1780), §§26, 43–4, 48–50; *MAS* (1784), §§v, xiv, xviii, xx–xxvii; Gillmor, *Coulomb* (1971), 18–41, 140–5, 175–81.

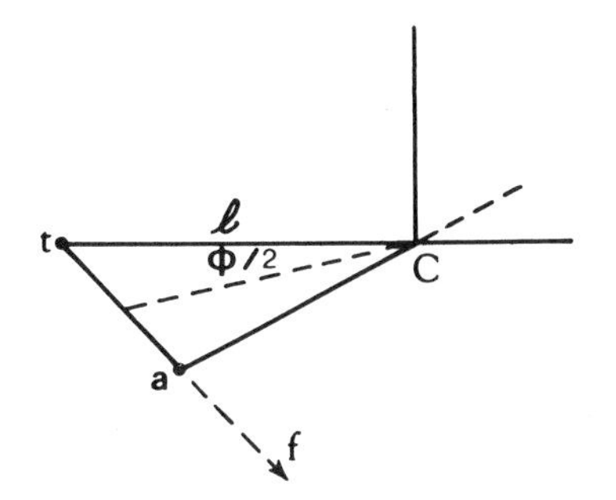

3.29 Schema of Coulomb's experiment: t, a, and C have the same significance as in fig. 3.28

rotation of q. Coulomb adjusted the knob b (fig. 3.28c) until a touched t without twisting the wire. He then electrified the probe (fig. 3.28d), and inserted it through the hole m to touch t, which divided its charge with a. Ball a withdrew counter-clockwise to a distance x, where the torque exerted by the electric force f balanced that of torsion, or where $f\, l \cos\phi/2 = k\phi$ (fig. 3.29). He twisted the wire further by turning the knob b clockwise, increasing the angle of torsion to θ, say, and reducing ϕ. Should the force be inverse-square and the diameter of the balls negligible in comparison with l, $\theta = \text{const.} (\cos\phi/2)/(\sin^2\phi/2)$; and if ϕ is small, $\theta = \text{const.}/\phi^2$. Coulomb reported two measurements that satisfied this last relation very closely. 'It follows,' he wrote, 'that the mutual repulsive action between two balls charged with the same kind of electricity follows the inverse ratio of the square of the distances.'[63]

Coulomb's colleagues at the Paris Academy did not doubt that his 'law,' and the Newtonian propositions that had provided the rationale for his experiments, constituted the backbone of the theory of electricity. They became both dualists and instrumentalists; 'whatever be the cause of electricity, one can explain all the phenomena, and calculation will be found agreeable to experiment, if one supposes two electrical fluids, whose particles attract and repel one another inversely as the square of the distance.'[64] Outside France applause came slower. With few exceptions, Coulomb's work was not reported in British and German textbooks until after 1800. Two authorities who did notice it, Deluc and Volta, wrongly dismissed it as concerned with a special geometry that did not permit deduction of universal law.[65] Unable themselves to move from force to *force*, they could not recognize another's success.

Progress with foreigners began in 1801, with Robison's treatise in the *Britannica*, the first modern exposition of electrostatic theory in any language. The better German textbooks began to include accounts of the torsion balance, its use and its results; Fischer, in particular, emphasized the resultant reduction of Symmer's ideas to 'an exact theory,' from which the phenomena follow by 'rigorous

63. Coulomb, *MAS* (1785), 569–77; *Mémoires* (1884), 107–15.
64. Coulomb, *MAS* (1786), 67–77; *Mémoires* (1884), 173–82; Haüy, *Exposition* (1787).
65. Deluc, *JP*, 36 (1790), 456; *VO*, V, 78–9, 81–3.

calculation.' No respectable physicist would settle for less, he wrote, embracing the French theory while warning his readers not to believe it. 'It is but a convenient way to explain the facts, and one can only conclude that the phenomena occur as if they were produced by two fluids endowed with the preceding properties [the law of squares], for the true nature of electricity lies hidden.'[66]

The growing enthusiasm for Coulomb was both cause and consequence of the penetration of the new French physics. Considered alone, the results of the torsion experiments did not compel. Coulomb had had to take many precautions to obtain the few results he reported; the wire would often twist when unstressed and the zero point migrate two or three degrees in a typical run. Many who tried to perform on his 'all too unsteady twisting machine,' as the Berlin physicist, P. L. Simon, called it, had trouble. Professor G. F. Parrot of Dorpat, for example, got errors of 12 percent or more, and Simon, emboldened by news of Volta's disbelief in Coulomb's law, announced that measurements made on his new gravity balance constantly gave $1/r$, not $1/r^2$. As late as the 1830s a respectable physicist, William Snow Harris, F.R.S., could call the law of squares into question. He bamboozled Whewell, who allowed that it was not yet supported by 'that complete evidence . . . which the precedents of other permanent sciences have led us to look for.'[67]

There are other important sources of error, leak for example, and the effects of mutual induction between the balls, and of segregation of charge in the glass housing. Coulomb later investigated these sources but did not entirely or successfully correct for them.[68] Nonetheless, none of his confrères and, after, 1800, few electricians outside France (and almost no historians) called his measurements into question. It appears that all were prepared to believe before Coulomb announced the signs of the faith.

Principia electricitatis Henry Cavendish, recluse and genius, one of the wealthiest men in England, 'so unsociable and cynical that he could stand honorably in the same tub as Diogenes,'[69] had the unusual acquirement of an eighteenth-century physicist of a Cambridge education. There it appears he encountered a problem that was to direct much of his later study: overcoming such difficulties in Newton's theory of interparticulate *forces* as occur in the derivation of Boyle's law.

The *Principia* constitutes air of stationary particles, evenly spaced and repelling one another with a *force* diminishing as their separation. The cardinal property of such a *force*, as of all *forces* decreasing more slowly than the inverse cube of the distance, is that particles far away from a given one will, in the ag-

66. Robison, *System* (1822), IV, 1–204; E. G. Fischer, *Physique* (1806), 257–61.

67. Simon, *Ann. Phys.*, 27 (1807), 325–7, and 28 (1808), 277–98; Parrot, *ibid.*, 60 (1818), 22; Harris, *PT*, 124:1 (1834), 239–41, and 126:1 (1836), 431–7; Whewell, *Hist.* (1858^3), II, 210.

68. Coulomb, *MAS* (1785), 616–38, *MAS* (1787), 421–67, and *MAS* (1788), 617–705; *Mémoires* (1884), 123–4, 147–52.

69. Landriani to Volta, 7 Oct. 1784, *VE*, III, 10–11; cf. G. Wilson, *Cavendish* (1851), 165–70.

gregate, exercise a greater force upon it than those close to hand. Newton's air would drive itself to the walls of rooms. To ease the derivation of Boyle's law and perhaps also to avoid this inconvenience, the *Principia* allows air particles to act only on nearest neighbors. This arbitrary and implausible restriction put off Cavendish.[70] In his first investigations, which concerned gases, he hoped to improve on Newton. He did not succeed and turned to another elastic fluid, the electrical, expecting its law of *force* to be more tractable.[71]

In his classic paper of 1771 Cavendish deepened and extended the approach of Aepinus, whose priority he granted while asserting his own independence. In Aepinus' manner he explained Canton's experiments, predicted plus-plus attraction, emphasized that segregation must precede electrical motions, etc.; but he went much further, introducing an advanced concept of potential and estimating the aggregate force of various distributions of electric fluids characterized by different laws of repulsion. He made shrewd guesses at the approximate behavior of elastic fluids not regulated by the inverse-square and deduced from Newton's theorem about the vanishing of force within a gravitational shell that those that are so regulated will crowd into the shallowest possible depth beneath the surface of a conducting sphere. He tested his deduction in the following way, which he saw no occasion to publish.[72]

A metal globe G mounted on waxed glass sticks SS stands within the frame BCADbc, which opens like a book (fig. 3.30) and carries the pasteboard hemispheres H, h. When it shuts, H and h enclose G leaving a space of about one-half inch all around. Tt is the electrometer. Cavendish closes the frame, runs a short wire W (not shown) from H to G, and electrifies the pasteboard with a Leyden jar. Then, with an elaborate system of strings, he disconnects the jar, removes W, opens the frame, and applies the electrometer. G shows no electricity. Cavendish infers the law of squares: the electrical fluid, contrary to the aereal, distributes itself in a manner entirely consistent with its elementary *force*, without the help of ad hoc restrictions of its range. Moreover, in a masterful analysis of experimental error, perhaps the earliest of its kind, he shows that, assuming the *force* to decrease as r^{-n}, his null result implies that n cannot differ from two by more than one part in fifty.

The high points of Cavendish's paper of 1771, indeed of the mathematical electrostatics of the eighteenth century, were considerations respecting narrow

70. Newton, *Princ.*, ed. Cajori, 214–15; Cavendish, *PT*, 61 (1771), 586–8, 647–8; *El. Res.* (1879), 5, 43, 411.

71. Wilson, *Cavendish* (1851), 16–27; McCormmach, *Isis*, 60 (1969), 299–301, and *El. Res.* (1967), 68–9, 168–73, 241–6.

72. Cavendish, *PT*, 61 (1771), 584–98, 608–11; *El. Res.* (1879), 3–11, 18–19, 48–51, 104–13. Suppose that the electrical matter is frozen in a uniform distribution within a conducting sphere. Now imagine a thaw: by Newton's theorem, any particle will experience a centrifugal force from the aggregate of electrical matter closer to the center than itself and no force at all from that further away. It will therefore be driven to the surface.

resistanceless 'canals' in which the compressible electrical matter is assumed to behave as an incompressible fluid. These considerations anchored a distinction that most electricians felt intuitively but failed to recognize in Cavendish's rigorous form and regrettable terminology. 'Though the terms positively and negatively electrified are much used [he wrote], yet the precise sense in which they are to be understood seems not well ascertained.' No earlier electrician, not even Nollet, had doubted the operational significance of Franklin's words: a body is plus when it possesses the same sort of electricity as a glass rod or the hook of a Leyden jar charged by the globe. Cavendish proposed another operational definition. Procure yourself a standard test body B and an infinite canal C; place B an infinite distance from A (the body under investigation) and from all other electrified objects; join A and B *via* C; if A becomes plus ('overcharged' in Cavendish's terminology), A is 'positively electrified.'

The canals represent wires. The assumed incompressibility of the fluid within them is an artifice introduced to preclude accumulations or deficiencies Cavendish could not calculate; in the case of long, thin wires, as he expected, the idealization creates unimportant errors. 'Positive electrification' corresponds exactly to positive potential with respect to ground. Consider a system of conductors A, A′, A″ . . . , joined by canals. Touch A with the tube: all bodies electrify positively to the same *degree*, that is, each will confer the same amount of superfluous fluid on the test body B, although no two need contain the same charge. Cavendish observed that a body can be both positively electrified and undercharged or, in modern terms, can have a positive potential and a net negative charge. Bring an insulated positively charged body D up to one of the conductors, say A. If D's electricity is sufficiently strong, it will drive all of A's surplus, and some of its natural fluid, into its interconnected fellows A′, A″, A‴,

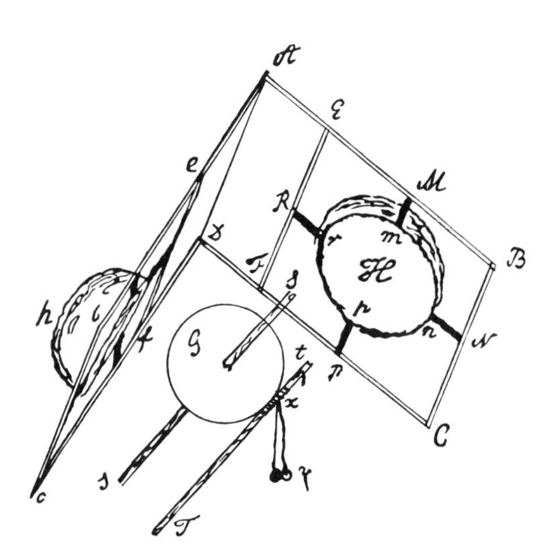

3.30 Cavendish's null experiment to show that electricity obeys the same law as gravity. From Cavendish, El. Res. *(1879), 104.*

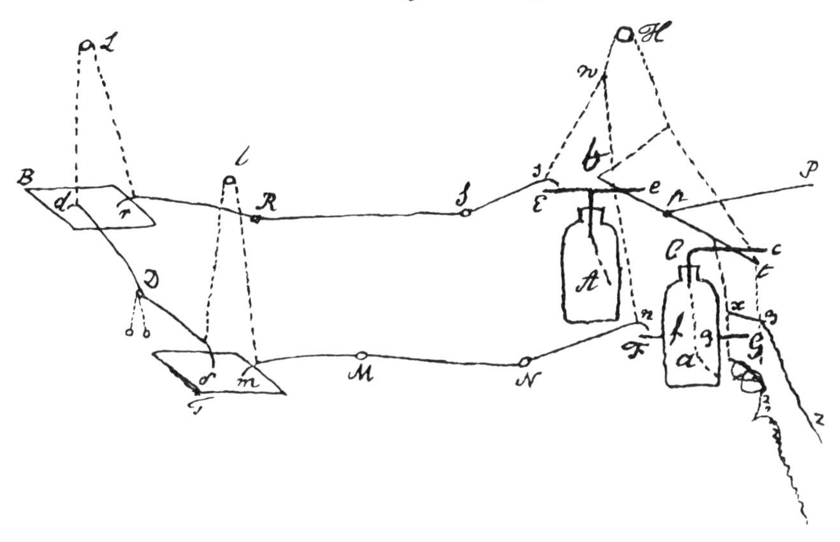

3.31 Cavendish's arrangement for measuring relative capacities. The purpose of the strings on the left is explained in the text; those on the right are to connect the jars to the prime conductor and to ground for charging, and to disconnect them for measuring. From Cavendish, El. Res. *(1879), 117.*

increasing the 'electrification' (potential) of the system. Although now minus, A will convey a larger plus charge to B than it did before.[73]

The artifice of the canal enabled Cavendish to unravel the old conundrum of the power of points. He showed that the smaller of two charged spheres connected by a long canal repels electrical matter at its surface more vigorously than the larger in the inverse ratio of their radii. In so far as a spiked PC communicating with the globe can be compared to two spheres, one much smaller than the other and connected by a canal, one understands immediately the apparent paradox that portions of the same surface, electrified to the same degree, can act very differently on the electrical fluid.[74] Cavendish was the first electrician to explain the power of points without recourse to ad hoc hypotheses. For a decade he continued to do work of unprecedented scope and originality, most of which he did not publish. Three sets of his experiments show how far it was possible to go in physics using the concepts, mathematics, and apparatus available in the late eighteenth century.

The first set employed an extensible tin plate T to test the relative capacities of bodies B (fig. 3.31). Two similar Leyden jars A and a charge B and T in opposite senses through rRSs and mMNn, respectively, while the wire dDδ is kept from A and B by silk strings passing over pullies L and *l*. An intricate arrangement of

73. *PT*, 61 (1771), 628–9, 650–3; *El. Res*. (1879), 31, 45–6, 95–6.
74. *PT*, 61 (1771), 660–4; *El. Res*. (1879), 52–4.

cords allowed Cavendish simultaneously to lift rR and mM and to drop dδ. If the electroscope at D registered zero, the capacities of T and B were equal; if not, he adjusted T and tried again. One of his most striking results made the ratio of the capacities of a disk and globe of the same diameter about 0.654; an approximate model, in which a canal connects a sphere to a disk, part of whose charge is spread evenly and the remainder placed at the circumference, gave 0.647; classical electrostatics prefers 0.640.[75] These experiments are models of method as well as of accuracy. Cavendish not only tried to minimize or to correct for leak and the perturbations introduced by his own body considered as a conductor; he also took into account the contribution of the walls of the laboratory to the measured capacitance of the globe. Although others, for example Barletti and Lichtenberg, had begun to appreciate these fine points,[76] no contemporary, not even Coulomb, approached Cavendish in subtlety, complexity, and precision of electrical measurement.

In the second set of experiments, Cavendish extended the method to compare capacities of Franklin squares armed with disks of tin foil. Using the fiction of the canal, he computed what amounts to the potential at the center of the positive plate, $V = (2d/a^2)Q$, where a is the radius of the disks and d the thickness of the glass. Comparison with Volta's relation, $Q = CT$, made $C = a^2/2d$, a proposition Cavendish confirmed using glass of different thicknesses.[77] But when he compared the capacities of the squares to that of a simple conductor, the computed values were eight times too small. Moreover, the amount of the discrepancy depended on the nature of the dielectric. These results, which point towards the concept of specific inductive capacity, greatly surprised and perhaps also disappointed Cavendish. The discovery that each insulator has an electrical property besides its poor conductivity limited the generality of his theory and may have checked his hope of preparing a *Principia electricitatis*.[78]

The third set of experiments concerned relative conductivities. In the winter of 1773, intrigued by accounts of an electrical fish, or torpedo, Cavendish began to measure the 'resistance' of solutions, placing himself in series with the test object and a Leyden jar and estimating the length x of each fluid required to diminish the explosion to a standard sensation. The technique established that fresh water resisted one hundred times better than salt but failed when applied to metals, for the resistance of Cavendish's body swamped that of any practicable length of wire. He therefore put himself in parallel with the test body and received the same stroke across 5.1 inches of saturated solution of salt—note the astounding accuracy—as across 2540 inches of a certain iron wire.[79]

The little fish that inspired these investigations provided the occasion for

75. *El. Res.* (1879), 30–6, 64, 114–37, 166–8, 447–8.

76. Barletti, Soc. ital., *Mem. mat. fis.*, 4 (1788), 306–7, and *Dubbî* (1776), letter v; Lichtenberg to Wolff, 3 Feb. 1785, in *Briefe*, II, 194–5.

77. *El. Res.* (1879), 77–80, 144–53, 161.

78. *Ibid.*, 172–82; McCormmach, *El. Res.* (1967), 469–76, 494–6.

79. Cavendish, *El. Res.* (1879), 262, 293–5.

Cavendish's second and last publication on electricity, an answer to those who refused to believe the torpedo electrical. And why not electrical? Because, as physicists had slowly and painfully learned, electrical matter manifests itself only after being collected upon an insulated body. 'When a Gentleman can so give up his reason as to believe in the possibility of an accumulation of electricity among conductors [that is, the ocean] sufficient to produce the effects ascribed to the Torpedo, he need not hesitate a moment to embrace *as truths* the greatest contradictions that can be laid before him.'[80] Cavendish replied that the fish need not maintain a charge and direct it towards a particular victim; it is only necessary for the torpedo to create electricity and for the electrician to understand the nature of divided circuits.

In discharging a Leyden jar through a wire held in the hands, the electrician opens two paths to the current, which, according to Cavendish, divides between the circuits inversely as their resistances. With a short, thick wire the experimenter feels no shock, not, as many electricians held, because no fluid traverses his body but because the quantity that does cannot amount to a millionth of the whole. When placed in parallel with a long, fine wire, however, he receives a sensible stroke, in exact analogy to the working of the torpedo. Here the parallel circuits are closed paths in the water, which carry less fluid the greater their lengths; if the experimenter, who conducts almost as well as the ocean, occupies a portion of one of the smaller circuits, he will pass most of its electricity.[81]

Another objection to ascribing the torpedo's prowess to electricity exploited its incompetence to throw sparks or attract chaff. How could it be electrical, lacking all the usual signs? Cavendish's answer returns us to our chief theme: one must distinguish carefully, he said, between quantity of electricity and its intensity, as measured by the spread of a Henley or the gap of a Lane electrometer, and recognize that the shock depends upon quantity and intensity conjointly. Although the stroke from a large battery greatly exceeds that of a single jar charged to the same degree, the battery will not discharge across a greater Lane gap than will any of its members taken separately. By multiplying the number of elements, one can preserve the stroke while decreasing the electrification; Cavendish designed a battery of forty-nine jars that struck like the fish when so weakly electrified that he needed a microscope to discern its spark. To amuse himself and convince doubters, he built torpedos of wood and leather, placed them in baths of sea water, hooked them up to his battery, and invited Lane and Priestley to play with them. They got shocks to their complete satisfaction.[82]

5. EPILOGUE

The work of Volta, Cavendish, and Coulomb brought electrostatics to the point that, within a few years, it could suffer its definitive quantification at the hands

80. Henley to Canton, 21 May 1775, quoting Thomas Ronayne, Canton Papers, II (RS).
81. Cavendish, *PT*, 66 (1776), 197–9, 210–11; *El. Res.* (1879), 195–6, 205, 301.
82. *PT*, 66 (1776), 200–9, 216–22; *El. Res.* (1879), 197–206, 210–13.

of the mathematical physicists of the Ecole Polytechnique. The step permanently removed higher electrical theories from the reach of the Gilberts, Franklins, and Voltas who had prepared it. In the late 1780s and early 1790s electricians without mathematics switched their attention to animal and medical electricity and to electrochemistry. 'Foreigners attend much more to [electro-] chemistry than electricity,' a traditionalist Briton told his students in 1794; 'therefore, you will discover nothing to reward the toil of reading [their journals], if you except Volta's papers.'[1] The nature and magnitude of the switch may be collected from Table 3.1, which shows the percentages of papers concerned with various aspects of electricity reviewed or abstracted in the *Commentarii de rebus in scientia naturali et medicina gestis* from 1752 to 1797. 'Traditional Electricity' signifies theories of electrical action, descriptions of apparatus, demonstrations of 'artificial' or laboratory electricity; 'Medical Electricity' includes accounts of therapy and experiments of electricity on animals; and 'Natural Electricity' means primarily the electricity of the atmosphere.

Perhaps the impression of some old-timers, that (to quote the *Britannica* for 1797) 'the science of electricity seems to be at a standstill,' arose from the diversion of electrical energy into new channels. Van Marum, unable as usual to design work for his big machine and now unguided by others, announced in 1795 that he was changing his line of research.[2] The machine remained idle for five years. Then, in 1800, Volta made public a discovery of the first importance, which, by uniting work on animal electricity with traditional concerns and the new electrochemistry, not only reawakened the interest of van Marum, but marked an epoch in the history of electricity.

ELECTROSTATICS AND THE PILE

In 1791 Luigi Galvani, professor of anatomy at the University of Bologna and of obstetrics at the Institute of Sciences of the Bologna Academy, published his now famous study of the electrical excitation of disembodied frog legs. He explained the jerking of a leg on completing a circuit through the relevant nerve and muscle as the consequence of the discharge of a 'nerveo-electrical fluid,' an 'animal electricity' *sui generis*, a new natural agent.[3]

Volta rejected Galvani's explanation of his grisly experiments: ordinary electricity sufficed for torpedos and should be good enough for frogs and anatomists.[4] He guessed that the 'electromotive force' (his words) arose at the junctions between the various conductors used to close the circuit or between them and the stripped animal that served as electroscope. He tried to maximize the twitch by composing the circuit of moist conductors (like frogs) and dry ones

1. Morgan, *Lectures* (1794), I, lxii.

2. *Encycl. Brit.* (1797[3]), VI, 424; Hackmann in Forbes, *Marum*, III, 329–30. The perception of standstill did not arise from a decline in the quantity of literature.

3. Galvani, *Commentary* (1953), 2–81, esp. 45.

4. *VO*, I, 10–11, 21–3, 26; *VE*, III, 143–5.

TABLE 3.1

Distribution of Articles on Electricity by Field[a] (percent)

	1752/61	*1762/74*	*1769*[b]	*1775/88*	*1789/97*
Traditional Electricity	50	60	60	60	10
Medical Electricity	45	35	30	30	70
Natural Electricity	5	5	10	5	10
Electrochemistry	0	0	0	5	10

[a] All data from the *Commentarii* except those for 1769. The editors may have been biased towards medical electricity in the 90s.
[b] From Krünitz, *Verz*. (1769). Since Krünitz' bibliography is retrospective, the agreement of the figures with those extracted from the *Commentarii* is a striking confirmation of the validity of the concept 'traditional electricity.'

(like metals) in various ways. The results, in order of decreasing power, expressed in Volta's notation (where capital and small letters signify dry and moist conductors, respectively, and r=rana=frog): rABr; raAr; rabr; rAr and rar, both zero. What about the all-dry circuit ABCA? Volta expected that a weak current would be generated, but he had no way to detect it because the only electroscope sensitive enough to register Galvanic electricity was itself a moist conductor.[5] The difficulty instanced a much more serious one: Volta's claim of the identity of Galvanic and common electricity rested on experiments in which pieces of animals played an indispensable part.[6]

The contact of zinc and silver develops about 0.78 volt. Volta's most sensitive straw electrometer marked about 40 volt/degree.[7] By the summer of 1796 he had managed to multiply the charges developed by touching dissimilar metals together enough to spread the straws.[8] He first succeeded with a Nicholson doubler and then with an unaided Bennet electroscope; and he later rendered contact electricity easily sensible by a 'condensing electroscope,' a straightforward combination of the condensatore and the straw electrometer.[9] It remained to find a way to multiply galvanic electricity directly. Volta discovered soon enough that piling metal disks on one another (say aABAB . . . a) did not help and that a circuit made only of metals gave no emf. These results led to the useful rule that the emf of a pile of disks is equal to what its extreme disks would generate if put into immediate contact.[10] How or when Volta hit on the far from obvious artifice

5. *VO*, I, 230, 371–82, 396–7, 401–6, 411–13.
6. *VO*, I, 490, 540–5.
7. One degree of Volta's best straw electroscope equalled 0.1 degree of a Henley (*VO*, I, 486, and V, 37, 52, 81), 35 degrees of which marked about 13,350 volt (Polvani, *Volta* [1942], 145). Hence, 1 degree straw indicated about 40 volt. Volta later estimated the tension between zinc and silver at 1/60 degree straw (*VO*, II, 39), or about 0.7 volt.
8. *VO*, I, 525; *VE*, III, 349, 359.
9. *VO*, I, 420–4, 435–6; *VE*, III, 438.
10. *VO*, II, 61; cf. *VO*, I, 326–7.

of repeating the apparently unimportant moist conductors in his generator is not known. The definitive pile, AZaAZaAZa . . . AZ, consisting of pairs of silver and zinc disks separated by pieces of moist cardboard, was first made public in 1800 in a now famous letter addressed to the president of the Royal Society of London.[11]

In explaining the operation of the pile, Volta appealed to an analogy to the torpedo, using precisely those concepts of tension and quantity of electricity that had been worked out in the 1770s and 1780s. The analogy was not hard to grasp: a medium-size pile, one with forty or fifty metallic pairs, gave anyone who touched its extremities about the same sensation he could enjoy holding an electric fish. In both cases, Volta said, a constant current running externally from top to toe of the electromotor passed through the arms and breast and agitated the sense of touch. Were it directed at the senses of vision, taste, or hearing, the current would cause light, taste, or sound instead.[12] Neither the pile nor the torpedo give electrostatic signs because, as Cavendish had argued long before, they operate at too low a tension; their effects derive rather from the quantity of electrical matter they move. As for the cause and continuance of the electricity generated by the contact of dissimilar conductors, Volta feigned no hypothesis: 'this perpetual motion may appear paradoxical, perhaps inexplicable; but it is nonetheless true and real, and can be touched, as it were, with the hands.'[13]

The pile was the last great discovery made with the instruments, concepts, and methods of the eighteenth-century electricians. It opened up a limitless field. It was immediately applied to chemistry, notably to electrolysis, and soon brought forth, for the first time, the shy elements sodium and potassium from fused soda and potash. Its steady current provided the long-sought means for establishing a relationship between electricity and magnetism. The consequent study of electromagnetism transformed our civilization.

THE QUANTIFICATION OF ELECTROSTATICS

As Galvani was examining the electricity of decapitated frogs, Coulomb was exploiting the newly demonstrated law of electrical force in researches of equal originality. After stressing, and confirming, that the charge on a conductor pressed itself within a layer of insensible thickness at its surface,[14] he set out to discover the distribution of the charge on the surfaces of bodies more interesting than isolated spheres. He represented the distribution as the varying surface density of one or the other electrical fluid, which he probably pictured as spread one molecule thick at the geometrical boundary of the conductor. The distribution can also be pictured as a variation in the depth of a fluid of uniform density.

Coulomb's methods, simple in principle, required great manual dexterity and

11. *PT*, 90:2 (1800), 403–31; *VO*, I, 563–82.
12. *VO*, I, 556, 578–82. 13. *VO*, I, 576.
14. Coulomb, *MAS* (1786), §§vii–xi; *MAS* (1787), §xviii; *Mémoires* (1884), 178–82, 205.

conceptual clarity. In one series of experiments he placed a charged sphere in his torsion balance, touched it with another of different radius, measured the decreased force of the first, compensated for leaks during the operation, and calculated the uniform densities of the separated partners. In another series, he used an invention of his now known as a proof-plane, a thin small disk of gilded paper attached to an insulating handle, which picks up an electricity proportional to the charge density of the place on a conductor to which it is touched. Coulomb measured the electricity of the plane in his torsion balance. This method, 'the easiest, the simplest, and perhaps the most exact,' and, one might add, the most general, way to compare charge densities, also allowed a convenient compensation for leak. To compare the densities at two places, A and B, Coulomb would at equal intervals of time measure that at A, at B, and again at A; assuming leak proportional to the time, he took the average of the measurements at A as the value of the density there at the time of the measurement at B.[15]

Coulomb compared such measurements with approximations deduced from analogies to the gravitational theory. One example will indicate the strength and limitation of his method. He required the distribution of charge on three equal, collinear, touching, insulated spherical conductors of radius a. Call the distributions on the extreme spheres ρ, that on the middle one ρ', and assume, what is not true, that charge spreads evenly over each ball. At either point of contact a unit charge is pushed toward the center ball by the nearest extreme one with a force f_1 and away from the center by forces f_2 and f_3, arising, respectively, from the middle and furthest ball. For equilibrium, $f_1 = f_2 + f_3$, which comes to $\rho = \rho' + 2\rho/9$. Hence $\rho/\rho' = 1.29$; measurement with the proof plane away from the points of contact made it 1.34.[16]

Coulomb deduced the equation among the densities as follows. From Newton's theorem about gravitating spheres, $f_3 = (4\pi a^2 \rho)/(3a)^2 = 4\pi\rho/9$. To get f_1 and f_2 he invoked an argument like the following. Consider points P and Q respectively inside and outside a charged spherical shell and distant from it by a small amount x; call s the part of the force at Q arising from a bit of surface g surrounding the intersection of PQ with the shell, and S the part arising from the rest of the surface G. At Q, $s + S = 4\pi a^2/(a+x)^2$. At P the force arising from g is opposite, and that from G identical, to what it is at Q, whence $s = S = 2\pi a^2/(a+x)^2$. In the case under discussion $x = 0, f_1 = 2\pi\rho, f_2 = 2\pi\rho'$. This pretty argument is credited by Poisson to Laplace. But it is already implicit in Coulomb's writings.[17]

The assumption of the uniformity of ρ and ρ', and similar rough but serviceable approximations for the case of unequal touching spheres, were the artifices

15. Coulomb, *MAS* (1787), §§ii–iii; *Mémoires* (1884), 187–90.
16. Coulomb, *MAS* (1787), §§xiii–xvi, xxii–xxiii; *Mémoires* (1884), 200–4, 210–11.
17. Coulomb, *MAS* (1786), §xi; *MAS* (1787), §xix; *Mémoires* (1884), 181–2, 207; Poisson, *MAS* (1811:1), 5–6.

of a resourceful engineer. Coulomb worked directly with the quantity he could measure, force, and with the theoretical object he could handle, a uniformly charged sphere. The definitive quantification of electrostatics came when mathematicians not limited to forces or to homogeneous spheres took up his problem of the distribution of electricity on conducting spheres. These mathematicians, or mathematical physicists, succeeded by applying Laplace's reformulation of the theory of gravity to the case of electricity.

The reformulation centered on a certain function V, later called 'potential,' introduced by Laplace around 1785. The function is the integral of the quotients of the gravitational masses dm by their respective distances from the point $P(x,y,z)$ at which V is to be computed. If the mass dm is situated at the point $P'(x',y',z')$, $V(x,y,z) = \int\int\int dm(x',y',z')/PP'$. Laplace observed that the negative partial derivatives of V are proportional to the components of the gravitational force.[18]

The function V allowed the computer to work with a scalar, additive quantity rather than with force, whose direction must be taken into account. In adopting this convenience, mathematizing physicists took up with an abstraction still further from experience than microscopic *force*. V proved exceptionally useful in electrostatics. The electrical fluid, as opposed to ponderable mass, is free to distribute itself over the surface of a closed conductor in such a way that, irrespective of the presence of other bodies however electrified, there is no electrical force within or tangent to the surface of the conductor. This complicated distribution can be represented easily by imposing conditions on the potential: V must be constant within all conductors and $\partial V/\partial s$ must be zero, where s signifies any direction in any plane tangent to the conductor's surface.

Laplace began to encourage others to apply the reformulated gravitational theory to electricity about the time Volta visited Paris to promote the pile and to help cement relations between the Napoleonic regimes of France and Italy. Among the earliest to look where Laplace pointed was Jean-Baptiste Biot, one of the first graduates of the new, high-powered Ecole Polytechnique. His brilliance and his offer to read proofs of the *Mécanique céleste* brought him the attention and the patronage of Laplace. While enjoying one of the fruits of this association, a professorship of mathematical physics at the former Collège Royal, Biot solved a difficult problem in electrostatics by a simple method suggested by his patron.[19] The problem: the distribution of electricity on the surface of an ellipsoid of revolution. The answer: the fluid arranges itself so that its depth y (assumed to be insensible) beneath any point P on the surface is proportional to the distance from P to the corresponding point on the surface of a similar and similarly placed ellipsoid within, and infinitesimally distant from, the original body. The proof: Laplace's expression for the force or potential of a gravitating ellipsoid implies

18. Cf. Todhunter, *Figure* (1873), II, 25–7, 31.
19. Crosland, *Society* (1967), 254–6; *DSB*, II, 133–40; Picard, *Eloges* (1931), 226–8, 231–3.

that no tangential force exists on a particle of electric fluid located anywhere between the similar ellipsoids.[20]

At this point all the ingredients of classical electrostatics had been identified: the law of squares, Coulomb's measurements of distribution, the Laplacian machinery and its demonstrated utility in problems of electricity, the potential function. Siméon-Denis Poisson, a precocious mathematical genius, a graduate of the Ecole Polytechnique, tenant of professorships at the Ecole and the University, disciple of Lagrange and Laplace, put them together.[21] He showed in great generality that the depth y, proportional to the quantity of redundant vitreous or resinous fluid in unit volume just beneath the surface of a conductor, is also proportional to the force just outside it; and that since this force must be perpendicular to the surface, $y \sim -\partial V/\partial n$, where n is the direction of the normal. Since the quantity in Poisson's electrostatic theory corresponding to the element of mass dm of the gravitational theory is $y\rho dS'$, where ρ is presumed constant and dS' is an element of surface, y satisfies the integral equation

$$y \sim -\frac{\partial}{\partial n}[\int (y\rho/\mathrm{PP'})dS'].$$

The function V must be constant within conductors, and its derivatives in their surfaces must vanish. Poisson managed to find V and y for the case of touching spheres by expressing the integrands as series, a trick at which he was a master.[22] His numerical results agreed better with Coulomb's measurements than Coulomb's theory did.

Poisson's V is the analytic form of Cavendish's 'electrification' and Volta's 'tension.' It is more supple than either, for it permits the statement of the classic problems of electrostatics—finding the distribution of electricity and the resultant ponderomotive forces—in full generality. To be sure, Poisson had to approximate to obtain numerical results; but his approximations, which could be made as close as desired, were approximations in the evaluation of mathematical forms, not substitutions of counterfactual physical systems for the ones under consideration.

Poisson shared the instrumentalism of the French school of mathematical physicists. A dualist for convenience, it was all the same to him whether electricity came in one fluid or two. He did not care whether these hypothetical fluids pressed themselves into a layer one molecule thick at the surface of the conductor to make a charge of varying surface density, as in Coulomb's representation, or whether, as in his and Biot's, they occupied an infinitesimal volume of constant surface density and variable depth. The problem of the mechanism of electrical action did not interest him. As for the vexed question of the agency that pre-

20. B[iot], Soc. phil., *Bull.*, 3 (1805), 21–3; cf. Poisson, *MAS* (1811:1), 27–8n.

21. Anon., *Rev. deux mondes*, 23:3 (1840), 410–37; Crosland, *Society* (1967), 127–8, 212–13, 318–19.

22. Poisson, *MAS* (1811:1), 1–92.

vented the escape of the fluids from the surface of conductors—a question by no means resolved by the elimination of atmospheres—Poisson was content to suppose with Biot that the 'pressure' of the air clamped electricity to its carrier. That, of course, was to ignore the evidence that vacuum insulates. But to Poisson the mechanism did not matter: it was enough that electricity remained on conductors and distributed itself in accordance with his calculations.

The chief moral of this history may be that when confronted with a choice between a qualitative model deemed intelligible and an exact description lacking clear physical foundations, the leading physicists of the Enlightenment preferred exactness. It was not a choice peculiar to them. 'The only object of theoretical physics is to calculate results that can be compared with experiment,' writes the dean of contemporary theoretical physicists; 'it is quite unnecessary that any satisfactory description of the whole course of the phenomena should be given.'[23]

23. Dirac, *Principles of Quantum Mechanics* (1930), 7.

Bibliography

The Bibliography lists only works cited in the notes; other early writings about electricity may be found in the catalogs of Ekelöf, Frost, Gartrell, Mottelay, Rosetti and Cantoni, and Weaver. Information sufficient for retrieval is given with each entry except for articles in a collaborative volume cited more than once and for references to reprints in collected works. In such cases referral to a fuller entry is given in short form, e.g., editor's last name, brief title, date (for collected works, brief title and date only). Names of learned societies and citations to their publications conform as closely as possible to the usage of the Library of Congress.

Titles of some articles, particularly in *PT*, have been shortened by dropping irrelevant opening phrases; e.g., 'An Extract of a Letter from Samuel Pickwick, Esq., to Lord So & So, F.R.S. on the Theory of Tittlebats,' would appear as 'The Theory of Tittlebats.' Book titles always begin with their original word or phrase; where necessary they are shortened internally by ellipsis. Subtitles, and the final phrase(s) of lengthy ones, are usually omitted.

Manuscripts do not appear in the Bibliography because citations to them in the notes usually give enough information to locate them. The few exceptions are 'Add. Ms.,' for items in the British Library, London; 'Burndy,' for the Burndy Library, Smithsonian Institution, Washington, D.C.; and 'Nollet, Journal,' for his 'Journal du voyage de Piédmont et d'Italie en 1749,' Bibliothèque Municipale, Soissons.

No distinction is made between primary and secondary sources or between books and articles. Russian words are transliterated according to the system of the Library of Congress. Accents are omitted from capital letters in French titles. A few special abbreviations besides those used in the notes occur:

AIHS.	*Archives internationales d'histoire des sciences.*
Ann. Sci.	*Annals of Science.*
BJHS.	*British Journal for the History of Science.*
CIHS.	Congrès International d'Histoire des Sciences.
IIET.	Akademiia nauk S.S.S.R. Institut istorii estestvoznaniia i tekhniki.

Abat, Bonaventure. *Amusemens philosophiques*. Amsterdam, 1763.

Achard, Franz Karl. 'Expériences sur l'électrophore avec une théorie de cet instrument.' *MAS*/Ber (1776), 122–34.

———. 'Sur l'analogie qui se trouve entre la production et les effects de l'électricité et de la chaleur.' *MAS*/Ber (1779), 27–35. (*Sammlung*, 141–53.)
———. *Chemisch-physische Schriften*. Berlin, 1780.
———. 'Mémoire sur la mesure de la force de l'électricité, et sa comparaison avec celle de la gravité.' *JP*, 21 (1782), 190–209.
———. *Sammlung physikalischer und chemischer Abhandlungen*. Berlin, 1784.
———. *Vorlesungen über die Experimentalphysik*. 4 parts. Berlin, 1791.
Aepinus, Franz Ulrich Theodor. 'Quelques nouvelles expériences électriques remarquables.' *MAS*/Ber (1756), 105–21.
———. 'Descriptio ac explicatio novorum quorundam experimentorum electricorum.' *NCAS*, 7 (1758–9), 277–302.
———. *Tentamen theoriae electricitatis et magnetismi*. St. Petersburg, 1759.
———. *Essay on the Theory of Electricity and Magnetism*. [1759] Home and Connor, *Aepinus's Essay* (1979), 227–478.
———. *Recueil de différents mémoires sur la tourmaline*. St. Petersburg, 1762.
———. 'De electricitate barometrorum disquisitio.' *NCAS*, 12 (1766–7), 303–24.
———. 'Examen theoriae magneticae a celeberr. Tob. Mayero propositae.' *Ibid*., 325–50.
Agrippa, Henry Cornelius. *The Vanity of Arts and Sciences* [1530], London, 1676.
Aiton, E. J. *The Vortex Theory of Planetary Motions*. London and New York, 1972.
Akademie der Wissenschaften, Göttingen. *Erster Säcularfeier*. Göttingen, 1852.
Akademiia Nauk, Leningrad. *Materialy dlia istorii imperatorskoi akademii nauk [1716–1750]*. 10 vols. St. Petersburg, 1885–1900.
Alembert, Jean Le Rond d'. *Oeuvres philosophiques, historiques et littéraires*. Jean François Bastien, ed. 18 vols. Paris, 1805.
———. *Oeuvres*. A. Belin, ed. 5 vols. Paris, 1821–2.
———. *Recherches sur différens points importans du système du monde*. 3 vols. Paris, 1754–6.
———. 'Essai sur les élémens de philosophie, ou sur les principes des connoissances humaines.' *Mélanges de littérature, d'histoire, et de philosophie*. 5 vols. Amsterdam, 1759–67. IV (1760), 1–292. (*Oeuvres* [1805], II, 3–476; *Oeuvres* [1821], I, 115–348.)
———. *Preliminary Discourse to the Encyclopedia of Diderot*. Richard N. Schwab, tr. Indianapolis, 1963.
Alexander, H. G. *The Leibniz-Clarke Correspondence*. Manchester, 1956.
Allamand, Jean Nicolas Sébastien. 'Mémoire contenant diverses expériences d'électricité.' *Bibliothèque britannique*, 24 (1746), 406–37.
———. 'Histoire de la vie et des ouvrages de M. 'sGravesande.' Gravesande, *Oeuvres*, I, ix–lix.
Allen, Phyllis. 'Scientific Studies in the English Universities of the 17th Century.' *JHI*, 10 (1949), 219–53.
Amodeo, Federico. *Vita matematica napoletana: studio storico*. 2 vols. Naples, 1905–24.
Anderson, P. J., ed. *Studies in the History and Development of the University of Aberdeen*. Aberdeen, 1906.
Anon. 'Mémoire historique.' Accademia delle scienze, Turin. *Mémoires*, 1 (1784–5), i–lvi.
———. 'Lettres à un Américain sur l'état des sciences en France. III. Poisson.' *Revue des deux mondes*, 23:3 (1840), 410–37.

———. 'Documents sur le Collège de France à la veille de la Révolution. *Revue internationale de l'enseignement*, 5 (1883), 406–12.

———. 'Aaresredogörelse.' Uppsala. University. *Aarsskrift* (1910), 30–50.

Anthiaume, Albert. *Le Collège du Havre*. 2 vols. Le Havre, 1905.

Antinori, Vincenzo. 'Notizie istoriche.' Accademia del Cimento. *Saggi dell'esperienze naturali fatte nell'Accademia del Cimento*. Florence, 1841³. Pp. 3–133.

Atkinson, Philip. *Elements of Static Electricity*. New York, 1887.

Bacon, Francis. *Works*. J. Spedding, R. L. Ellis and D. D. Heath, eds. 15 vols. New York, 1863–9.

———. *The Advancement of Learning*. G. W. Kitchin, ed. London, 1915.

Bacon, Roger. *Opus maius*. J. H. Bridges, ed. 2 vols. London, 1900.

Baker, Henry. 'Medical Experiments of Electricity.' *PT*, 45 (1748), 270–5.

Baker, Keith. 'Les débuts de Condorcet au secrétariat de l'Académie royale des sciences (1773–1776).' *RHS*, 20 (1967), 229–80.

Baldi, Francesco. *De igne, luce et fluido electrico propositiones physicae*. Florence, 1790.

Barber, William Henry. *Leibniz in France from Arnauld to Voltaire*. Oxford, 1955.

Barbosa Machado, Diogo. *Biblioteca lusitana*. 4 vols. Lisbon, 1930².

Barletti, Carlo. *Physica specimina*. Milan, 1772.

———. *Dubbî e pensieri sopra la teoria degli elettrici fenomeni*. Milan, 1776.

———. 'Introduzione a nuovi principi della teoria elettrica dedotti dall'analisi de' fenomeni delle elettriche punte.' Società italiana delle scienze, Rome. *Memorie di matematica e di fisica*, 1 (1782), 1–54.

———. 'Della supposta eguaglianza di contrarie elettricità nelle due opposte faccie del vetro.' *Ibid.*, 4 (1788), 304–9.

———. 'Della legge d'immutabile capacità e necessaria contrarietà di eccesso e difetto di elettricità sugli opposti lati del vetro.' *Ibid.*, 7 (1794), 444–61.

Barrière, Pierre. *L'Académie de Bordeaux . . . au XVIIIe siècle*. Bordeaux, 1951.

Bartholmess, Christian Jean Guillaume. *Histoire philosophique de l'Académie de Prusse depuis Leibniz jusqu'à Schelling*. 2 vols. Paris, 1850–1.

Bayle, François. *Institutiones physicae ad usum scholarum accomodatae*. 3 vols. Toulouse, 1700.

Bayle, Pierre. *Dictionnaire historique et critique*. 4 vols. Amsterdam, 1740⁵.

Beccaria, Giambatista. *Dell'elettricismo artificiale e naturale libri due*. Turin, 1753.

———. *Dell'elettricismo. Lettere . . . al chiarissimo sig. Giacomo Bartolomeo Beccari*. Turin, 1758.

———. 'Experiments in Electricity.' *PT*, 51 (1760), 514–25.

———. 'Novorum quorundam in re electrica experimentorum specimen.' *PT*, 56 (1766), 105–18; 57 (1767), 297–311.

———. *Experimenta, atque observationes, quibus electricitas vindex late constituitur, atque explicatur*. Turin, 1769.

———. *Elettricismo artificiale*. Turin, 1772.

———. *A Treatise upon Artificial Electricity . . . To which is added An Essay on the Mild and Slow Electricity which prevails in the Atmosphere during Serene Weather*. London, 1776.

Beckman, Anna. 'Två svenska experimentalfysiker på 1700-talet: Mårten Triewald och Nils Wallerius.' *Lychnos* (1967–8), 186–214.

Bednarski, Stanislas. 'Déclin et renaissance de l'enseignement des Jésuites en Pologne.' *Archivum historicum Societatis Jesu*, 2 (1933), 199–223.

Benjamin, Park. *A History of Electricity (The Intellectual Rise in Electricity) from Antiquity to the Days of Benjamin Franklin*. New York, 1898.

Bennet, Abraham. 'Description of a new Electrometer.' *PT*, 76 (1786), 26–34.

Bennett, J. A. 'Robert Hooke as Mechanic and Natural Philosopher.' RS, *Not. Rec.*, 35 (1980), 33–48.

Berger, Peter. 'Johann Heinrich Lamberts Bedeutung in der Naturwissenschaft des 18. Jahrhunderts.' *Centaurus*, 6 (1959), 157–254.

Bergman, Torbern Olof. 'Elektrische Versuche mit Seidenbande von unterschiedlicher Farbe.' *AKSA*, 25 (1763), 344–52.

———. *Torbern Bergman's Foreign Correspondence*. G. Carlid and J. Nordstrom, eds. I. *Letters from Foreigners to Torbern Bergman*. Stockholm, 1965.

Bernoulli, Daniel, and Jean [II] Bernoulli: 'Nouveaux principes de mécanique et de physique, tendans à expliquer la nature et les propriétés de l'aiman.' AS. *Pièces qui ont remporté le prix* [*en 1743 et 1744/6*]. Paris, 1748. Pp. 117–44. (AS. *Recueil des pièces . . .* , 5.)

Bernoulli, Jean [III]. *Lettres astronomiques*. Berlin, 1771.

———. *Lettres sur différens sujets écrites pendant le cours d'un voyage par l'Allemagne, la Suisse, la France méridionale et l'Italie, en 1774 et 1775*. 3 vols. Berlin, 1777–9.

Berthé de Besaucèle, Louis. *Les Cartésiens d'Italie . . . aux XVIIe et XVIIIe siècles*. Paris, 1920.

Bierens de Haan, J. A. *De Hollandsche Maatschappij der Wetenschappen, 1752–1952*. Haarlem, 1952.

Bioche, Charles. 'Les savants et le gouvernement aux XVIIe et XVIIIe siècles.' *Revue des deux mondes*, 107:2 (1937), 174–89.

B[iot], Jean Baptiste. 'Sur un problème de physique, relatif à l'électricité.' Société philomathique, Paris. *Bulletin*, 3 (1805), 21–3.

Birch, Thomas. *History of the Royal Society of London*. 4 vols. London, 1756–7.

Biwald, Leopold. *Institutiones physicae in usum philosophiae auditorum*. 2 vols. Vienna, 1779[2].

Boas, Marie. *See* Hall, Marie Boas.

Böhm, Wilhelm. *Die Wiener Universität*. Vienna, 1952.

Boerhaave, Herman. *A New Method of Chemistry*. Peter Shaw, tr. 2 vols. London, 1741[2].

Börner, Friedrich. *Nachrichten von den vornehmsten Lebensumständen und Schriften jetztlebender berümter Aerzte und Naturforscher in und um Deutschland*. 3 vols. Wolfenbüttel, 1749–54.

Boffito, Giuseppe. 'Un *qui pro quo* degli storici galileiani,' *Bibliofilia*, 44 (1942), 176–84.

Bopp, Karl. 'Leonhard Eulers und Johann Heinrich Lamberts Briefwechsel.' Akademie der Wissenschaften, Berlin. Phys.-Math. Kl. *Abhandlungen* (1924:2). 45 p.

———. 'J. H. Lamberts und H. G. Kaestners Briefe.' Akademie der Wissenschaften, Heidelberg. *Sitzungsberichte* (1928:18). 34 p.

Borgeaud, Charles. *Histoire de l'Université de Genève*. I. *L'Académie de Calvin, 1559–1798*. Geneva, 1900.

Bortolotti, Ettore. *La storia della matematica nella Università di Bologna*. Bologna, 1947.

Boscovich, Roger Joseph. *De viribus vivis*. Rome, 1745.

———. *A Theory of Natural Philosophy* [1763²]. J. M. Child, tr. Cambridge, Mass., 1961.

Bose, Georg Mathias. *Tentamina electrica*. Wittenberg, 1744. (This contains three essays, designated in the notes as follows:
I. i: 'De attractione et electricitate,' 1738;
I. ii: 'De electricitate,' 1742;
I. iii: 'De electricitate et beatificatione,' 1744.)

———. *Recherches sur la cause et sur la véritable théorie de l'électricité*. Wittenberg, 1745.

———. *Tentamina electrica tandem aliquando hydraulicae chymiae et vegetabilibus utilia*. Wittenberg, 1747. (This contains two essays, as follows:
II. i: 'De electricitate commentarius novus,' 1746;
II. ii: 'De electricitate commentarius epistolaris,' n.d.)

———. *L'électricité son origine et ses progrès. Poème en deux livres*. Leipzig, 1754.

Bosmans, Henri. 'Grégoire de St. Vincent.' *Biographie nationale de Belgique*, XXI (1911), cols. 141–71.

Boss, Valentin. *Newton and Russia; the Early Influence, 1698–1796*. Cambridge, Mass., 1972.

Bossut, Charles. *Traité théorique et expérimental d'hydrodynamique*. 2 vols. Paris, an IV².

Bouguer, Pierre. *Entretiens sur la cause de l'inclinaison des orbites des planètes*. Paris, 1748². (*AS. Recueil des pièces qui ont remporté les prix*. 2:7 [Paris, 1752]. 140 p.)

Bouillier, François. *Histoire de la philosophie cartésienne*. 2 vols. Paris, 1854.

Boutroux, Pierre. 'L'enseignement de la mécanique en France au XVIIᵉ siècle.' *Isis*, 4 (1921–2), 276–94.

Bowles, Geoffrey. 'John Harris and the Powers of Matter.' *Ambix*, 22 (1975), 21–38.

Boyle, Robert. *The Excellency and Grounds of the Corpuscular or Mechanical Philosophy* [annexed to *The Excellency of Theology*]. London, 1674. (*Works*, IV, 67–78; Fulton, *Bibl.*, no. 116; M. B. Hall, *Boyle* [1965], 187–209.)

———. *The Excellency of Theology compar'd with Natural Philosophy*. London, 1674. (*Works*, IV, 1–66; Fulton, *Bibl.*, no. 116.)

———. *Some Considerations about the Reconcileableness of Reason and Religion*. London, 1675. (*Works*, IV, 151–91; Fulton, *Bibl.*, no. 122.)

———. *The Works . . . To which is prefixed The Life of the Author*. Thomas Birch, ed. 6 vols. London, 1772². (Fulton, *Bibl.*, no. 241.)

Brandes, Ernst. *Betrachtungen über den Zeitgeist in Deutschland in den letzten Decennien des vorigen Jahrhunderts*. Hannover, 1808.

Briggs, J. Morton. 'D'Alembert: Philosophy and Mechanics in the 18th Century.' Colorado University. *Studies in History*, no. 3 (1964), 38–56.

Brischar, Karl. *P. Athanasius Kircher. Ein Lebensbild*. Würzburg, 1877.

Brisson, Mathurin-Jacques. *Pésanteur spécifique des corps*. Paris, 1787.

Browne, Thomas. *Pseudodoxia epidemica: or, Enquiries into very many Received Tenets and Commonly Presumed Truths*. London, 1646.

Browning, John. 'Fray Beneto Jerónimo de Feijóo and the Sciences in 18th-Century Spain.' Hughes and Williams, eds. *The Varied Pattern*, 353–71.

Brunet, Pierre. *Les physiciens hollandais et la méthode expérimentale en France au XVIIIᵉ siècle*. Paris, 1926.

———. *Maupertuis*. 2 vols. Paris, 1929.
———. *L'introduction des théories de Newton en France au XVIIIe siècle*. Paris, 1931.
———. 'Les premières recherches expérimentales sur la foudre et l'électricité atmosphérique.' *Lychnos*, 10 (1946–7), 117–48.
———. *La vie et l'oeuvre de Clairaut (1713–1765)*. Paris, 1952. (*RHS*, 4 [1951], 13–40, 109–53; 5 [1952], 334–49; 6 [1953], 1–17.)
Bryan, Margaret. *Lectures on Natural Philosophy*. London, 1806.
Bryden, D. J. *Scottish Scientific Instrument Makers 1600–1900*. Edinburgh, 1972.
Buchdahl, Gerd. 'History of Science and Criteria of Choice.' *Historical and Philosophical Perspectives of Science*. R. H. Stuewer, ed. Minneapolis, 1970. Pp. 204–30. (*Minnesota Studies in the Philosophy of Science, V*.)
Buffon, Georges-Louis Leclerc, Comte de. *Oeuvres complètes*. 6 vols. Paris, 1835–6. (*Oeuvres* [1954], Bibl., no. 177.)
———. 'Traité de l'aimant.' *Oeuvres complètes*. III (1836), pp. 76–113. (*Histoire naturelle des minéraux*, V [1788]; *Oeuvres* [1954], Bibl., no. 143.)
———. *Oeuvres philosophiques*. Jean Pivateau, ed. Paris, 1954. (*Corpus général des philosophes français*, XLI:1.)
———. 'Seconde vue de la nature.' *Oeuvres* (1954), 35–41. (*Histoire naturelle*, XIII [1765].)
Bugge, Thomas. *Science in France in the Revolutionary Era*. Maurice P. Crosland, ed. Cambridge, Mass., 1969.
Buonanni, Filippo. *Rerum naturalium historia nempe quadrupedum . . . existentium in Museo Kircheriano*. J. A. Battarra, ed. Rome, 1773.
Burtt, Edwin Arthur. *The Metaphysical Foundations of Modern Physical Science*. London, 1932^2.

Cabeo, Niccolò. *Philosophia magnetica*. Cologne, 1629.
———. *Meteorologicorum Aristotelis commentarii*. 4 vols. Rome, 1646.
Calinger, Ronald S. 'Frederick the Great and the Berlin Academy of Sciences (1740–1766).' *Ann. Sci.*, 24 (1968), 239–49.
Campo, Mariano. *Cristiano Wolff e il razionalismo precritico*. 2 vols. Milan, 1939. (Università cattolica del S. Cuore. *Pubblicazioni*, ser. 1, 30.)
[Canton, John.] 'Electrical Experiments, with an Attempt to account for Their several Phaenomena, together with Observations on Thunder Clouds.' *PT*, 48 (1753), 350–8. (*BFP, V*, 149–54; *EO*, 293–9.)
———. 'New Electrical Experiments.' *PT*, 48 (1754), 780–5.
Caramuel, Juan. *Rationalis et realis philosophia*. Louvain, 1642.
Cardano, Girolamo. *De subtilitate libri XXI*. Lyon, 1550.
Carpenter, Nathanael. *Geographie Delineated Forth in Two Bookes*. Oxford, 1635^2.
Carré, Jean Raoul. *La philosophie de Fontenelle*. Paris, 1932.
Carteron, H. 'L'idée de force mécanique dans le système de Descartes.' *Revue philosophique de la France et de l'Etranger*, 47 (1922), 243–77, 483–511.
Carvalho, Joaquim de. *Correspondência científica dirigida a João Jacinto de Magalhães*. Coimbra, 1952. (Coimbra. University. Faculdade de ciências. *Revista*, 20 [1951], 93–283.)
Casini, Paolo. *L'universo-macchina*. Bari, 1969.

Cassirer, Ernst. *The Philosophy of the Enlightenment*. Princeton, 1951.

Catania. University. *Storia della Università di Catania*. Catania, 1934.

Caullet de Veaumorel, L., tr. *Description de la machine électrique négative et positive de M. Nairne*. Paris, 1784.

Caut, Ronald J. *The College of St. Salvator*. Edinburgh and London, 1950.

Cavallo, Tiberius. *A Complete Treatise on Electricity, in Theory and Practice; with Original Experiments*. London, 1777; 1782[2]; 3 vols., 1786–95.

Cavendish, Henry. 'An Attempt to explain some of the Principal Phaenomena of Electricity by Means of an Elastic Fluid.' *PT*, 61 (1771), 584–677. (Cavendish. *Electrical Researches*, 3–63.)

———. 'Some Attempts to imitate the Effects of the *Torpedo* by Electricity.' *PT*, 66 (1776), 196–225. (Cavendish. *Electrical Researches*, 194–215.)

———, *et al*. 'The best Method of adjusting the fixed Points of Thermometers; and of the Precautions necessary to be used in making Experiments with those Instruments.' *PT*, 67 (1777), 816–57.

———. *The Electrical Researches of Henry Cavendish*. J. C. Maxwell, ed. Cambridge, 1879.

Ceñal, Ramon. 'Emmanuel Maignan: su vida, su obra, su influencia.' *Revista de estudios políticos*, 46 (1952), 111–49.

———. 'Juan Caramuel. Su epistolario con Atanasi Kircher, S. J.' *Revista de filosofía*, 12 (1953), 101–47.

———. 'La filosofía de Emmanuel Maignan.' *Ibid.*, 13 (1954), 15–68.

Ceyssens, L. 'L'action antijanséniste de Brunon Neusser, O.F.M.' *Franziskanische Studien*, 35 (1953), 401–11.

Chaldecott, J. A. *Handbook of the King George III Collection of Scientific Instruments*. London, 1951.

Chapin, S. L. 'The Academy of Sciences During the 18th Century: An Astronomical Appraisal.' *French Historical Studies*, 5 (1968), 371–404.

Charles le géomètre. 'Essai sur les moyens d'établir entre les thermomètres une comparabilité.' *MAS* (1787), 567–82.

Cheyne, George. *Philosophical Principles of Religion: Natural and Reveal'd*. 2 vols. in 1. London, 1715–6[2].

Chipman, R. A. 'An Unpublished Letter of Stephen Gray on Electrical Experiments, 1707–1708.' *Isis*, 45 (1954), 33–40.

Chossat, Marcel. *Les Jésuites et leurs oeuvres à Avignon 1553–1768*. Avignon, 1896.

Christie, John R. R. 'The Origins and Development of the Scottish Scientific Community, 1680–1760.' *History of Science*, 12 (1974), 122–41.

Cigna, Gianfrancesco. 'De novis quibusdam experimentis electricis.' *Ibid.*, 3 (1762–5), 31–72.

———. 'De electricitate.' *Ibid.*, 5 (1770–3), 97–108.

———. 'Esperienze elettriche.' *Scelte di opuscoli interessanti tradotti di varie lingue*, 9 (1775), 68–83.

Clairaut, Alexis Claude. 'Sur les explications cartésienne et newtonienne de la réfraction de la lumière.' *MAS* (1739), 259–75.

Cohen, Ernst, and W. A. T. Cohen-de Meester. 'Daniel Gabriel Fahrenheit.' *Chemisch Weekblad*, 33 (1936), 374–92; 34 (1937), 727–30.

Cohen, I. Bernard. *Some Early Tools of American Science*. Cambridge, 1950.

———. 'Prejudice against the Introduction of Lightning Rods.' *Journal of the Franklin Institute*, 253 (1952), 393–440.

———. *Franklin and Newton*. Philadelphia, 1956.

Collegium conimbricense. *Commentarii . . . in octo libros Physicorum*. 2 vols. Lyon, 1602; Cologne and Frankfurt, 1609.

———. *Commentarii in quattuor libros de Coelo*. Lyon, 1603².

———. *Commentarii in duos libros de Generatione et Corruptione*. Lyon, 1606².

———. *Commentarii in universam dialecticam Aristotelis* [I]. Lyon, 1607.

Condorcet, Marie-Jean-Antoine-Nicolas Caritat, Marquis de. 'Eloge de Fouchy.' *HAS* (1788), 37–49.

Convegno internazionale celebrativo del 250° anniversario della nascità di R. G. Boscovich e del 200° anniversario della fondazione dell'Osservatorio di Brera. *Atti*. Milan, 1963.

Cornelio, Tommaso. *Epistola, qua motuum illorum, qui vulgo ob fugam vacui fieri dicuntur, vera causa per circumpulsionem ad mentem Platonis explicatur*. Rome, 1648.

Corson, D. W. 'Pierre Polinière, Francis Hauksbee and Electroluminescence: A Case of Simultaneous Discovery.' *Isis*, 59 (1968), 402–13.

Cosentino, Giuseppe. 'L'insegnamento delle matematiche nei collegi gesuitici nell'Italia settentrionale. Nota introduttiva.' *Physis*, 13 (1971), 205–17.

Costa, Gustavo. 'Il rapporto Frisi-Boscovich alla luce di lettere inedite di Frisi, Boscovich, Mozzi, Lalande e Pietro Verri.' *Rivista storica italiana*, 79 (1967), 819–76.

Costabel, Pierre. 'Le *De viribus vivis* de R. Boscovich ou De la vertu des querelles de mots.' *AIHS*, 14 (1961), 3–21.

———. 'La correspondance Le Sage-Boscovich.' Convegno Boscovich. *Atti*, 205–16.

Costello, William Thomas. *The Scholastic Curriculum at Early 17th Century Cambridge*. Harvard, 1958.

Coulomb, Charles-Augustin. 'Recherches sur la meilleure manière de fabriquer les aiguilles aimantées.' AS. *Mémoires de mathématique et de physique, présentés . . . par divers savans*, 9 (1780), 166–264.

———. 'Recherches théoriques et expérimentales sur la force de torsion et sur l'élasticité des fils de métal.' *MAS* (1784), 229–69.

———. 'Sur l'électricité et le magnétisme.'

[I] 'Construction et usage d'une balance électrique.' *Ibid.*, 569–77.

[II] 'Où l'on détermine suivant quelles lois le fluide magnétique ainsi que le fluide électrique agissent.' *Ibid.*, 578–611.

[III] 'De la quantité d'électricité qu'un corps isolé perd dans un temps donné.' *Ibid.*, 616–38.

[IV] 'Où l'on démontre deux principales propriétés du fluide électrique.' *MAS* (1786), 67–77.

[V] 'Sur la manière dont le fluide électrique se partage entre deux corps conducteurs mis en contact.' *MAS* (1787), 421–67.

[VI] 'Suite des recherches sur la distribution du fluide électrique entre plusieurs corps conducteurs.' *MAS* (1788), 617–705.

———. *Mémoires*. Paris, 1884. (Société française de physique. *Collection de mémoires relatifs à la physique*, I.)

Cousin, Jean. 'L'Académie des sciences, belles lettres et arts de Besançon au XVIIIe siécle et son oeuvre scientifique.' *RHS*, 12 (1959), 327–44.

Cousin, Victor. *Fragments philosophiques pour servir à l'histoire de la philosophie*. 5 vols. Paris, 1865–6[5].

Coutts, James. *A History of the University of Glasgow from its Foundation in 1451 to 1909*. Glasgow, 1909.

Crane, V. W. 'The Club of Honest Whigs: Friends of Science and Liberty.' *William and Mary Quarterly*, 23 (1966), 210–33.

Crino, Anna Maria. *Fatti e figure del seicento anglo-toscano*. Florence, 1957.

Cromaziano, Agatopisto. *Della restaurazione di ogni filosofia ne' secoli XVI, XVII e XVIII*. 3 vols. Venice, 1785–9.

Crombie, Alistair C. *Robert Grosseteste and the Origins of Experimental Science, 1100–1700*. Oxford, 1962[2].

Crommelin, Claude-August. 'Die holländische Physik im 18. Jahrhundert mit besonderer Berücksichtigung der Entwicklung der Feinmechanik.' *Sudhoffs Archiv*, 28 (1935), 129–42.

———. 'L'ascension du Mont-Blanc par H. B. de Saussure en 1787 et un relief du Mont-Blanc de 1790.' Nederlands aardrijkskundig genootschap. *Tijdschrift*, 66 (1949), 327–31.

———. *Descriptive Catalogue of the Physical Instruments of the 18th Century in the Rijksmuseum voor de Geschiedenis der Natuurwetenschappen*. Leyden, 1951. (Leyden. Rijksmuseum voor de Geschiedenis der Natuurwetenschappen. *Communications*, no. 81.)

Crosland, Maurice P. *The Society of Arcueil*. Cambridge, Mass., 1967.

Cuthbertson, John. *Abhandlung von der Elektricität*. Leipzig, 1786.

———. *Practical Electricity*. London, 1807.

Cuvier, Georges-Frédéric. 'Eloge historique de M. Banks.' *HAS*, 5 (1821–2), 204–31.

Dähnert, Johann Carl, ed. *Sammlung gemeiner und besonderer Pommerscher und Rügischer Landes-Urkunden*. II. Stralsund, 1767.

Dainville, François de. *La géographie des humanistes*. Paris, 1940.

———. 'L'enseignement des mathématiques dans les collèges Jésuites de France du XVIe au XVIIIe siècle.' *RHS*, 7 (1954), 6–21, 109–23.

———. 'L'enseignement des mathématiques au XVII siècle.' *XVIIe siècle*, no. 30 (1956), 62–8.

———. 'L'enseignement scientifique dans les collèges des Jésuites.' Taton, ed. *Enseignement*, 27–65.

Dalibard, Thomas François. 'Histoire abrégée de l'électricité.' *EO* (Dal)[1], i–lxx.

Dalzel, Andrew. *History of the University of Edinburgh from its Foundation*. 2 vols. Edinburgh, 1862.

Daujat, Jean. *Origines et formation de la théorie des phénomènes électriques et magnétiques*. 3 vols. Paris, 1945.

Daumas, Maurice. *Les instruments scientifiques aux XVIIe et XVIIIe siècles*. Paris, 1953.

———. 'La vie scientifique au XVII siècle.' *XVIIe siècle*, no. 30 (1956), 110–33.

———. 'Precision Mechanics.' Charles Singer et al., eds. *A History of Technology*. IV. *The Industrial Revolution*. Oxford, 1958. Pp. 379–416.

———. 'Precision of Measurement and Physical and Chemical Research in the 18th Century.' Alistair C. Crombie, ed. *Scientific Change*. New York, 1963. Pp. 418–30.
Debus, Allen G. 'Robert Fludd and the Use of Gilbert's *De magnete* in the Weapon-Salve Controversy.' *Journal of the History of Medicine*, 19 (1964), 389–417.
———. 'Mathematics and Nature in the Chemical Texts of the Renaissance.' *Ambix*, 15 (1968), 1–28.
———. *Science and Education in the 17th Century. The Webster-Ward Debate*. London and New York, 1970.
Dee, John. *Propaedeumata aphoristica*. London, 1568 [2].
Dekker, T. 'De popularisering der natuurwetenschap in Nederland in de achttiende eeuw.' *Geloof en wetenschap*, 53 (1955), 173–88.
Delandine, Antoine François. *Couronnes académiques*. 2 vols. Paris, 1787.
———. *Manuscrits de la bibliothèque de Lyon*. 3 vols. Paris and Lyon, 1812.
Delattre, Pierre. *Les établissements des Jésuites en France depuis quatre siècles*. 5 vols. Paris, 1939–57.
Delfour, J. *Les Jésuites à Poitiers (1604–1702)*. Paris, 1902.
Deluc, Jean André. *Recherches sur les modifications de l'atmosphère*. 2 vols. Geneva, 1772.
———. 'Cinquième lettre . . . sur le fluide électrique.' *JP*, 36 (1790), 450–69.
———. *Précis de la philosophie de Bacon et des progrès qu'ont fait les sciences naturelles par ses préceptes et son exemple*. 2 vols. Paris, 1802.
Desaguliers, Jean Théophile, *Physico-Mechanical Lectures*. London, 1717.
———. 'An Account of a Book entitul'd *Vegetable Staticks*.' *PT*, 34 (1728), 264–91; 35 (1729), 323–31. (*Course* [1763 [3]], II, 403–11.)
———. 'An Attempt to Solve the Phenomenon of the Rise of Vapours, Formation of Clouds, and Descent of Rain.' *PT*, 36 (1729), 6–22.
———. *A Course of Experimental Philosophy*. 2 vols. London, 1734–44; 1745 [2]; 1763 [3].
———. 'Some Thoughts and Conjectures Concerning the Cause of Electricity.' *PT*, 41 (1739–40), 175–85.
———. 'Some Thoughts and Experiments Concerning Electricity.' *Ibid.*, 186–93.
Descartes, René. *Oeuvres*. D. Adam and P. Tannery, eds. 13 vols. Paris, 1897–1913; 1964–74 [2].
———. *Principes de la philosophie* [1644]. *Oeuvres*, IX:2.
———. *Correspondence of Descartes and Constantyn Huygens, 1635–1647*. L. Roth, ed. Oxford, 1926.
———. *Correspondence*. C. Adam and G. Milhaud, eds. 8 vols. Paris, 1936–63.
———. *Discourse on Method, Optics, Geometry and Meteorology*. Paul J. Olscamp, tr. Indianapolis, Ind., 1965.
Desplat, Christian. *Un milieu socio-culturel provincial: L'Académie royale de Pau au XVIIIe siècle*. Pau, 1971.
Dibner, Bern. *Early Electrical Machines*. Norwalk, Conn., 1957.
———. 'The Great van Marum Electrical Machine.' *Natural Philosopher*, 2 (1963), 8–103.
Dictionnaire de physique. Gaspard Monge et al., eds. 4 vols. Paris, 1793–1822. (*Encyclopédie méthodique, ou par ordre des matières*. Vols. 186–9.)
Digby, Kenelm. *Two Treatises. In the One of which the Nature of Bodies; in the Other, the Nature of Man's Soul, is looked into*. London, 1644.

———. *A Late Discourse . . . touching the Cure of Wounds by the Powder of Sympathy*. R. White, tr. London, 1658.
Dijksterhuis, E. J. *The Mechanization of the World Picture*. C. Dikshoorn, tr. Oxford, 1961.
Dobbs, Betty Jo. 'Studies in the Natural Philosophy of Sir Kenelm Digby.' *Ambix*, 18 (1971), 1–25.
Doerberl, Michael. *Entwicklungsgeschichte Bayerns*. 3 vols. Munich, 1916–31[3].
Doppelmayr, Johann Gabriel. *Neu-entdeckte Phaenomena von bewunderswürdigen Würkungen der Natur*. Nuremburg, 1774.
Doublet, E. 'L'abbé Bossut.' *Bulletin des sciences mathématiques*, 38:1 (1914), 93–6, 121–5, 158–9, 186–90, 220–4.
Droysen, H. 'Die Marquise du Châtelet, Voltaire und der Philosoph Christian Wolff.' *Zeitschrift für französiche Sprache und Literatur*, 35 (1910), 226–48.
Du Châtelet-Lomont, Gabrielle-Emilie, Marquise. *Institutions de physique*. Paris, 1740.
———. *Les lettres de la marquise du Châtelet*. Theodore Besterman, ed. 2 vols. Geneva,1958.
Dufay, Charles François de Cisternay. 'Mémoire sur les baromètres lumineux.' *MAS* (1723), 295–306.
———. 'Mémoire sur un grand nombre de phosphores nouveaux.' *MAS* (1730), 524–35.
———. 'A Letter from Mons. Du Fay . . . concerning Electricity.' *PT*, 38 (1733–4), 258–66.
———. 'Mémoires sur l'électricité. 1er. L'histoire de l'électricité.' *MAS* (1733), 23–35.
'2e. Quels sont les corps qui sont susceptibles d'électricité.' *Ibid.*, 73–84.
'3e. Des corps qui sont les plus vivement attirés par les matières électriques, et de ceux qui sont les plus propres à transmettre l'électricité.' *Ibid.*, 233–54.
'4e. L'attraction et la répulsion des corps électriques.' *Ibid.*, 457–76.
'5e. Des nouvelles découvertes sur cette matière, faites depuis peu par M. Gray; et où l'on examine quelles sont les circonstances qui peuvent apporter quelque changement à l'électricité pour l'augmentation ou la diminution de sa force, comme la température de l'air, la vide, l'air comprimé, etc.' *MAS* (1734), 341–61.
'6e. Quel rapport il y a entre l'électricité et la faculté de rendre de la lumière, qui est commune à la plupart des corps électriques, et ce qu'on peut inférer de ce rapport.' *Ibid.*, 503–26.
'7e. Quelques additions aux mémoires précédents.' *MAS* (1737), 86–100.
'8e.' *Ibid.*, 307–25.
Duhem, Pierre. 'L'optique de Malebranche.' *Revue de métaphysique et de morale*, 23 (1916), 37–91.
———. *To Save the Phenomena. An Essay on the Idea of Physical Theory from Plato to Galileo*. E. Doland and C. Maschler, trs. Chicago, 1969.
Duhr, Bernhard. *Geschichte der Jesuiten in den Ländern deutscher Zunge in der zweiten Hälfte des XVII. Jahrhunderts*. 4 vols. Munich and Regensberg, 1921.
Dulieu, Louis. 'La contribution montpelliéraine aux Recueils de l'Académie royale des sciences.' *RHS*, 11 (1958), 250–62.
———. 'Le mouvement scientifique montpelliérain au XVIIIe siècle.' *Ibid.*, 227–49.
Dunken, Gerhard. *Die Deutsche Akademie der Wissenschaften zu Berlin in Vergangenheit und Gegenwart*. Berlin, 1960.
Dupont-Ferrier, Gustave. *La vie quotidienne d'un collège parisien pendant plus de 350*

ans: du collège de Clermont au lycée Louis-le-Grand, 1563–1920. 3 vols. Paris, 1921–5.

Dutour, Etienne François. 'Essai sur l'aiman.' AS. *Pièces qui ont remporté le prix* [*en 1743 et 1744/6*]. Paris, 1748. [Prix de 1744/6], 51–114. (AS. *Recueil des pièces . . .* , 5.)

Edleston, J. *Correspondence of Sir Isaac Newton and Professor Cotes*. London and Cambridge, 1850.

Ekelöf, Stig, ed. *Catalogue of Books and Papers relating to the History of Electricity in the Library of the Institute for Theoretical Electricity, Chalmers University of Technology*. 2 vols. Göteborg, 1964–6. (Chalmers Tekniska Högskola. Institution för elektricitetslära och elektrisk mätteknik. *Meddelande*, 11, 20.)

Ellicott, John. 'Of Weighing the Strength of Electrical Effluvia.' *PT*, 44 (1746), 96–9.

———. 'Several Essays toward Discovering the Laws of Electricity.' *PT*, 45 (1748), 195–224.

Elvius, Pehr. 'Geschichte der Wissenschaften, vom Ausdünsten des Wassers.' *AKSA*, 10 (1748), 3–10.

Encyclopédie méthodique. See *Dictionnaire de physique*.

Encyclopédie, ou dictionnaire raisonné des sciences, des arts, et des métiers [1751–65]. 36 vols. Geneva, 1778[3].

Erxleben, Johann Christian Polykarp. *Anfangsgründe der Naturlehre* [1772]. G. C. Lichtenberg, ed. Göttingen, 1784[3]; 1787[4]; 1791[5]; 1794[6].

Eschenberg, Johann Joachim. *Lehrbuch der Wissenschaftskunde. Ein Grundriss encyclopädischer Vorlesungen*. Berlin and Stettin, 1792.

Eschweiler, K. 'Die Philosophie der spanischen Spätscholastik auf den deutschen Universitäten des siebzehnten Jahrhunderts.' Görresgesellschaft. *Spanische Forschungen*, 1 (1928), 251–324.

———. 'Roderigo de Arriaga, S. J. Ein Beitrag zur Geschichte der Barockscholastik.' *Ibid*., 3 (1931), 253–85.

Eulenburg, Franz. *Die Frequenz der deutschen Universitäten von ihrer Gründung bis zur Gegenwart*. Leipzig, 1904. (Akademie des Wissenschaften, Leipzig. Phil.-Hist. Kl. *Abhandlungen*, 24:2.)

Euler, Johann Albrecht, et al. *Dissertationes selectae quae ad imperialem scientiarum petropolitanam academiam an. 1755 missae sunt*. St. Petersburg and Lucca, 1757.

———. 'Disquisitio de causa physica electricitatis.' *Ibid*., 1–40.

———. 'Recherches sur la cause physique de l'électricité.' *MAS*/Ber (1757), 125–59.

Euler, Leonhard. 'Dissertatio de magnete.' AS. *Pièces qui ont remporté le prix* [*en 1743 et 1744/6*]. Paris, 1748. [Prix de 1744/6], 3–47. (AS. *Recueil des pièces . . .* , 5.)

———. *Lettres . . . sur différentes questions de physique et de philosophie*. M. J. A. Condorcet and S. F. de Lacroix, eds. 3 vols. Paris, 1787–9.

———. *Die Berliner und die Petersburger Akademie der Wissenschaften im Briefwechsel Leonhard Eulers*. I. *Der Briefwechsel L. Eulers mit G. F. Müller*. A. P. Juškević and E. Winter, eds. Berlin, 1959.

[Fabri, Honoré.] *Metaphysica demonstrativa sive Scientia rationum universalium*. Lyon, 1648. (Issued under pseudonym, Petrus Mousnerius.)

———. *Dialogi physici in quibus de motu terrae disputatur, marini aestus nova causa proponitur, necnon acquarum & mercurii supra libellam elevatio examinatur*. Lyon, 1655.

———. *Physica id est Scientia rerum corporearum in decem tractatus distributa*. 5 vols. Lyon, 1669–71.

———. *Epistolae tres de sua hypothesi philosophica*. Mainz, 1674.

Fabroni, Angelo. *Lettere inedite di uomini illustri*. 2 vols. Florence, 1773–5.

Farrell, Allan P. *The Jesuit Code of Liberal Education*. Milwaukee, Wis., 1938.

Faucillon, Jean-Marcellin-Ferdinand. *Le Collège de Jésuites de Montpellier (1629–1762)*. Montpellier, 1857.

Favaro, Antonio. 'Per la storia della Accademia del Cimento.' Istituto veneto di scienze, lettere ed arti. *Atti*, 71 (1912), 1173–8.

———. *L'Università di Padova*. Venice, 1922.

Fawcett, Trevor. 'Popular Science in 18th-Century Norwich.' *History Today*, 22 (1972), 590–5.

Fellmann, Emil A. 'Die mathematischen Werke von Honoratus Fabry (1607–1688).' *Physis*, 1 (1959), 6–29, 73–102.

Ferguson, Alan, ed. *Natural Philosophy through the 18th Century and Allied Topics*. London, 1948.

Fernández Diéguez, D. 'Un matemático español del siglo XVIII: Juan Caramuel.' *Revista matemática hispano-americana*, 1 (1919), 121–7, 178–89, 203–12.

Ferretti-Torricelli, A. 'Padre Francesco Lana nel III centenario della nascità.' Ateneo di Brescia. *Commentari* (1931), 321–90.

Ferrner, Bengt. *Resa i Europa 1758–1762*. S. G. Lindberg, ed. Uppsala, 1956.

Fester, Richard. *'Der Universitäts-Bereiser.' Friedrich Gedike und sein Bericht an Friedrich Wilhelm II*. Berlin, 1905. (Archiv für Kulturgeschichte. *Ergänzungsheft*, 1.)

Fichter, Joseph Henry. *Man of Spain, Francis Suarez*. New York, 1940.

Finn, Bernard S. 'An Appraisal of the Origins of Franklin's Electrical Theory.' *Isis*, 60 (1969), 362–9.

———. 'Output of 18th Century Electrical Machines.' *BJHS*, 5 (1971), 289–91.

Fisch, Max H. 'The Academy of the Investigators.' E. Ashworth Underwood, ed. *Science, Medicine and History. Essays . . . [for] Charles Singer*. 2 vols. Oxford, 1953. I, 521–63.

Fischer, Ernst Gottfried. *Physique mécanique*. J. B. Biot, tr. Paris, 1806.

Fischer, Johann Carl. *Geschichte der Physik seit der Wiederherstellung der Künste und Wissenschaften bis auf die neuesten Zeiten*. 8 vols. Göttingen, 1801–8.

Fischer, Johann Nepomuk. *Beweis dass das Glockenlaeuten bey Gewittern mehr schaedlich als nuetzlich sey*. Munich, 1784.

Fitzpatrick, Edward Augustus, ed. *St. Ignatius and the Ratio Studiorum*. New York and London, 1933.

Fletcher, Harris Francis. *The Intellectual Development of John Milton*. 2 vols. Urbana, Ill., 1956–61.

Fletcher, John E. 'A Brief Survey of the Unpublished Correspondence of Athanasius Kircher, S. J. (1602–1680).' *Manuscripta*, 13 (1969), 150–60.

———. 'Medical Men and Medicine in the Correspondence of Athanasius Kircher.' *Janus*, 56 (1969), 259–77.

Flourens, Pierre. *Fontenelle ou De la philosophie moderne relativement aux sciences physiques*. Paris, 1847.

Förster, Johann Christian. *Übersicht der Geschichte der Universität zu Halle in ihrem ersten Jahrhunderte*. Halle, 1799.

Fogolari, Gino. 'Il museo Settala. Contributo per la storia della coltura in Milano nel secolo XVII.' *Archivio storico lombardo*, 14 (1900), 58–126.

Fontana, Felice. 'Description abrégée du cabinet de physique et d'histoire naturelle du grand-duc de Toscane, à Florence.' *JP*, 9 (1777), 41–8, 104–12, 194–203, 267–72, 377–9.

Fontenelle, Bernard le Bovier de. 'Doutes sur le système physique des causes occasionnelles' [1686]. *Oeuvres*, IX, 37–68.

———. *The Elogium of Isaac Newton*. London, 1728. (Newton. *Papers and Letters*, 444–74.)

———. 'Sur l'attraction newtonienne.' *HAS* (1732), 112–17.

———. 'Préface sur l'utilité des mathématiques et de la physique' [1733]. *Oeuvres*, V, 1–14.

———. 'Eloge de M. Saurin.' *HAS* (1737), 110–20. (*Oeuvres*, VI, 331–42.)

———. *Oeuvres*. 12 vols. Amsterdam, 1764.

Forbes, Eric, and I. K. Kopelevich. 'Neiznestnye pisma L. Eĭlera k T. Maĭery.' *Istorikoastronomicheskie issledovaniia*, 10 (1969), 285–310.

Forbes, Eric G. 'Tobias Mayer, Zur Wissenschaftsgeschichte des 18. Jahrhunderts.' *Jahrbuch für Geschichte der oberdeutschen Reichsstädte*, 16 (1970), 132–67.

Forbes, Robert J., ed. *Martinus van Marum, Life and Work*. 4 vols. Haarlem, 1969–73.

———. 'A Member of the First Class.' Forbes, ed. *Martinus van Marum*. III, 1–21.

Formey, Jean Henri Samuel. *Abrégé de physique*. 2 vols. Berlin, 1770–2.

———. *Souvenirs d'un citoyen*. 2 vols. Paris, 1797[2].

Fourier, J. J. 'Eloge historique de M. Charles.' *MAS*, 8 (1829), lxxiii–lxxxviii.

Fracastoro, Girolamo. *De sympathia & antipathia rerum*. Lyon, 1550.

France, Anatole. *L'Elvire de Lamartine. Notes sur M. et Mme Charles*. Paris, 1893.

Frank, Robert G., Jr. 'Science, Medicine and the Universities of Early Modern England: Background and Sources.' *History of Science*, 11 (1973), 236–69.

Franklin, Alfred. *Les origines du Palais de l'Institut. Recherches historiques sur le Collège des Quatre-Nations*. Paris, 1862.

———. *Histoire de la Bibliothèque Mazarine et du Palais de l'Institut*. Paris, 1901[2].

———. *Dictionnaire historique des arts, métiers et professions*. Paris and Leipzig, 1906.

Franklin, Benjamin. *Experiments and Observations on Electricity*.

EO[1] London, 1751–4.

EO[2] London, 1754.

EO[3] London, 1760–5.

EO[4] London, 1769.

EO[5] London, 1774.

EO (Dal)[1] Paris, 1752. (Tr. by T. Dalibard of the first part of *EO*[1].)

EO (Dal)[2] 2 vols. Paris, 1756. (Tr. of the complete *EO*[1].)

EO (Wilcke) Leipzig, 1758. (Tr. of *EO*[1] by J. C. Wilcke.)

———. *The Writings of Benjamin Franklin*. A. H. Smyth, ed. 10 vols. New York, 1907. (Cited as *BFP* [Smyth].)

———. *Benjamin Franklin's Experiments. A New Edition of Franklin's Experiments and Observations on Electricity*. I. B. Cohen, ed. Cambridge, Mass., 1941. (Taken from *EO*[5]; cited as *EO*.)

———. *Papers*. L. W. Labaree, et al., eds. New Haven, 1959+. (Cited as *BFP*.)
Freind, John. *Praelectiones chymicae*. Amsterdam, 1710.
French, Peter J. *John Dee. The World of an Elizabethan Magus*. London, 1972.
Freudenthal, Gad. 'Early Electricity between Chemistry and Physics.' HSPS, 11:2 (1981), 203–29.
Freyberg, Bruno von. *Johann Gottlob Lehmann*. Erlangen, 1955.
Friedländer, Paul. 'Athanasius Kircher und Leibniz; ein Beitrag zur Geschichte der Polyhistorie im XVII. Jahrhundert.' Pontificia accademia romana di archeologia. *Rendiconti*, 13 (1937), 229–47.
Frost, A. J. *Catalogue of Books and Papers relating to Electricity, Magnetism, and the Electric Telegraph, etc., including the Ronalds Library*. London, 1880.
Fulton, John Farquhar. *A Bibliography of the Honourable Robert Boyle, Fellow of the Royal Society*. Oxford, 1961^2.

Gabrieli, Vittorio. *Sir Kenelm Digby*. Rome, 1957.
Galvani, Luigi. *Commentary on the Effect of Electricity on Muscular Motion*. Robert Montraville Greca, tr. Cambridge, Mass., 1953.
García Villoslada, Riccardo. *Storia del Collegio romano del suo inizio (1551) alla soppressione della Compagnia di Gesù (1773)*. Rome, 1954. (*Analecta Gregoriana, 66* [1953]. Series Facultatis historiae ecclesiasticae, sec. A, no. 2.)
Garin, Eugenio. 'Antonio Genovesi e la sua introduzione storica agli *Elementa physicae* di Pietro van Musschenbroek.' *Physis*, 11 (1969), 211–22.
Gartrell, Ellen G. *Electricity, Magnetism and Animal Electricity. A Checklist of Printed Sources, 1600–1850*. Wilmington, Del., 1975.
Gassendi, Pierre. *Opera omnia*. 6 vols. Lyon, 1658.
Gaubil, Antoine. *Correspondance de Pékin, 1722–1759*. Renée Simon, ed. Geneva, 1970.
Gaullieur, Ernst. *Collège de Guyenne d'après un grand nombre de documents inédits*. Paris, 1874.
Gehler, Johann Samuel Traugott. *Physikalisches Wörterbuch oder Versuch einer Erklärung der vornehmsten Begriffe und Kunstwörter der Naturlehre*. 4 vols. Leipzig, 1787–91; 11 vols. in 22, 1825–45^2.
Gelbart, Nina Rattner. 'The Intellectual Development of Walter Charleton.' *Ambix*, 18 (1971), 149–68.
Genovese, Antonio. 'Disputatio [later Dissertatio] physico-historica de rerum corporearum origine et constitutione.' Petrus van Musschenbroek. *Elementa physicae conscripta in usus academicos*. 2 vols. Naples, 1745. I, 1–79.
Gerland, Ernst, and Friedrich Traumüller. *Geschichte der physikalischen Experimentierkunst*. Leipzig, 1899.
Geymonat, Ludovico. *Galileo Galilei*. Stillman Drake, tr. New York and London, 1965.
Gibbs, F. W. 'Boerhaave's Chemical Writings.' *Ambix*, 6 (1958), 117–35.
———. 'Itinerant Lecturers in Natural Philosophy.' *Ambix*, 8 (1961), 111–17.
Gibson, James. *Locke's Theory of Knowledge and its Historical Relations*. Cambridge, 1917.
Gilbert, William. *De magnete, magneticisque corporibus, et de magno magnete tellure; physiologia nova*. London, 1600. P. F. Mottelay, tr. New York, 1893; S. P. Thompson, tr. London, 1900.

———. *De mundo nostro sublunari philosophia nova*. Amsterdam, 1651.

Gilen, Leonhard. 'Über die Beziehungen Descartes' zur zeitgenössichen Scholastik.' *Scholastik*, 32 (1957), 41–66.

Gillmor, C. Stewart. *Coulomb and the Evolution of Physics and Engineering in 18th-Century France*. Princeton, 1971.

Gilson, Etienne Henri. *Index scolastico-cartésien*. Paris, 1912.

Glick, Thomas F. 'On the Influence of Kircher in Spain.' *Isis*, 62 (1971), 379–81.

Gliozzi, Mario. 'L'elettrologia nel secolo XVII.' *Periodico di matematiche*, 13 (1933), 1–14.

———. 'Giambatista Beccaria nella storia dell'elettricità.' *Archeion*, 17 (1935), 15–47.

———. *L'elettrologia fino al Volta*. 2 vols. Naples, 1937.

———. 'La costituzione della materia nella concezione di Boscovich e di Faraday.' Convegno Boscovich. *Atti*, 115–19.

———. 'Consonanze e dissonanze tra l'elettrostatica di Cavendish e quella di Volta.' *Physis*, 11 (1969), 231–48.

Godley, Alfred Denis. *Oxford in the 18th Century*. London, 1908.

Gordon, Andreas. *Versuch einer Erklärung der Electricität*. Erfurt, 1745; 1746[2].

Gottsched, J. C. *Historische Lobschrift des weiland hoch- und wohlgeborenen Herrn Christians, des H.R.R. Freyherrn von Wolff*. Halle, 1755.

Grabmann, M. 'Die disputationes metaphysicae des Franz Suarez in ihrer methodischen Eigenart und Fortwirkung.' *Mittelalterliches Geistesleben*. 3 vols. Munich, 1926–56. I, 525–60.

Gralath, Daniel. 'Geschichte der Electricität.' [Erster Abschnitt.] Naturforschende Gesellschaft, Danzig. *Versuche und Abhandlungen*, 1 (1747), 175–304.
'Zweiter Abschnitt.' *Ibid*., 2 (1754), 355–460.
'Dritter Abschnitt.' *Ibid*., 3 (1757), 492–556.

———. 'Nachricht von einigen electrischen Versuchen.' *Ibid*., 1 (1747), 506–41. Grandjean de Fouchy, Jean-Paul. 'Eloge de Nollet.' *HAS* (1770), 121–37.

Grant, Alexander. *The Story of the University of Edinburgh during its first Three Hundred Years*. 2 vols. London, 1884.

Gravesande, Willelm Jacob van 's. *Mathematical Elements of Natural Philosophy confirmed by Experiments, or an Introduction to Isaac Newton's Philosophy*. J. T. Desaguliers, tr. 2 vols. London, 1721; 1731[4].

———. *Oeuvres philosophiques et mathématiques*. J. N. S. Allamand, ed. 2 vols. Amsterdam, 1774.

Gray, Stephen. 'An Account of Some New Electrical Experiments.' *PT*, 31 (1720–1), 104–7.

———. 'Several Experiments concerning Electricity.' *PT*, 37 (1731–2), 18–44.

———. 'The Electricity of Water.' *Ibid*., 227–30.

———. 'Farther Account of his Experiments concerning Electricity.' *Ibid*., 285–91; 397–407.

———. 'Experiments and Observations upon the Light that is produced by communicating Electrical Attraction to Animate or Inanimate Bodies.' *PT*, 39 (1735–6), 16–24.

———. 'Some Experiments relating to Electricity.' *Ibid*., 166–70.

Great Britain. Parliament. House of Commons. 'Evidence, Oral and Documentary, Taken and Received by the Commissioners for Visiting the Universities of Scotland.' *Sessional Papers*, 1837: Edinburgh (92), *35;* Glasgow (93), *36;* St. Andrews (94), *37;* Aberdeen (95), *38*. (Command no. in parentheses, vol. in italics.)

Greaves, Richard L. *The Puritan Revolution and Eductional Thought*. New Brunswick, N.J., 1969.

Greene, Robert. *The Principles of the Philosophy of the Expansive and Contractive Forces*. Cambridge, 1727.

Grew, Nehemiah. *Museum regalis societatis*. London, 1681.

Grimsley, Ronald. *Jean d'Alembert (1717–*83). Oxford, 1963.

Grosier, Jean Baptiste Gabriel Alexandre. *Mémoires d'une société célèbre considérée comme corps littéraire . . . ou Mémoires des Jésuites sur les sciences, les belles-lettres et les arts*. 3 vols. Paris, 1792.

Günther, Siegmund. 'Die mathematischen und Naturwissenschaften an der Nürnbergischen Universität Altdorf.' Verein für Geschichte der Stadt Nürnberg. *Mitteilungen*, Heft 3 (1881), 1–36.

Guericke, Otto von. *Experimenta nova (ut vocantur) magdeburgica de vacuo spatio*. Amsterdam, 1672.

———. *Neue (sogenannte) magdeburger Versuche über den leeren Raum*. Hans Schimank et al., trs. and eds. Düsseldorf, 1968.

Guerlac, Henry. 'A Note on Lavoisier's Scientific Education.' *Isis*, 47 (1956), 211–16.

———. 'Newton's Changing Reputation in the 18th Century.' Raymond O. Rockwood, ed. *Carl Becker's Heavenly City Revisted*. Ithaca, 1958. Pp. 3–26.

———. 'Francis Hauksbee: expérimentateur au profit de Newton.' *AIHS*, 16 (1963), 113–28.

———. *Newton et Epicure*. Paris, 1963. (Paris. Palais de la Découverte. *Conférences*. Series D, no. 91.)

———. 'Sir Isaac and the Ingenious Mr. Hauksbee.' *Mélanges Alexandre Koyré*. I, 228–53.

———. 'Where the Statue Stood: Divergent Loyalties to Newton in the 18th Century.' Earl R. Wasserman, ed. *Aspects of the 18th Century*. Baltimore, 1965. Pp. 317–34.

———. 'An Augustan Monument: The Opticks of Sir Isaac Newton.' Hughes and Williams, eds. *Varied Pattern*, 131–63.

———. 'Chemistry as a Branch of Physics: Laplace's Collaboration with Lavoisier.' *Historical Studies in the Physical Sciences*, 7 (1976), 193–276.

———. *Essays and Papers in the History of Modern Science*. Baltimore, 1977.

Guitton, Georges. *Les Jésuites à Lyon sous Louis XIV et Louis XV*. Lyon, 1954.

Gunther, Robert T. *Early Science at Oxford*. 14 vols. London, 1920–45.

Gurlt, Ernst Julius, and A. Hirsch. *Biographisches Lexikon der hervorragenden Ärzte aller Zeiten und Völker*. 6 vols. Vienna and Leipzig, 1884–8.

Gutmann, Joseph. *Athanasius Kircher, 1602–1680, und das Schöpfungs- und Entwicklungsproblem*. Fulda, 1938.

Habertzettl, Hermann. *Die Stellung der Exjesuiten in Politik und Kulturleben Österreichs zu Ende des 18. Jahrhunderts*. Vienna, 1973. (Vienna. University. *Dissertationen*, 94.)

Hackmann, Willem D. 'Electrical Researches.' Forbes, ed. *Martinus van Marum*. III, 329–78.
———. *John and Jonathan Cuthbertson. The Invention and Development of the 18th Century Plate Electrical Machine*. [Leyden], 1973. (Leyden. Rijksmuseum voor de Geschiedenis der Natuurwetenschappen. *Communications*, no. 142.)
———. *Electricity from Glass*. Alphen aan den Rijn, 1978.
Hagelgans, Johann Georg. *Orbis literatus academicus germanico-europaeus*. Frankfurt a.M., 1737.
Hahn, Paul. *Georg Christoph Lichtenberg und die exakten Wissenschaften*. Göttingen, 1927.
Hahn, Roger. *The Anatomy of a Scientific Institution. The Paris Academy of Sciences 1666–1803*. Berkeley, 1971.
———. 'Scientific Careers in 18th Century France.' Maurice P. Crosland, ed. *The Emergence of Science in Western Europe*. London, 1975. Pp. 127–38. (*Minerva*, 13 [1975], 501–13.)
———. 'Scientific Lecturers in Paris, 1777–1789.' Unpublished ms.
Hales, Stephen. *Vegetable Staticks . . . Also, a Specimen of an Attempt to Analyse the Air*. London, 1727.
Hall, Alfred Rupert, and Marie Boas Hall. *Unpublished Scientific Papers of Isaac Newton*. Cambridge, 1962.
Hall, Marie Boas. *Robert Boyle on Natural Philosophy. An Essay with Selections from his Writings*. Bloomington, Ind., 1965.
———, and Alfred Rupert Hall. 'Newton's Electric Spirit: Four Oddities.' *Isis*, 50 (1959), 473–6.
[Haller, Albrecht von.] 'Histoire des nouvelles découvertes faites, depuis quelques années en Allemagne, sur l'électricité; mémoire sur les nouvelles découvertes qu'on a faites par rapport à l'électricité.' *Bibliothèque raisonnée des ouvrages des savans de l'Europe*, 34 (1745), 3–20.
Hamberger, Georg Christoph, and Johan Georg Meusel. *Das gelehrte Teutschland*. 23 vols. Lemgo, 1796–1834.
Hamberger, Georg Erhard. *Elementa physices methodo mathematica in usum auditorii conscripta*. Jena, 1735²; 1741³.
Hammermayer, Ludwig. *Gründungs- und Frühgeschichte der bayerischen Akademie der Wissenschaften*. Kallmünz/Opf., 1959. (*Münchener historische Studien. Abteilung bayerische Geschichte*, 4.)
Hankins, Thomas L. 'The Influence of Malebranche on the Science of Mechanics during the 18th Century.' *JHI*, 28 (1967), 193–210.
———. *Jean d'Alembert. Science and the Enlightenment*. Oxford, 1970.
Hanna, Blake T. 'Polinière and the Teaching of Experimental Physics at Paris 1700–1730.' P. Gay, ed. *18th Century Studies Presented to Arthur M. Wilson*. Hanover, N.H., 1972. Pp. 15–39.
Hans, Nicholas A. *New Trends in Education in the 18th Century*. London, 1951.
Hansteen, Christopher. *Untersuchungen über den Magnetismus der Erde*. Christiania, 1819.
Hardin, Clyde L. 'The Scientific Work of the Reverend John Michell.' *Ann. Sci.*, 22 (1966), 27–47.

Harding, Diana. 'Mathematics and Science Education in 18th-Century Northamptonshire.' *History of Education*, 1 (1972), 139–59.

Harnack, Adolf von. *Geschichte der Königlich Preussischen Akademie der Wissenschaften*. 3 vols. in 4. Berlin, 1900.

Harris, John. *Lexicon technicum*. 2 vols. London, 1704–10.

Harris, William Snow. 'On Some Elementary Laws of Electricity.' *PT*, 124:1 (1834), 213–45.

———. 'Inquiries concerning the Elementary Laws of Electricity. Second series.' *PT*, 126:1 (1836), 417–52.

———. *Rudimentary Magnetism*. Henry M. Noad, ed. London, 1872².

Hartsoeker, Nicolas. *Conjectures physiques*. 3 vols. Amsterdam, 1706–12.

———. *Cours de physique*. The Hague, 1730.

Harvey, Edmund Newton. *A History of Luminescence from the Earliest Times until 1900*. Philadelphia, 1957.

Hauksbee, Francis. 'Experiments on the Production and Propagation of Light from the *Phosphorus in Vacuo*. *PT*, 24 (1704–5), 1865–6.

———. 'Several Experiments on the Mercurial Phosphorus.' *Ibid.*, 2129–35.

———. 'The Production of a Considerable Light upon a slight Attrition of the Hand on a Glass Globe Exhausted of its Air: with other Remarkable Occurrences.' *PT*, 25 (1706–7), 2277–82.

———. 'The Extraordinary Elistricity [!] of Glass, produceable on a smart Attrition of it; with a Continuation of Experiments on the same subject, and other *Phaenomena*.' *Ibid.*, 2327–31.

———. 'A Continuation of the Experiments on the Attrition of Glass.' *Ibid.*, 2332–5.

———. 'Several Experiments shewing the Strange Effects of Glass, produceable on the Motion and Attrition of it.' *Ibid.*, 2372–7.

———. 'The Production of Light, by the Effluvia of one Glass falling on another in Motion.' *Ibid.*, 2412–13.

———. 'An Experiment touching Motion given Bodies included in a Glass, by the Approach of a Finger near its outside: with other Experiments on the Effluvia of Glass.' *PT*, 26 (1708–9), 82–6.

———. *Physico-Mechanical Experiments on Various Subjects*. London, 1709; 1719².

———. 'An Account of Experiments concerning the Proportion of the Power of the Loadstone at different Distances.' *PT*, 27 (1710–2), 506–11.

Hausen, Christian August. *Novi profectus in historia electricitatis*. Leipzig, 1743.

Haüy, René Just. *Exposition raisonnée de la théorie de l'électricité et du magnétisme d'aprés les principes de M. Aepinus*. Paris, 1787.

———. *Traité élémentaire de physique*. 2 vols. Paris, 1803.

Heathcote, N. H. de V. 'Guericke's Sulphur Globe.' *Ann. Sci.*, 6 (1948–50), 293–305.

———. 'The Early Meaning of Electricity: Some Pseudodoxia Epidemica.' *Ann. Sci.*, 23 (1967), 261–75.

Heilbron, J. L. 'G. M. Bose: The Prime Mover in the Invention of the Leyden Jar?' *Isis*, 57 (1966), 264–7.

———. 'A propos de l'invention de la bouteille de Leyde.' *RHS*, 19 (1966), 133–42.

———. 'Honoré Fabri and the Accademia del Cimento.' CIHS, 1968. *Actes*, III:B (1971), 45–9.

———. *H. G. J. Moseley. The Life and Letters of an English Physicist, 1887–1915*. Berkeley, 1974.

———. 'Franklin, Haller, and Franklinist History.' *Isis*, 68 (1977), 539–49.

———. 'Introductory Essay.' Wayne Shumaker and J. L. Heilbron. *John Dee on Astronomy*. Berkeley, 1978. Pp. 1–99.

———. 'Experimental Natural Philosophy.' G. S. Rousseau and Roy Porter, eds. *The Ferment of Knowledge. Studies in the Historiography of 18th-Century Science*. Cambridge, 1980. Pp. 357–87.

Heimann, P. M., and J. E. McGuire, 'Newtonian Forces and Lockean Powers: Concepts of Matter in 18th Century Thought.' *Historical Studies in the Physical Sciences*, 3 (1971), 233–306.

Heinrichs, Norbert. 'Scientia magica.' Diemer, ed. *Der Wissenschaftsbegriff*, 30–46.

Helbig, Herbert. *Universität Leipzig*. Frankfurt a.M., 1961.

Henderson, Ebenezer. *Life of James Ferguson*. Edinburgh, 1867.

[Henley, William.] 'Experiments and Observations on a New Apparatus called, A Machine for exhibiting perpetual Electricity.' *PT*, 66 (1776), 513–22.

Henley, W., et al. 'Report of the Committee appointed by the Royal Society, for examining the Effect of Lightning, May 15, 1777, on the Parapet-Wall of the House of the Board of Ordnance, at Purfleet.' *PT*, 68 (1778), 236–8.

Herbert, Josephus Nobilis de. *Theoriae phaenomenorum electricorum quae seu electricitatis ex redundante corpore in deficiens trajectu, seu sola atmosphaerae electricae actione gignuntur*. Vienna, 1772; 1778².

Hermann, Armin. 'Die Anfänge der "akademischen" Physik in München.' *Physikalische Blätter*, 22 (1966), 388–96.

Hermelink, Heinrich, and S. A. Kähler. *Die Philipps-Universität zu Marburg, 1527–1927*. Marburg, 1927.

Herrmann, Dieter B. "George Christoph Lichtenberg als Herausgeber von Exlebens Werk *Anfangsgründe der Naturlehre*.' *NTM*, 6:1 (1969), 68–81; 6:2 (1969), 1–12. (*Mitteilungen der Archenhold-Sternwarte Berlin-Treptow*, no. 102.)

Hesse, Mary B. *Forces and Fields*. London, 1961.

Hesse, Werner. *Beiträge zur Geschichte der früheren Universität in Duisburg*. Duisburg, 1875.

Hildebrand, Bengt. *Kungl. Svenska Vetenskapsakademien. Förhistoria, grundläggning och första organisation*. 2 vols. Stockholm, 1939.

Hildebrandsson, Hugo Hildebrand. *Samuel Klingenstiernas levnad och verk*. I. *Levnadsteckning*. Uppsala, 1919.

Hilgers, Joseph. *Der Index der verbotenen Bücher*. Freiburg, 1904.

Hill, Christopher. *Intellectual Origins of the English Revolution*. Oxford, 1965.

Hoadley, Benjamin, and Benjamin Wilson. *Observations on a Series of Electrical Experiments*. London, 1756; 1759².

Hofmann, Josef Ehrenfried. *Die Mathematik an altbayerischen Hochschulen*. Munich, 1954. (Akademie der Wissenschaft, Munich. Math.-Wiss. Kl. *Abhandlungen*, 62.)

Hollmann, Samuel Christian. 'Succincta attractionis historia.' Akademie der Wissenschaften, Göttingen. *Commentarii*, 4 (1754), 215–44.

Home, Roderick W. *The Effluvial Theory of Electricity*. Ph.D. thesis. University of Indiana, 1967.

———. 'Francis Hauksbee's Theory of Electricity.' *Archive for History of Exact Sciences*, 4 (1967–8), 203–17.

———. 'The Reception Accorded Aepinus's Theory of Electricity.' CIHS, 1971. *Actes*, VI (1974), 287–94.

———. 'Aepinus and the English Electricians: The Dissemination of a Scientific Theory.' *Isis*, 63 (1972), 190–204.

———. 'Franklin's Electrical Atmospheres.' *BJHS*, 6 (1972), 131–51.

———. 'Electricity in France in the Post-Franklin Era.' CIHS, 1974. *Actes*, II (1975), 269–72.

———. 'Aepinus, the Tourmaline Crystal, and the Theory of Electricity and Magnetism.' *Isis*, 67 (1976), 21–30.

———. 'Introduction.' Home and Connor, *Aepinus's Essay* (1979), 3–224.

———, and P. J. Connor. *Aepinus's Essay on the Theory of Electricity and Magnetism*. Princeton, 1979.

Hooke, Robert. *Micrographia*. London, 1665.

———. *Diary*. Henry W. Robinson and Walter Adams, eds. London, 1935.

Hoppe, Edmund. *Geschichte der Elektrizität*. Leipzig, 1884.

Horvath, Joannes Baptist. *Physica particularis auditorum usibus accommodata*. Venice, 1782.

Hoskin, M. A. '"Mining all within": Clarkes notes to Rohault's *Traité de physique*.' *The Thomist*, 24 (1951), 353–63.

Huber, Paulus. *Der Parnassus Boicus. Ein Beitrag zur Kulturgeschichte Baierns während der ersten Hälfte des 18. Jahrhunderts*. Munich, 1868. (Munich. Ludwigs-Gymnasium. *Program*, [1867–8]. 20 p.)

Hughes, Arthur. 'Science in English Encyclopedias, 1704–1875. I.' *Ann. Sci.*, 7 (1951), 340–70.

'II. Theories of the Elementary Composition of Matter.' *Ann. Sci.*, 8 (1952), 323–67.

Hughes, Peter, and David Williams, ed. *The Varied Pattern: Studies in the 18th Century*. Toronto, 1971. (MacMaster University. Association for 18th Century Studies. *Publications*, I.)

Hume, David. *A Treatise of Human Nature* [1739–40]. L. A. Selby-Bigge, ed. 3 vols. Oxford, 1888.

———. *An Enquiry concerning the Human Understanding* [1748]. L. A. Selby-Bigge, ed. Oxford, 1894.

Hunter, Michael. 'The Social Bias and Changing Fortunes of an Early Scientific Institution: an Analysis of the Membership of the Royal Society, 1660–1685.' *RS. Notes and Records*, 31(1976), 9–114.

Huter, Franz. *Die Fächer Mathematik, Physik und Chemie an der Philosophischen Fakultät zu Innsbruck bis 1945*. Innsbruck, 1971. (Innsbruck. University. *Veröffentlichungen*, 66; *Forschungen zur Innsbrucker Universitätsgeschichte*, 10.)

Hutton, Charles. *Philosophical and Mathematical Dictionary*. 2 vols. London, 1815[2].

Huygens, Christian. *Treatise on Light* [1690]. Sylvanus P. Thompson, tr. Chicago, 1945.

———. *Oeuvres complètes*. 22 vols. The Hague, 1888–1950.

Imhof, Maximus. *Grundriss der öffentlichen Vorlesungen über die Experimental-Naturlehre*. 2 vols. Munich, 1794–5.

Index librorum prohibitorum. Vatican City, 1948.

Ingenhousz, Jan. 'Electrical Experiments, to explain how far the Phaenomena of the Electrophorus may be accounted for by Dr. Franklin's Theory of positive and negative Electricity.' *Ibid.*, 1027–48.

———. 'Improvements in Electricity.' *PT*, 69 (1779), 661–73.

———. 'Exposition de plusieurs loix qui paroissent s'observer constamment dans les divers mouvemens du fluide électrique & auxquelles les physiciens n'avoient fait une suffisante attention.' *JP*, 16 (1780), 117–26.

Irsay, Stephen d'. *Histoire des universités françaises et étrangères*. 2 vols. Paris, 1933–5.

Italy. Ministero della pubblica istruzione. *Monografie delle università e degli istituti superiori*. 2 vols. Rome, 1911–3.

Jacquet de Malzet, Louis Sébastien. *Précis de l'électricité ou Extrait expérimental & théorétique des phénomènes électriques*. Vienna, 1775.

[———.] 'Sur l'électrophore perpétuel de M. Volta.' *JP*, 7 (1776), 501–8.

Jacquot, Jean. 'Sir Hans Sloane and French Men of Science.' RS. *Notes and Records*, 10 (1953), 85–98.

Jansen, B. 'Die scholastische Philosophie des 17. Jahrhunderts.' Görres-Gesellschaft. *Philosophisches Jahrbuch*, 50 (1937), 401–44.

———. 'Die Pflege der Philosophie im Jesuitenorden während des 17./18. Jahrhunderts.' *Ibid.*, 51 (1938), 172–215, 344–66, 435–56.

Joachim, Johannes. *Die Anfänge der königlichen Sozietät der Wissenschaften zu Göttingen*. Berlin, 1936. (Akademie der Wissenschaft, Göttingen. Math.-Phys. Kl. *Abhandlungen*, 19.)

Jöcher, Christian Gottlieb. *Allgemeines Gelehrten-Lexikon*. 4 vols. Leipzig, 1750–1.

———. *Fortsetzung und Ergänzungen*. Johann Christoph Adelung et al., eds. 7 vols. Leipzig, 1784–1897.

Johnson, Francis Rarick. 'Gresham College: Precursor of the Royal Society.' *JHI*, 1 (1940), 413–38.

Jourdain, Charles. *Histoire de l'Université de Paris au XVIIe et XVIIIe siècle*. 2 vols. Paris, 1888.

Jurin, James. 'An Account of some Experiments . . . with an Enquiry into the Cause of the Ascent and Suspension of Water in Capillary Tubes.' *PT*, 30 (1717–9), 739–47.

———. 'Disquisitiones physicae de tubulis capillaribus.' *CAS*, 3 (1728), 281–92.

Kästner, Abraham Gotthelf. 'Über die Verbindung der Mathematik und Naturlehre.' [1768]. *Vermischte Schriften*, II, 358–64.

———. *Vermischte Schriften*. 2 parts. Althenburg, 1783³.

———. *Selbstbiographie und Verzeichnis seiner Schriften. Nebst Heynes Lobrede auf Kästner*. Rudolf Eckart, ed. Hanover, [1909?].

———. *Briefe aus sechs Jahrzehnten, 1745–1800*. Berlin, 1912.

[Kangro, Hans.] 'Zur Geschichte der Physik an der Universität Freiburg i. Br.' Zentgraf, ed. *Geschichte der Naturwissenschaften*, 9–22.

Kargon, Robert Hugh. *Atomism in England from Hariot to Newton*. Oxford, 1966.

Karsten, W. C. G. *Physisch-chemische Abhandlungen durch neuere Schriften von hermetischen Arbeiten und andre neuere Untersuchungen veranlasset*. 2 Hefte. Halle, 1786.

Kauffeldt, Alfons. 'Otto von Guericke on the Problem of Space.' CIHS, 1965. *Actes*, III (1968), 364–8.

———. *Otto von Guericke. Philosophisches über den leeren Raum*. Berlin, 1968.

Keill, John. 'Epistola . . . in qua leges attractionis aliaque physices principia traduntur.' *PT*, 26 (1708), 97–110.

———. *An Introduction to Natural Philosophy; or, Philosophical Lectures read in the University of Oxford, Ann. Dom. 1700*. London, 1720; 1745[4].

Kernkamp, G. W. *De Utrechtsche Universiteit, 1636–1936*. I. *De Utrechtsche Academie 1636–1815*. Utrecht, 1936.

King, William. 'Useful Transactions.' [1709] *The Original Works*. John Nichols, ed. 3 vols. London, 1776. II, 57–178.

Kink, Rudolf. *Geschichte der kaiserlichen Universität zu Wien*. 2 vols. Vienna, 1854.

Kircher, Athanasius. *Magnes sive De arte magnetica libri tres*. Rome, 1641; Cologne, 1643[2].

———. *Mundus subterraneus in XII libros digestus*. 2 vols. Amsterdam, 1665.

Kirchvogel, Paul Adolf. 'Das astronomisch-physikalische Kabinett des hessischen Landesmuseums.' *Physikalische Blätter*, 9 (1953), 259–63.

Klee, Friedrich. *Die Geschichte der Physik an der Universität Altdorf bis zum Jahre 1650*. Erlangen, 1908.

Klingenstierna, Samuel. *Tal om de nyaste rön vid elektriciteten*. Stockholm, 1755.

Knappert, L. 'Les relations entre Voltaire et 'sGravesande.' *Janus*, 13 (1908), 249–57.

Knight, Gowin. *An Attempt to demonstrate that all the Phaenomena in Nature may be explained by Two Simple Active Principles, Attraction and Repulsion*. London, 1748.

Knight, Isabel F. *The Geometric Spirit. The abbé de Condillac and the French Enlightenment*. New Haven and London, 1968.

Körber, Hans-Günther. 'Die Berliner Instrumentenmacher im 18. Jahrhundert.' CIHS, 1971. Actes, VI (1974), 271–6.

Kopelevich, I. K. 'Perepiska Leonarda Eĭlera i Tobiasa Maĭera.' *Istoriko-astronomicheskie issledovaniia*, 5 (1959), 271–444.

Koyré, Alexandre. 'Les Queries de l'optique.' *AIHS*, 13 (1960), 15–29.

———. *Newtonian Studies*. London, 1965.

———. *Metaphysics and Measurement. Essays in Scientific Revolution*. Cambridge, Mass., 1968.

———, and I. B. Cohen. 'Newton's "Electric and Elastic Spirit".' *Isis*, 51 (1960), 337.

Krafft, Georg Wolfgang. 'De calore ac frigore experimenta varia.' *CAS*, 14 (1744–6), 218–39.

Kratzenstein, Christian Gottlieb. *Theoria electricitatis more geometrico explicata*. Halle, 1746.

Kronick, D. A. 'Scientific Journal Publication in the 18th century.' Bibliographical Society of America. *Papers*, 59 (1965), 28–44.

Krüger, Johann Gottlob. *Geschichte der Erde*. Halle, 1746.

Kühn, Karl Gottlob. *Die neuesten Entdeckungen in der physikalischen und medizinischen Elektrizität*. 2 vols. Leipzig, 1796–7.

Kunik, Ernst-Edward. 'Einleitung.' Wolff. *Briefe* (1860), vii–xxxv.

Lacoarret, M., and Mme Ter-Menassian. 'Les universités.' Taton, ed. *Enseignement*, 125–68.

Laer, Pierre Henri van. *Philosophico-Scientific Problems*. Henry J. Koren, tr. Pittsburgh, 1953. (*Duquesne Studies, Philosophical Series*, 3.)

Lagrange, Joseph Louis. *Oeuvres*. J. A. Serret, ed. 14 vols. Paris, 1867–92.

Lallamand, P. J. *Histoire de l'éducation dans l'ancien Oratoire de France*. Paris, 1888.

Lambert, Johann Heinrich. *Photometrie (Photometria, sive De mensura et gradibus luminis, colorum et umbrae, 1760)*. E. Anding, tr. 3 vols. Leipzig, 1892. (*Ostwalds Klassiker der exakten Wissenschaften*, nos. 31–3.)

———. 'Analyse de quelques expériences faites sur l'aiman.' *MAS*/Ber (1766), 22–48.

———. 'Sur la courbure du courant magnétique.' *Ibid*., 49–77.

———. *Deutscher gelehrter Briefwechsel*. Johann Bernoulli, ed. 5 vols. Berlin, 1781–7.

Lamontagne, Roland. 'Lettres de Bouguer à Euler.' *RHS*, 19 (1966), 225–46.

Lamprecht, S. P. 'The Role of Descartes in 17th Century England.' Columbia University. Department of Philosophy. *Studies in the History of Ideas*. 3 vols. New York, 1918–35. III, 178–240.

Lana Terzi, Francesco. *Prodromo; overo Saggio di alcune inventioni nuove premesso all'Arte maestra*. Brescia, 1670. (The 'Proemio' to the *Prodromo* is reprinted in M. L. A. Biagi, ed. *Scienziati del seicento*. Milan, 1969. Pp. 711–29.)

———. *Magisterium naturae et artis*. 3 vols. Brescia, 1684–92.

[———]. 'Excerpta actis Novae academiae brixiensis philo-exoticorum naturae et artis.' *Acta eruditorum* (1686), 556–65.

———. 'Nova methodus construendae pyxidis magneticae.' *Ibid*., 560–2.

Lane, T. 'Description of an Electrometer.' *PT*, 57 (1767), 451–60. (*BFP*, XII, 459–65.)

Lantoine, H. E. *Histoire de l'enseignement secondaire en France au XVIIe siècle*. Paris, 1874[4].

Lasswitz, Kurd. *Geschichte der Atomistik vom Mittelalter bis Newton*. 2 vols. Hamburg and Leipzig, 1890.

Laudan, Laurens, 'Comment [on a paper by G. Buchdahl].' Roger H. Stuewer, ed. *Historical and Philosophical Perspectives of Science*. Minneapolis, 1970. Pp. 230–8. (*Minnesota Studies in the Philosophy of Science*, 5.)

Lavoisier, A. L., and P. S. Laplace. 'Mémoire sur la chaleur.' *MAS* (1780), 355–408.

Lecot, Victor Lucien Sulpice. *L'abbé Nollet de Pimprez*. Noyon, 1856.

Leeuwen, Henry G. van. *The Problem of Certainty in English Thought 1630–1690*. The Hague, 1970[2]. (*International Archives of the History of Ideas*, 3.)

Leibniz, Gottfried Wilhelm, freiherr von. *Die philosophischen Schriften*. C. I. Gerhardt, ed. 7 vols. Berlin, 1875–90.

———. *Briefwechsel zwischen Leibniz und Christian Wolf*. C. I. Gerhardt, ed. Halle, 1860.

Leide, Arvid. *Fysiska institutionen vid Lunds universitet*. Lund, 1968. (Lund. University. *Acta*. Sectio 1: *Theologica, juridica, humaniora*, no. 8.)

Leipzig. University. 'Das physikalische und das theoretisch-physikalische Institut.' *Festschrift zur Feier des 500 jährigen Bestehens*. 4 vols. Leipzig, 1909. IV:2, 24–69.

Le Monnier, Louis-Guillaume. 'Recherches sur la communication de l'électricité.' *MAS* (1746), 447–64.

———. 'Extract of a Memoir concerning the Communication of Electricity.' *PT*, 44:1 (1746–7), 290–5.

Lenoble, Robert. *Mersenne; ou, La naissance du mécanisme*. Paris, 1943.

Léotaud, Vincent. *Magnetologia in qua exponitur nova de magnetis philosophia*. Lyon, 1668.

Le Roy, Jean Baptiste. 'Mémoire sur l'électricité. [I] Où l'on montre . . . qu'il y a deux espèces d'électricités . . . & qu'elles ont chacune des phénomènes particuliers qui les caractérisent parfaitement.' *MAS* (1753), 447–59.
[II] 'Où l'on rapporte les expériences qui confirment l'existence des deux électricités *par condensation & par raréfaction*.' *Ibid.*, 459–68.
[III] 'Supplément au mémoire précédent, où l'on fait voir . . . que tous les feux que l'on observe aux extrémités des corps présentés à ceux qui sont électrisés *par condensation* . . . sont formés *par l'entrée du fluide ou feu électrique dans ces corps, & non par sa sortie*.' *Ibid.*, 468–74.

[———]. 'Electromètre.' *Encyclopédie, ou dictionnaire raisonné*. Geneva, 1778[3]. XII, 52–60.

Levere, T. H. 'Martinus van Marum and the Introduction of Lavoisier's Chemistry in the Netherlands.' Forbes, ed. *Martinus van Marum*. I, 158–286.

———. 'Friendship and Influence. Martinus van Marum, F.R.S.' RS, *Notes and Records*, 25 (1970), 113–20.

Libanori, Antonio. *Ferrara d'oro imbrunito*. 3 parts. Ferrara, 1665–74.

Lichtenberg, Georg Christoph. 'De nova methodo naturam ac motum fluidi electrici investigandi.' Akademie der Wissenschaften, Göttingen. Phys.-Math. Kl. *Novi commentarii*, 8 (1777), 168–80. (German tr.: Lichtenberg. *Über eine neue Methode*, 17–28.)

———. 'Super nova methodo motum ac naturam fluidi electrici investigandi.' Akademie der Wissenschaften, Göttingen. Math. Kl. *Novi commentarii*, 1 (1778), 65–79. (German tr.: Lichtenberg. *Über eine neue Methode*, 31–45.)

———. *Briefe*. Albert Leitzmann and Carl Schüddekopf, eds. 3 vols. Leipzig, 1901–4.

———. *Aphorismen*. Albert Leitzmann, ed. 5 vols. Berlin and Leipzig, 1902–8.

———. *Über eine neue Methode, die Natur und die Bewegung der elektrischen Materie zu erforschen*. Herbert Pupke, ed. Leipzig, 1956. (*Ostwalds Klassiker der exakten Wissenschaften*, no. 246.)

Lilley, S. 'Nicholson's Journal, 1797–1813.' *Ann. Sci.*, 6 (1948), 78–101.

Lindborg, Rolf. *Descartes i Uppsala. Striderna om 'nya filosofien' 1663–1689*. Stockholm, 1965. (*Lychnos-Bibliotek*, 22.)

Lindroth, Sten. *Kungl. Svenska Vetenskapsakademiens historia 1739–1818*. 2 vols in 3. Stockholm, 1967.

Lipski, A. 'The Foundation of the Russian Academy.' *Isis*, 44 (1953), 349–54.

Lissa, Giuseppe. *Cartesianismo e anticartesianismo in Fontenelle*. Naples, 1971. (Università degli studi, Salerno. *Collana di studi e testi*, 6.)

Locke, John. *An Essay Concerning Human Understanding* [1690]. Alexander Campbell Fraser, ed. 2 vols. Oxford, 1894.

Loeb, Leonard Benedict. *Fundamental Processes of Electrical Discharges in Gases*. New York, 1939.

Lohne, J. 'Newton's "Proof" of the Sine Law and his Mathematical Theory of Colours.' *Archive for History of Exact Sciences*, 1 (1961), 389–406.

Lomonosov, Mikhail Vasil'evich. *Ausgewählte Schriften*. 2 vols. Berlin, 1961.

Lorey, Wilhelm. 'Die Physik an der Universität Giessen im 17. und 18. Jahrhundert.' Giessener Hochschulgesellschaft. *Nachrichten*, 14 (1940), 14–39.

———. 'Die Physik an der Universität Giessen im 19. Jahrhundert.' *Ibid.*, 15 (1941), 80–132.

McClellan, James Edward, III. *The International Organization of Science and Learned Societies in the 18th Century*. Ph.D. thesis. Princeton, 1975.

McCormmach, Russell K. *The Electrical Researches of Henry Cavendish*. Ph.D. dissertation. Case Institute of Technology, 1967.

———. 'Henry Cavendish: A Study of Rational Empiricism in 18th-Century Natural Philosophy.' *Isis*, 60 (1969), 293–306.

McDonald, John F. 'Properties and Causes: An Approach to the Problem of Hypotheses in the Scientific Methodology of Sir Isaac Newton.' *Ann. Sci.*, 28 (1972), 217–33.

McGuire, J. E. 'Body and Void and Newton's *De mundi systemate:* Some New Sources.' *Archive for History of Exact Sciences*, 3 (1966), 206–48.

———. 'Transmutation and Immutability: Newton's Doctrine of Physical Qualities.' *Ambix*, 14 (1967), 69–95.

———. 'Force, Active Principles and Newton's Invisible Realm.' *Ambix*, 15 (1968), 154–208.

———. 'The Origin of Newton's Doctrine of Essential Qualities.' *Centaurus*, 12 (1968), 233–60.

———. 'Atoms and the "Analogy of Nature": Newton's Third Rule of Philosophizing.' *Studies in History and Philosophy of Science*, 1 (1970), 3–58.

———. 'Boyle's Conception of Nature.' *JHI*, 33 (1972), 523–42.

McKie, Douglas. 'Mr. Warltire, a good Chymist.' *Endeavor*, 10 (1951), 46–9.

———, and N. H. de V. Heathcote. *The Discovery of Specific and Latent Heats*. London, 1935.

Mackie, J. D. *The University of Glasgow, 1451–1951*. Glasgow, 1954.

Maclaurin, Colin. *An Account of Sir Isaac Newton's Philosophical Discoveries*. London, 1748; 1750².

Madeira Arrais, Duarte. *Novae philosophiae et medicinae de qualitatibus occultis a nemine unquam excultae pars prima philosophis, et medicis, pernecessaria, theologis vero apprime utilis*. 2 vols. Lisbon, 1650.

[Magalotti, Lorenzo, Conte]. *Saggi dell'esperienze naturali fatte nell'Accademia del Cimento*. Florence, 1667; 1841³.

[———]. *Essays of Natural Experiments Made in the Academie del Cimento Under the Protection of His Most Serene Prince Leopold of Tuscany*. R. Waller, tr. London, 1684. (Reprinted, New York, 1964, with introduction by A. R. Hall.)

Magrini, Silvio. "Il *De magnete* del Gilbert e i primordi della magnetologia in Italia alla lotta intorno ai massimi sistemi.' *Archivio di storia della scienza*, 8 (1927), 17–39.

Maheu, Gilles. 'Introduction à la publication des lettres de Bouguer à Euler.' *RHS*, 19 (1966), 206–24.

Mahieu, l'abbé Léon. *François Suarez, sa philosophie et les rapports qu'elle a avec sa théologie*. Paris, 1921.

Maignan, Emanuel. *Perspectiva horaria*. Rome, 1648.
———. *Cursus philosophicus*. 4 vols. Toulouse, 1653; Lyon, 1673 [2].
Maindron, Ernest. 'Les fondations de prix à l'Académie des sciences (1714–1880).' *Revue scientifique*, 18 (1880), 1107–17.
———. *L'Académie des sciences*. Paris, 1888.
Mairan, Jean-Baptiste Dortous de. 'Eloge de Molières.' *HAS* (1742), 195–205.
Malebranche, Nicolas. *Oeuvres complètes*. André Robinet, ed. 20 vols. in 18. Paris, 1958–68.
———. 'Reflexions sur la lumière et les couleurs et la generation du feu.' *MAS* (1699), 22–36. (Reprinted, with additions, in the 'XVI. Eclaircissement' to the *Recherche de la vérité*, in *Oeuvres*, III, 255–69.)
Mallet, Sir Charles Edward. *A History of the University of Oxford*. 3 vols. London, 1924–7.
Mandelbaum, Maurice. *Philosophy, Science and Sense Perception: Historical and Critical Studies*. Baltimore, 1964.
Manegold, K. H., ed. *Wissenschaft, Wirtschaft und Technik*. Munich, 1969.
Marcović, Zeljko. 'Boscovich's *Theoria*.' Whyte, ed. *Roger Joseph Boscovich*, 127–52.
Marek, J. 'Un physicien tchèque du XVIIe siècle: Joannes Marcus Marci de Kronland (1595–1667).' *RHS*, 21 (1968), 109–30.
———. 'Zu der Entwicklung der Physik im postrudolphinischen Prag.' *Bohemia*, 16 (1975), 98–109.
Marsak, Leonard M. 'Bernard de Fontenelle: The Idea of Science in the French Enlightenment.' American Philosophical Society. *Transactions*, 49:7 (1959). 64 p.
Martin, Benjamin. *A Supplement: containing Remarks on a Rhapsody of Adventures of a Modern Knight-Errant in Philosophy*. Bath, 1746.
Martin, Geneviève. 'Documents de l'Académie de Rouen concernant l'enseignement des sciences au XVIIIe siècle.' *RHS*, 11 (1958), 207–26.
Martin, T. H. 'Observations et théories des anciens sur les attractions et les répulsions magnétiques et sur les attractions électriques.' Pontificia accademia delle scienze, Rome. *Atti*, 18 (1865), 17–32, 97–123.
Martini, Angelo. *Manuale di metrologia, ossia misure, pesi e monete*. Turin, 1883.
Maupertuis, Pierre Louis Moreau de. 'Sur les loix de l'attraction.' *MAS* (1732), 343–62.
———. *Oeuvres*. 4 vols. Lyon, 1756.
Mautner, Franz H., and Franklin Miller, Jr. 'Remarks on G. C. Lichtenberg, Humanist-Scientist.' *Isis*, 43 (1952), 223–31.
Mayer, Johann Tobias. *Anfangsgründe der Naturlehre zum Behulf der Vorlesungen über die Experimental-Physik*. Göttingen, 1801; 1804 [2]; 1812 [3].
Mayer, Tobias. *The Unpublished Writings of Tobias Mayer*. III. *The Theory of the Magnet and its Application to Terrestrial Magnetism*. Eric G. Forbes, ed. Göttingen, 1972. (Göttingen. Niedersächsische Staats- und Universitätsbibliothek. *Arbeiten*, 11.)
Mayor, J. E. B., ed. *Cambridge under Queen Anne*. Cambridge, 1911.
Mazières, Jean Simon. *Traité des petits tourbillons de la matière subtile*. Paris, 1727.
Meister, Richard. 'Geschichte des Doktorates der Philosophie an der Universität Wien.' Akademie der Wissenschaften, Vienna. Phil.-Hist. Kl. *Sitzungsberichte*, 232:2 (1958). 158 p.
Mélanges Alexandre Koyré, publiés à l'occasion de son soixante-dixième anniversaire. 2 vols. Paris, 1964.

Mendham, Joseph. *The Literary Policy of the Church of Rome Exhibited*. London, 1830[2].
Mercier, Barthélemy. *Notice raisonnée des ouvrages de Gaspar Schott*. Paris, 1785.
Merland, Constant. 'Mathurin-Jacques Brisson.' *Biographies vendéennes*, II. Nantes, 1883, Pp. 1–47.
Mersenne, Marin. *Correspondance du P. Marin Mersenne*. M. Tannery and C. de Waard, eds. 12 vols. Paris, 1932–72.
Metzger, Hélène. *Newton, Stahl, Boerhaave et la doctrine chimique*. Paris, 1930.
Meyer, Friedrich Albert. 'Petrus van Musschenbroek. Werden und Werk und seine Beziehungen zu Daniel Gabriel Fahrenheit.' *Duisburger Forschungen*, 5 (1961), 1–51.
Michell, John. *A Treatise of Artificial Magnets*. London, 1750.
———. *Traité sur les aimans artificiels*. A. Rivoire, tr. Paris, 1752.
Middleton, William Edgar Knowles. *The History of the Barometer*. Baltimore, 1964.
———. *A History of the Thermometer and its Use in Meteorology*. Baltimore, 1966.
———. *The Experimenters. A Study of the Accademia del Cimento*. Baltimore, 1971.
———. 'Science in Rome, 1675–1700, and the Accademia Fisicomatematica of Giovanni Giustino Ciampini.' *BJHS*, 8 (1975), 138–54.
Millburn, John R. *Benjamin Martin, Author, Instrument-Maker, and 'Country Showman.'* Leyden, 1976.
Molières, Joseph Privat de. *Leçons de physique*. 4 vols. Paris, 1734–9.
Moll, Gerard. 'A Biographical Account of J. H. van Swinden.' *Edinburgh Journal of Science*, 1 (1824), 197–208.
Monchamp, Georges. *Histoire du cartésianisme en Belgique*. Brussels, 1886.
Monconys, Balthasar de. *Voyages*. Lyon, 1665–6; Paris, 1695[2].
Montandon, Cléopâtre. *Le développement de la science à Genève aux XVIIIe et XIXe siècles. Le cas d'une communauté scientifique*. Vevey, 1975.
Montesquieu, Charles Louis de Secondat, Baron de la Brède et de. *Oeuvres complètes*. E. Laboulaye, ed. 7 vols. Paris, 1875–9.
Montzey, Charles de. *Histoire de la Flèche et de ses seigneurs*. 3 vols. Le Mans and Paris, 1877.
Mor, Carlo Guido. *Storia della Università di Modena*. Modena, 1953.
Morgan, Alexander. *Scottish University Studies*. Oxford, 1933.
Morgan, George Cadogan. *Lectures on Electricity*. 2 vols. Norwich, 1794.
Morhof, Daniel Georg. *Polyhistor literarius, philosophicus, et practicus*. 2 vols. Lübeck, 1747[4].
Mornet, Daniel. *Les origines intellectuelles de la révolution française*. Paris, 1947.
Morrell, J. B. 'The University of Edinburgh in the Late 18th Century: Its Scientific Eminence and Academic Structure.' *Isis*, 62 (1971), 158–71.
Mousnerius, Petrus, *pseud*. *See* Fabri, Honoré.
Mouy, Paul. *Le développement de la physique cartésienne, 1646–1712*. Paris, 1934.
Müller, Conrad Heinrich. 'Studien zur Geschichte der Mathematik an der Universität Göttingen.' *Abhandlungen zur Geschichte der mathematischen Wissenschaften*, 18 (1904), 51–143.
Mumford, Alfred Alexander. *The Manchester Grammar School, 1515–1915*. London, 1919.
Muntendam, Alida M. 'Dr. Martinus van Marum (1750–1837).' Forbes, ed. *Martinus van Marum*. I, 1–72.

Murray, David. *Memories of the Old College of Glasgow*. Glasgow, 1927.

Musschenbroek, Pieter van. 'De viribus magneticis.' *PT*, 33 (1724–5), 370–7.

———. *Elementa physicae conscripta in usus academicos*. Leyden, 1734; 1741[2]. (The second and third editions of *Epitome elementorum*.)

———. *Essai de physique . . . avec une Description de nouvelles sortes de machines pneumatiques et un recueil d'expériences par Mr. J. V. M.* Pierre Massuet, tr. 2 vols. Leyden, 1739; 1751[2].

———. *The Elements of Natural Philosophy*. John Colson, tr. 2 vols. London, 1744.

———. *Dissertatio physica experimentalis de magnete*. Vienna, 1754[2].

———. *Cours de physique expérimentale et mathématique*. J. A. Sigaud de la Fond, tr. 3 vols. Paris, 1769. (Tr. of *Introductio* [1762].)

Musson, Albert Eduard, and Eric Robinson. *Science and Technology in the Industrial Revolution*. Toronto, 1969.

Nairne, Edward. 'An Account of some Experiments made with an Air-pump on Mr. Smeaton's Principle; together with some Experiments with a common Air-pump.' *PT*, 67 (1777), 614–48.

———. 'Experiments on Electricity, being an Attempt to shew the Advantage of elevated pointed Conductors.' *PT*, 68 (1778), 823–60.

———. *See* Caullet de Veaumorel, L.

Nauck, E. T. 'Die Vertretung der Naturwissenschaften durch Freiburger Medizinprofessoren.' *Beiträge zur Freiburger Wissenschafts- und Universitätsgeschichte*, no. 4 (1954).

Neave, E. W. J. 'Chemistry in Rozier's Journal.' *Ann. Sci.*, 6 (1950), 416–21; 7 (1951), 101–6, 144–8, 284–99, 393–400; 8 (1952), 28–45.

Nedeljković, Dusan. *La philosophie naturelle et relativiste de R. J. Boscovich*. Paris, 1922.

Needham, J. T. 'Some new Electrical Experiments lately made at Paris.' *PT*, 44 (1746), 247–63.

Newton, Isaac. *Philosophiae naturalis principia mathematica*. Thomas Le Seur and François Jacquier, eds. 3 vols. Geneva, 1739–42.

———. *Mathematical Principles of Natural Philosophy*. A. Motte, tr.; F. Cajori, ed. Berkeley, 1934.

———. *Opticks*. New York, 1952. (Reprint of London, 1730[4].)

———. *Isaac Newton's Papers and Letters on Natural Philosophy*. I. B. Cohen, ed. Cambridge, Mass., 1958.

———. *Correspondence*. H. W. Turnbull et al., eds. 6 vols. Cambridge, 1959–76.

Neyreneuf, F. V. 'Expériences sur le condensateur d'Aepinus.' *JP*, 1 (1872), 62–3.

Nicéron, Jean Pierre, ed. *Mémoires pour servir à l'histoire des hommes illustres dans la république des lettres, avec un catalogue raisonné de leurs ouvrages*. 43 vols. in 44. Paris, 1729–45.

Nicholson, William. *An Introduction to Natural Philosophy*. 2 vols. London, 1782; 1790[3]; 1797[4].

———. 'A Description of an Instrument which, by the turning of a Winch, produces the Two States of Electricity without Friction or Communication with the Earth.' *PT*, 78 (1788), 403–7.

———. 'Experiments and Observations on Electricity.' *PT*, 79 (1789), 265–88.

Nollet, Jean Antoine. *Programme, ou, Idée générale d'un cours de physique expérimentale, avec un catalogue raisonné des instrumens qui servent aux expériences*. Paris, 1738.

———. *Leçons de physique expérimentale*. 6 vols. Paris, 1743–8. (Often reprinted.)

———. 'Discours sur les dispositions et sur les qualités qu'il faut avoir pour faire du progrès dans l'étude de la physique expérimentale.' *Leçons de physique expérimentale*. I, xlv–xciv.

———. 'Conjectures sur l'électricité des corps.' *MAS* (1745), 107–51.

———. 'Observations sur quelques nouveaux phénomènes d'électricité.' *MAS* (1746), 1–23.

———. *Essai sur l'électricité des corps*. Paris, 1746; 1750².

———. 'Eclaircissemens sur plusieurs faits concernant l'électricité.' *MAS* (1747), 102–31.

'Second Mémoire. Des circonstances favorables ou nuisibles à l'électricité.' *Ibid.*, 149–99.

———. 'Comparaison raisonnée des plus célèbres phénomènes de l'électricité.' *MAS* (1753), 429–46.

———. 'Suite du mémoire dans lequel j'ai entrepris d'examiner si l'on est bien fondé à distinguer des électricités *en plus & en moins, résineuse & vitrée*, comme autant d'espèces différentes.' *MAS* (1755), 293–317.

———. 'Nouvelles expériences de l'électricité, faites à l'occasion d'un ouvrage publié depuis peu en Angleterre, par M. Robert Symmer.' *MAS* (1761), 244–58.

———. *Lettres sur l'électricité*. [I] *dans lesquelles on examine les dernières découvertes qui ont été faites sur cette matière, & les conséquences que l'on en peut tirer*. Paris, 1753.

[II] *dans lesquelles on soutient le principe des Effluences et Affluences simultanées contre la doctrine de M. Franklin, & contre les nouvelles prétensions de ses partisans*. Paris, 1760.

[III] *dans lesquelles on trouvera les principaux phénomènes qui ont été découverts depuis 1760, avec des discussions sur les conséquences qu'on en peut tirer*. Paris, 1767.

Nordin-Pettersson, B. S. 'K. Svenska Vetenskapsakademiens äldre prisfrågor och belöningar 1739–1820.' Svenska Vetenskapsakademie. *Årsbok* (1959), 435–516.

Occhialini, Augusto. *Notizie sull'Istituto di fisica sperimentale dello studio pisano*. Pisa, 1914.

Oldenburg, Henry. *Correspondence*. A. R. and M. B. Hall, eds. and trs. 10 vols. Madison, 1965–75.

Olson, Richard. 'The Reception of Boscovich's Ideas in Scotland.' *Isis*, 60 (1969), 91–103.

Ornstein, Martha. *The Rôle of Scientific Societies in the 17th Century*. Chicago, 1928.

Oseen, Carl Wilhelm. *Johan Carl Wilcke, experimentalfysiker*. Uppsala, 1939.

Ostertag, Heinrich. *Der philosophische Gehalt des Wolff-Manteuffelschen Briefwechsels*. Leipzig, 1910. (*Abhandlungen zur Philosophie und ihrer Geschichte*, Heft 13.)

Pacaut, ———. 'Le physicien Jacques Rohault. 1620–1672.' Académie des sciences, des lettres et des arts, Amiens. *Mémoires*, 8 (1881), 1–26.

Pacchi, Arrigo. *Cartesio in Inghilterra da More a Boyle*. Bari, 1973.

Pachtler, G. M. *Ratio studiorum et Institutiones scholasticae societatis Jesu per Germaniam olim vigentes*. 4 vols. Berlin, 1887–94. (*Monumenta germaniae paedegogica*, II, V, IX, XVI.)

Palter, Robert. 'Early Measurements of Magnetic Force.' *Isis*, 63 (1972), 544–58.

Parrot, Georg Friedrich. 'Über das Gesetz der elektrischen Wirkungen in der Entfernung.' *Annalen der Physik*, 60 (1818), 22–32.

Paulian, Aimé Henri. *L'électricité soumise à un nouvel examen, dans différentes lettres addressées à M. l'abbé Nollet*. Paris, 1768.

———. *Dictionnaire de physique*. 3 vols. Paris, 1773[2].

Paulsen, Friedrich. 'Wesen und geschichtliche Entwicklung der deutschen Universitäten.' Wilhelm H. R. A. Lexis, ed. *Die deutschen Universitäten*. 2 vols. Berlin, 1896. I, 1–168.

———. *Geschichte des gelehrten Unterrichts*. 3 vols. Leipzig, 1896–7.

———. 'Professorengehalt und Kollegenhonarat in geschichtlicher Beleuchtung.' *Preussische Jahrbücher*, 87 (1897), 136–44.

Pavia. University. *Contributi alla storia dell'Università di Pavia*. Pavia, 1925.

Pemberton, Henry. *A View of Sir Isaac Newton's Philosophy*. London, 1728.

Pernetti, l'abbé Jacques. *Recherches pour servir à l'histoire de Lyon*. 2 vols. Lyon, 1757.

Petersen, Peter. *Geschichte der aristotelischen Philosophie im protestantischen Deutschland*. Leipzig, 1921.

Petersson, Robert Torsten. *Sir Kenelm Digby*. Cambridge, Mass., 1956.

Phillips, Edward C. 'The Correspondence of Father Christopher Clavius, S. J., preserved in the Archives of the Pont. Gregorian University.' *Archivum historicum Societatis Jesu*, 8 (1939), 193–222.

Picard, Charles E. 'La vie et l'oeuvre de Jean Baptiste Biot.' *Eloges et discours académiques*. Paris, 1931. Pp. 221–87.

Pietro, Pericle di. *Lo studio pubblico di S. Carlo in Modena (1682–1772). Novant' anni di storia della Università di Modena*. Modena, 1970.

Playfair, John. 'Biographical Account of the late John Robison.' RS Edin. *Transactions*, 7 (1815), 495–539.

Poisson, Siméon D. 'Sur la distribution de l'électricité à la surface des corps conducteurs.' *MAS* (1811), 1–92, 163–274.

Polinière, Pierre. *Expériences de physique* [1709]. Paris, 1718[2].

Polvani, Giovanni. *Alessandro Volta*. Pisa, 1942.

Power, Henry. *Experimental Philosophy in Three Books . . . in Avouchment and Illustration of the now famous Atomical Hypothesis*. London, 1664.

Prandtl, Karl von. *Geschichte der Ludwig-Maximilians-Universität in Ingolstadt, Landshut, München*. 2 vols. Munich, 1872.

Prevost, Pierre. *Recherches physico-mécaniques sur la chaleur*. Geneva, 1792.

———. *Notice de la vie et des écrits de Georges-Louis Le Sage de Genève*. Geneva, 1805.

Priestley, Joseph. *The History and Present State of Electricity, with Original Experiments*. 2 vols. London, 1767; 1775[3], reprinted, New York, 1966.

———. 'An Account of a New Electrometer.' *PT*, 62 (1772), 359–64.

Pringle, John, *et al*. 'A Report of the Committee, appointed by the Royal Society, to consider of the most effectual method of securing the Powder Magazines at Purfleet against the Effects of Lightning.' *PT*, 68 (1778), 313–17.

Prinz, Hans. 'Nachdenkliches und Belustigendes über das Hochspannungsfeld.' Schweizerischer Elektrotechnischer Verein. *Bulletin*, 61:1 (1970), 8–18.

———. 'Dalla "commozione elettrica" alle affascinanti prospettive dell'elettricità immagazzinata nel condensatore statico.' *Museoscienza*, 11:5 (1971), 3–27 (Tr. from Schw. Elekt. Ver. *Bull.*, 62:2 [1971].)

———. 'I mirabili artifici sperimentali dell'elettricità scintillante.' *Museoscienza*, 12:5 (1972), 3–26. (Tr. from Schw. Elekt. Ver. *Bull.*, 63:1 [1972].)

Pütter, Johann Stephan. *Versuch einer academischen Gelehrten-Geschichte von der Georg-August Universität zu Göttingen*. 4 vols. Göttingen and Hanover, 1765–1838.

Pujoulx, Jean Baptiste. *Paris à la fin du 18e siècle*. Paris, 1801.

Putnam, George Haven. *The Censorship of the Church of Rome and Its Influence upon the Production and Distribution of Literature*. 2 vols. New York, 1906.

Quinn, Arthur J. *Evaporation and Repulsion: A Study of English Corpuscular Philosophy from Newton to Franklin*. Ph.D. thesis. Princeton, 1970.

Rackstrow, B. *Miscellaneous Observations, together with a Collection of Experiments on Electricity*. London, 1748.

Rait, Robert Sangster. *The Universities of Aberdeen. A History*. Aberdeen, 1895.

Randolph, Herbert. *Life of General Sir Robert Wilson*. 2 vols. London, 1862.

Ranke, Leopold. *The Popes of Rome: Their Ecclesiastical and Political History during the 16th and 17th Centuries*. Sarah Austin, tr. 2 vols. London, 1847[3].

Réaumur, René Antoine Ferchault de. *Lettres inédites de Réaumur*. G. Musset, ed. La Rochelle, 1886.

Régis, Pierre Sylvain. *Système de philosophie*. 7 vols. Lyon, 1691.

Regnault, Noël. *L'origine ancienne de la physique nouvelle*. 3 vols. Paris, 1734.

Reif, Sister Patricia. 'The Textbook Tradition in Natural Philosophy, 1600–1650.' *JHI*, 30 (1969), 17–32.

Reilly, Conor. 'A Catalogue of Jesuitica in the *Philosophical Transactions* of the Royal Society of London (1665–1715).' *Archivum historicum societatis Jesu*, 27 (1958), 339–62.

———. *Francis Line, S. J. An Exiled English Scientist, 1595–1675*. Rome, 1969. (*Bibliotheca instituti historici Societatis Jesu*, XXIX.)

Riccioli, Giovanni Battista. *Almagestum novum*. Bologna, 1651.

Richmann, G. W. 'De quantitate caloris, quae post miscelam fluidorum certo gradu calidorum, oriri debet, cogitationes.' *NCAS*, 1 (1747–8), 152–67.

———. 'Formulae pro gradu excessus caloris supra gradum [0°F] . . . post miscelam duarum massarum acquarum, diverso gradu calidarum, confirmatio per experimenta.' *Ibid.*, 168–73.

———. 'De argento vivo calorem celerius recipiente et celerius perdente quam multa fluida leviora experimenta et cogitationes.' *NCAS*, 3 (1750–1), 309–39.

———. 'Inquisitio in decrementa et incrementa caloris solidorum in aëre.' *NCAS*, 4 (1752–3), 241–70.

———. 'Tentamen rationem calorum respectivorum lentibus et thermometris definiendi.' *Ibid.*, 277–300.

———. 'De indice electricitatis et de eius usu in definiendis artificialis et naturalis electricitatis phaenomenis dissertatio.' *Ibid.*, 301–40.

Riedl, C. C. 'Suarez and the Organization of Learning.' Gerard Smith, ed. *Jesuit Thinkers of the Renaissance*. Milwaukee, 1939. Pp. 1–62.

Rigaud, Stephen Peter, ed. *Correspondence of Scientific Men of the 17th Century*. 2 vols. Oxford, 1846.

Ring, Walter. *Geschichte der Stadt Duisburg*. Ratingen, 1949².

Robinet, André. 'La vocation académicienne de Malebranche.' *RHS*, 12 (1959), 1–18.

———. 'La philosophie malebranchiste des mathématiques.' *RHS*, 14 (1961), 205–54.

———. 'Malebranche dans la pensée de Fontenelle.' *Revue de synthèse*, 82 (1961), 79–86.

———. *Malebranche de l'Académie des sciences: L'oeuvre scientifique, 1674–1715*. Paris, 1970.

Robinson, Bryan. *A Dissertation on the Aether of Sir Isaac Newton*. Dublin, 1743.

———. *Sir Isaac Newton's Account of the Aether*. Dublin, 1745.

Robinson, Eric. 'The Profession of Civil Engineer in the 18th Century: A Portrait of Thomas Yeoman, F. R. S., 1704?–1781.' *Ann. Sci.*, 18 (1962), 195–215.

———. 'Benjamin Donn (1729–1798), Teacher of Mathematics and Navigation.' *Ann. Sci.*, 19 (1963), 27–36.

———, and Douglas McKie, eds. *Partners in Science. Letters of James Watt and Joseph Black*. London, 1970.

Robison, John. *A System of Mechanical Philosophy*. David Brewster, ed. 4 vols. Edinburgh, 1822.

Roche, Daniel. 'Milieux académiques provinciaux et société des lumières. Trois académies provinciales au 18ᵉ siècle: Bordeaux, Dijon, Châlons-sur-Marne.' *Livre et société dans la France du XVIIIᵉ siècle*. Paris, 1965. Pp. 93–184. (Paris. University. *Civilisations et société*, 1.)

Rochemonteix, Camille de. *Un collège de Jésuites aux XVIIᵉ et XVIIIᵉ siècles: le Collège Henri IV de la Flèche*. 4 vols. Paris, 1889.

Rochot, Bernard. *La correspondance scientifique du père Mersenne*. Paris, 1966. (Paris. Palais de la Découverte. *Conférences*, series D, no. 110.)

Rodis-Lewis, Geneviève. *Nicolas Malebranche*. Paris, 1963.

Roger, Jacques. 'Diderot et Buffon en 1749.' *Diderot Studies*, 4 (1963), 221–36.

Rohault, Jacques. *Traité de physique*. 2 vols. Paris, 1671; 1692⁶.

Roller, D., and Duane Henry Du Bose Roller. 'The Development of the Concept of Electric Charge.' J. B. Conant, ed. *Harvard Case Histories in Experimental Science*. 2 vols. Cambridge, Mass., 1957. II, 543–639.

Rooseboom, Maria. *Bijdrage tot de geschiedenis der instrumentmakerkunst in de norrdelijke Nederlanden tot omstreeks 1840*. Leyden, 1950. (Leyden. Rijksmuseum voor de Geschiedenis der Natuurwetenschappen. *Mededeeling*, no. 74.)

Rosenberger, Ferdinand. 'Die erste Entwicklung der Elektrisiermaschine.' *Abhandlungen zur Geschichte der Mathematik*, no. 8 (1890), 69–88.

Rosenfeld, Léon. 'Newton and the Law of Gravitation.' *Archive for History of Exact Sciences*, 2 (1965), 365–86.

Rosenfeld, Leonora Cohen. 'Peripatetic Adversaries of Cartesianism in 17th Century France.' *Review of Religion*, 22 (1957), 14–40.

Rosetti, Francesco, and G. Cantoni. *Bibliografia italiana di elettricità e magnetismo*. Padua, 1881.

Rossi, Paolo. 'Hermeticism, Rationality and the Scientific Revolution.' M. L. Righini-Bonelli and William R. Shea, eds. *Reason, Experiment and Mysticism in the Scientific Revolution*. New York, 1975. Pp. 247–73.

Rota Ghibaudi, Silvia. *Ricerche su Ludovico Settala. Biografia, bibliografia, iconografia e documenti*. Florence, 1959.

Rouse Ball, W. W. *A History of the Study of Mathematics at Cambridge*. Cambridge, 1889.

Rowbottom, Margaret E. 'The Teaching of Experimental Philosophy in England, 1700–1730.' CIHS, 1965. *Actes*, IV (1968), 46–53.

———. 'John Theophilus Desaguliers (1683–1744).' Huguenot Society of London. *Proceedings*, 21 (1968), 196–218.

Royal Society of London. *The Record*. London, 1940[4].

Ruestow, Edward G. *Physics at 17th and 18th Century Leiden*. The Hague, 1973. (*Archives internationales d'histoire des idées*, series minor, no. 11.)

Sabra, A.I. *Theories of Light: From Descartes to Newton*. London, 1967.

Sadoun-Goupil, Michelle. 'La correspondance de Claude-Louis Berthollet et Martinus van Marum (1786–1805).' *RHS*, 25 (1972), 221–52.

Sanden, Heinrich von. *Dissertatio physico-experimentalis de succino electricorum principe*. Königsberg, 1714.

Sander, Franz. *Die Auffassungen des Raumes bei Emmanuel Maignan und Johannes Baptiste Morin*. Paderborn, 1934. (*Geschichtliche Forschungen zur Philosophie der Neuzeit*, 3.)

Sarton, George. 'Lagrange's Personality (1736–1813).' American Philosophical Society. *Proceedings*, 88 (1944), 456–96.

Saurin, Joseph. 'Examen d'une difficulté considérable proposée par M. Hughens contre le système cartésien sur la cause de la pésanteur.' *MAS* (1709), 131–48.

Saussure, Horace Bénédict de. *Voyages dans les alpes*. 4 vols. Geneva and Neuchâtel, 1779–96.

———. *Essais sur l'hygrométrie*. Neuchâtel, 1783.

Savérien, Alexandre. *Histoire des philosophes modernes avec leur portrait ou allégorie*. VI. *Histoire des physiciens*. Paris, 1768. Notices of Jacques Rohault, pp. 1–62; Pierre Polinière, pp. 165–216; Joseph Privat de Molières, pp. 217–48; Jean Théophile Desaguliers, pp. 249–88; Petrus van Musschenbroek, pp. 345–98.

Saxtorph, Friedrich. *Electricitets-Laere grundet paa Erfaring og Forsøg*. 2 vols. Copenhagen, 1802–3.

Scaduto, Mario. 'Il matematico Francesco Maurolico e i gesuiti.' *Archivum historicum Societatis Jesu*, 18 (1949), 126–41.

Schaff, Josef. *Geschichte der Physik an der Universität Ingolstadt*. Erlangen, 1912.

Schilling, Johann Jacob. 'Observationes et experimenta de vi electrica vitri aliorumque corporum.' *Miscellanea berolinensia*, 4 (1734), 334–43.

Schimank, Hans. 'Traits of Ancient Natural Philosophy in Otto von Guericke's World Outlook.' *Organon*, 4 (1967), 27–37.

———. 'Die Wandlung des Begriffs "Physik" während der ersten Hälfte des 18. Jahrhunderts.' Manegold, ed. *Wissenschaft*, 454–68.

———. 'Zur Geschichte der Physik an der Universität Göttingen vor Wilhelm Weber (1734–1830).' *Rete*, 2 (1974), 207–52.

Schimberg, André. *L'éducation morale dans les collèges de la Compagnie de Jésus en France sous l'ancien régime*. Paris, 1913.

Schmid, Christoph von. *Erinnerungen aus meinem Leben*. Hubert Schiel, ed. Freiburg, 1953.

Schmidt-Schönbeck, Charlotte. *300 Jahre Physik und Astronomie an der Kieler Universität*. Kiel, 1965.

Schofield, Robert E. 'Histories of Scientific Societies: Needs and Opportunities for Research.' *History of Science*, 2 (1963), 70–83.

———. *Mechanism and Materialism. British Natural Philosophy in an Age of Reason*. Princeton, 1970.

———. 'Benjamin Franklin: Natural Philosopher.' RS. *Proceedings*, 235:A (1956), 433–44.

Schott, Gaspar. *Thaumaturgus physicus sive Magiae universalis naturae et artis pars IV*. Würzburg, 1659.

———. *Pantometrum kircherianum*. Würzburg, 1660.

———. *Physica curiosa sive Mirabilia naturae et artis*. 2 vols. Würzburg, 1662.

———. *Magia optica, das ist, Geheime doch naturmässige Besicht- und Augen-Lehre*. Bamberg, 1671.

Schrader, Wilhelm. *Geschichte der Universität Halle*. 2 vols. Berlin, 1894.

Schreiber, Heinrich. *Geschichte der Stadt und Universität Freiburg im Breisgau*. 7 vols. in 2 Bände. Freiburg, 1857–60. (Band II: *Geschichte der Albert-Ludwigs-Universität*.)

Schur, Friedrich H. *Johann Heinrich Lambert als Geometer*. Karlsruhe, 1905.

Scolopio, Everado. *Storia dell'Università di Pisa dal 1737 al 1859*. Pisa, 1877.

Scott, J. F. *The Scientific Work of René Descartes (1596–1650)*. London, [1952].

Scriba, Christoph J. 'The Autobiography of John Wallis, F.R.S.' RS. *Notes and Records*, 25 (1970), 17–46.

Sédillot, M. L. A. 'Les professeurs de mathématiques et de physique générale au Collège de France.' *Bulletino di bibliografia e di storia delle scienze matematiche e fisiche*, 2 (1869), 343–68, 387–448, 461–510; 3 (1870), 107–70.

Segner, Johann Andreas von. *De mutationibus aeris a luna pendentibus*. Jena, 1733.

Selle, Götz von. *Die Georg-August Universität zu Göttingen, 1737–1937*. Göttingen, 1937.

Senebier, Jean. *Mémoire historique sur la vie et les écrits de Horace Bénédict Desaussure*. Geneva, an IX.

Senguerdius, Wolferdus. *Philosophia naturalis*. Leyden, 1680; 1685[2].

Shapin, Steven. 'Property, Patronage, and the Politics of Science: The Founding of the Royal Society of Edinburgh.' *BJHS*, 7 (1974), 1–41.

Shorr, Philip. *Science and Superstition in the 18th Century. A Study of the Treatment of Science in Two Encyclopedias of 1725–1750*. New York, 1932. (Columbia University. *Studies in History, Economics and Public Law*, no. 364.)

Shuckburgh-Evelyn, Sir George Augustus William, Bart. 'Observations made in Savoy, in order to ascertain the height of Mountains by means of the Barometer.' *PT*, 67 (1777), 513–97.

Sigaud de la Fond, Joseph Aignan. *Précis historique et expérimental des phénomènes électriques, depuis l'origine de cette découverte jusqu'à ce jour*. Paris, 1781.

Silliman, Robert H. 'Fresnel and the Emergence of Physics as a Discipline.' *Historical Studies in the Physical Sciences*, 4 (1974), 137–62.

Simeoni, Luigi. *Storia della Università di Bologna*. II. *L'età moderna (1500–1888)*. Bologna, 1940.

Simon, Paul Ludwig. 'Auszug aus einem Schreiben . . . an den Professor Gilbert.' *Annalen der Physik*, 27 (1807), 325–7.

———. 'Über die Gesetze, welche dem elektrischen Abstossen zum Grunde liegen.' *Ibid.*, 28 (1808), 277–98.

Skempton, A. W., and Joyce Brown. 'John and Edward Troughton, Mathematical Instrument Makers.' RS. *Notes and Records*, 27 (1973), 233–62.

Smeaton, John. 'Some Improvements . . . in the Air Pump.' *PT*, 47 (1751–2), 415–28.

Smend, Rudolph. 'Die Göttinger Gesellschaft der Wissenschaften.' Akademie der Wissenschaften, Göttingen. *Festschrift des zweihundertjährigen Bestehens*. Berlin, Göttingen and Heidelberg, 1951. Pp. x–xix.

Snorrason, Egill. *Kratzenstein, en flittig Dansk professor fra det 18. århunderde*. Copenhagen, 1967.

———. *C. G. Kratzenstein, Professor Physices Experimentalis Petropol. et Havn., and his Studies on Electricity during the 18th Century*. Odense, 1974. (*Acta historica scientiarum naturalium et medicinalium*, XXIX.)

Socin, Abel. 'Tentamina electrica in diversis morborum generibus, quibus accedunt levis electrometri bernoulliani adumbratio, et quorundam experimentorum instituendorum ratio.' *Acta helvetica*, 4 (1760), 214–30.

Södergren, Olaus. *De recentioribus quibusdam in electricitate detectis*. Uppsala, [1746].

Sommervogel, Carlos, ed. *Bibliothèque de la Compagnie de Jésus*. 11 vols. Paris, 1890–1932.

Sortais, Gaston. *Le cartésianisme chez les Jésuites français au XVIIe et au XVIIIe siècle*. Paris, 1929. (*Archives de philosophie*, IV:3.)

Spano, Nicola. *L'Università di Roma*. Rome, 1935.

Specht, Thomas. *Geschichte der ehemaligen Universität Dillingen (1549–1804)*. Freiburg i. Br., 1902.

Speziali, Pierre. 'Une correspondance inédite entre Clairaut et Cramer.' *RHS*, 8 (1955), 193–237.

Spiess, Otto. *Basel anno 1760. Nach den Tagebüchern der ungarischen Grafen Joseph und Samuel Teleki*. Basel, 1936.

Sprat, Thomas. *The History of the Royal Society of London* [1767]. J. I. Cope and H. W. Jones, eds. St. Louis and London, 1958.

[Squario, Eusebio]. *Dell'elettricismo: o sia delle forze elettriche dei corpi*. Venice, 1746.

Stanhope, Charles Stanhope, 3rd Earl. *Principles of Electricity*. London, 1779.

Stearns, Raymond Phineas. *Science in the British Colonies of America*. Urbana, Ill., 1970.

Steck, Max. *Bibliographia lambertiana*. Hildesheim, 1970^{2}.

Steffens, Henry John. *The Development of Newtonian Optics in England*. New York, 1977.

Steiglehner, Cölestin. 'Beantwortung der Preisfrage über die Analogie der Elektricität und des Magnetismus.' Akademie der Wissenschaften, Munich. *Neue philosophische Abhandlungen*, 2 (1780), 229–350.

Steinmetz, Max, et al., eds. *Geschichte der Universität Jena 1548/58–1958*. 2 vols. Jena, 1958.

Stieda, Wilhelm. 'Die Anfänge der kaiserlichen Akademie der Wissenschaften in St. Petersburg.' *Jahrbücher für Kultur und Geschichte der Slaven*, 2 (1926), 133–68.

———. 'Franz Karl Achard und die Frühzeit der deutschen Zuckerindustrie.' Akademie der Wissenschaften, Leipzig. Phil.-Hist. Kl. *Abhandlungen*, 39:3 (1928).

———. 'Die Übersiedlung Leonhard Eulers von Berlin nach St. Petersburg.' Akademie der Wissenschaften, Leipzig. Phil.-Hist. Kl. *Abhandlungen*, 83:3 (1931), 62 p.

———. *Erfurter Universitätsreformpläne im 18. Jahrhundert*. Erfurt, 1934. (Akademie Gemeinnütziger Wissenschaften, Erfurt. *Sonderschriften*, Heft 5.)

Stone, Lawrence, ed. *The University in Society*. 2 vols. Princeton, 1974.

———. 'The Size and Composition of the Oxford Student Body, 1580–1909.' Stone, ed. *University* (1974), I, 3–110.

Strong, Edward W. *Procedures and Metaphysics. A Study in the Philosophy of Mathematical-Physical Science in the 16th and 17th Centuries*. Berkeley, 1936.

———. 'Newtonian Explications of Natural Philosophy.' *JHI*, 18 (1957), 49–83.

Sturm, Johann Christoph. *Physica electiva sive hypothetica*. 2 vols. Nuremberg, 1697–1722.

———. *Kurzer Begriff der Physic*. Hamburg, 1713.

Swinden, Jan Hendrik van. *Dissertatio philosophica inauguralis de attractione*. Leyden, 1766.

———. *Oratio inauguralis de causis errorum in rebus philosophicis*. Freneker, 1767.

———. 'Dissertatio de analogia electricitatis et magnetismi.' Akademie der Wissenschaften, Munich. *Neue philosophische Abhandlungen*, 2 (1780), 1–226.

Symmer, Robert. 'New Experiments and Observations Concerning Electricity. I. Of the Electricity of the Human Body, and the Animal Substances, Silk and Wool.' *PT*, 51 (1759), 340–7.
'II. Of the Electricity of Black and White Silk.' *Ibid.*, 348–58.
'III. Of Electrical Cohesion.' *Ibid.*, 359–70.
'IV. Of Two Distinct Powers in Electricity.' *Ibid.*, 371–89.

Tabarroni, G. 'La torre dell'Università di Bologna e l'elettricità atmosferica.' *Coelum*, 34 (1966), 102–6, 127–33.

Tandberg, J. G. 'Die Triewaldsche Sammlung am Physikal. Institut der Universität zu Lund und die Original-luftpumpe Guerickes.' Lund. University. *Årsskrift*, avd. 2, 16:9 (1920). 31 p.

Targe, Maxime. *Professeurs et régents des collèges dans l'ancienne Université de Paris*. Paris, 1902.

Targioni-Tozzetti, Giovanni. *Notizie degli aggrandimenti delle scienze fisiche accaduti in Toscana nel corso di anni LX del secolo XVII*. 4 vols. in 3. Florence, 1780.

Taton, René, ed. *Enseignement et diffusion des sciences en France au XVIIIe siècle*. Paris, 1964.

Taylor, Brook. 'An Account of an Experiment made . . . in order to discover the Law of Magnetical Attraction.' *PT*, 29 (1714–6), 294–5.

———. 'An Account of an Experiment, made to ascertain the Proportion of the Expansion of the Liquor in the Thermometer, with Regard to the Degrees of Heat.' *PT*, 32 (1723), 291.

———. 'An Account of some Experiments relating to Magnetism.' *PT*, 31 (1720–1), 204–8.

Taylor, Eva Germaine Rimington. *The Mathematical Practitioners of Tudor and Stuart England*. Cambridge, 1954.

———. *The Mathematical Practitioners of Hanoverian England, 1714–1840*. London, 1966.

Taylor, F. S. 'Science Teaching at the End of the 18th Century.' Ferguson, ed. *Natural Philosophy*, 144–64.

Tega, Walter. 'Le *Institutiones in physicam experimentalem* di Giambattista Beccaria.' *Rivista critica di storia della filosofia*, 24 (1969), 179–211.

Tenca, Luigi. 'Guido Grandi, matematico cremonese (1671–1742).' Istituto lombardo di scienze e lettere, Milan. Classe di scienze matematiche e naturali. *Rendiconti*, 83 (1950), 493–510.

———. 'Relationi fra Gerolamo Saccheri e il suo allievo Guido Grandi.' Pavia. Collegio Ghislieri. *Studi matematici-fisici*, 1 (1952), 19–46.

Thackray, Arnold. *Atoms and Powers*. Cambridge, 1970.

Thomas Aquinas, Saint. *Commentary on Aristotle's Physics*. R. J. Blackwell, *et al.*, trs. London, 1963.

Thompson, Silvanus P. *Notes on the De Magnete of Dr. William Gilbert*. London, 1901.

Thorndike, Lynn. *A History of Magic and Experimental Science*. 8 vols. New York, 1923–58.

———. '*L'Encyclopédie* and the History of Science.' *Isis*, 6 (1924), 361–86.

Todhunter, Isaac. *A History of the Mathematical Theories of Attraction and the Figure of the Earth from the Time of Newton to that of Laplace*. 2 vols. London, 1873.

Tolnai, Gabriel, ed. *La cour de Louis XV; Journal de voyage du Comte Joseph Teleki*. Paris, 1943.

Tonelli, Giorgio. 'La nécessité des lois de la nature au XVIIIe siècle et chez Kant en 1762.' *RHS*, 12 (1959), 225–41.

Torlais, Jean. *Réaumur et sa société*. Bordeaux, 1933.

———. *Un esprit encyclopédique en dehors de "l'Encyclopédie". Réaumur, d'après les documents inédits*. Paris, 1935; 1961².

———. *Un Rochelais grand-maitre de la franc-maçonnerie et physicien au XVIIIe siècle, le révérend J. T. Désaguliers*. La Rochelle, 1937

———. *Un physicien au siècle des lumières, l'abbé Nollet*. Paris, 1954.

———. 'Un prestidigitateur célèbre, chef de service d'électrothérapie au XVIIIe siècle. Ledru dit Comus (1731–1807).' *Histoire de la médicine* (Feb. 1955). 13–25.

———. 'Une grande controverse scientifique au XVIIIe siècle. L'abbé Nollet et Benjamin Franklin.' *RHS*, 9 (1956), 339–49.

———. 'Réaumur philosophe.' *RHS*, 11 (1958), 13–33. (Centre international de synthèse. *La vie*, 144–65.)

———. 'Une rivalité célèbre. Réaumur et Buffon.' *Presse médicale*, 66:2 (1958), 1057–8.

———. 'L'Académie de la Rochelle et la diffusion des sciences au XVIIIe siècle.' *RHS*, 12 (1959), 111–25.

———. 'La physique expérimentale.' *Ibid.*, 619–46.

Tressan, H. A. G. de La Vergne, Marquis de. *Souvenirs du Comte de Tressan, Louis-Elisabeth de la Vergne*. Versailles, 1897.

Turner, Gerard L'E. 'The Auction Sales of the Earl of Bute's Instruments, 1793.' *Ann. Sci.*, 23 (1967), 213–43.

———. 'Van Marum's Scientific Instruments in Teyler's Museum. Descriptive Catalogue.' Forbes, ed. *Martinus van Marum*, IV (1973), 129–401.

———. 'Henry Baker, F.R.S., Founder of the Bakerian Lecture.' RS. *Notes and Records*, 29 (1974), 53–79.

Turner, R. Stephen. 'University Reform and Professorial Scholarship in Germany, 1760–1806.' Stone, ed. *University* (1974), II, 495–531.

Tweedie, Charles. *James Stirling. A Sketch of his Life and Works, along with his Scientific Correspondence*. Oxford, 1922.

Urbanitzky, Alfred von. *Elektrizität und Magnetismus in Alterthume*. Vienna, 1887.

Vallauri, Tommaso. *Storia delle Università degli Studi del Piemonte*. 3 vols. Turin, 1845–6.

Van Marum, Martinus. 'Description d'une très grande machine électrique placée dans le Muséum de Teyler, à Harlem, et des expériments faits par le moyen de cette machine.' Teyler's Tweede Genootschap. *Verhandelingen*, 3 (1785). (Excerpted in *JP*, 27 [1785], 145–55.)

Varićak, V. 'Drugi ulomak Boškovićeve Korrespondencije (sa 2 slike van teksta).' Jugoslavenska akademija znanosti i umjetnosti. *Rad*, 193 (1912), 163–338.

Vassalli-Eandi, A. M. 'Notizia storica di Giambattista Beccaria.' *Lo spettatore*, 5, parte italiana (Milan, 1816), 101–5, 117–22.

———. 'Memorie istoriche intorno alla vita ed agli studi di Gianfrancesco Cigna.' Accademia delle scienze, Turin. *Memorie*, 26 (1821), xiii–xxxvi.

Veratti, Giovan Giuseppe. 'De electricitate caelesti.' Accademia delle scienze dell'Istituto di Bologna. *Commentarii et opuscula*, 3 (1755), 200–4.

Vidari, Giovanni. *L'educazione in Italia dall'umanesimo al risorgimento*. Rome, 1930.

Viemeister, Peter E. *The Lightning Book*. New York, 1961.

Villoslada. *See* García Villoslada.

Visconti, Alessandro. *La storia dell'Università di Ferrara*. Bologna, 1950.

Vleeschauwer, H. J. de. 'La genèse de la méthode mathématique de Wolff. Contribution à l'histoire des idées au XVIII siècle.' *Revue belge de philologie et d'histoire*, 11 (1932), 651–77.

Volpati, Carlo. *Alessandro Volta nella gloria e nell'intimità*. Milan, 1927.

Volta, Alessandro. 'Sur l'électrophore perpétuel.' *JP*, 8 (1776), 21–4.

———. 'Osservazioni sulla capacità dei conduttori elettrici.' *Opuscoli scelti sulle scienze e sulle arti*, 1 (Milan, 1778), 273–80. (*VO*, III, 201–29; *JP*, 12 [1779], 249–77.)

———. 'Del modo di render sensibilissima la più debole elettricità sia naturale, sia artificiale.' *PT*, 72 (1782), 237–80. (*VO*, III, 271–300; English tr.: *PT*, 72 (1782), Appendix, xii–xxxiii.)

———. 'Sur la capacité des conducteurs coniugués.' *JP*, 22 (1783), 325–50; 23 (1783), 1–16, 81–99. (*VO*, III, 313–77.)

———. 'On the Electricity excited by the mere Contact of Conducting Substances of Different Kinds.' *PT*, 90 (1800), 403–31. (*VO*, I, 563–82.)

———. *Le opere*. 7 vols. Milan, 1918–29. (Cited as *VO*.)

———. *Epistolario*. 5 vols. Bologna, 1949–55. (Cited as *VE*.)

Volta, Zanino. 'Il primo viaggio di Alessandro Volta a Parigi [1781/2].' Istituto lombardo di scienze e lettere, Milan. *Rendiconti*, 15 (1882), 29–37.

Voltaire, François Marie Arouet de. *Correspondance*. 107 vols. Theodore Besterman, ed. Geneva, 1953–65.

———. *Lettres philosophiques* [1734]. *Edition critique*. Gustave Lanson, ed.; rev. by André M. Rousseau. 2 vols. Paris, 1964.

Vregille, P. de. 'Un enfant du Bugey. Le père Honoré Fabri.' *Bulletin de la société Gorini*, 9 (1906), 5–15.

Vries, Joseph de. 'Zur aristotelisch-scholastischen Problematik von Materie und Form.' *Scholastik*, 32 (1957), 161–85.

Vucinich, Alexander S. *Science in Russian Culture. A History to 1860*. Stanford. 1963.

Wahl, Richard. 'Professor Bilfingers Monadologie und prästabilierte Harmonie in ihrem Verhältniss zu Leibniz und Wolf.' *Zeitschrift für Philosophie und philosophische Kritik*, 85 (1884), 66–92, 202–23.

Waitz, Jacob Sigismund von, et al. *Abhandlungen von der Elektricität*. Berlin, 1745.

Walker, W. Cameron. 'The Detection and Estimation of Electric Charges in the 18th Century.' *Ann. Sci.*, 1 (1936), 66–99.

Wall, Samuel. 'Experiments of the Luminous Qualities of *Amber, Diamonds* and *Gum Lac*.' *PT*, 26 (1708–9), 69–76.

Wallerius, Niels. 'Versuche von Beschaffenheit der Dünste und dem Ursachen ihres Aufsteigens.' *AKSA*, 9 (1747), 272–81.

Watson, William. 'Experiments and Observations tending to illustrate the Nature and Properties of Electricity.' *PT*, 43 (1744–5), 481–501.

———. *A Sequel to the Experiments and Observations tending to illustrate the Nature and Properties of Electricity*. London, 1746. (Reprint [or preprint] of *PT*, 44 [1747], 704–49; cf., *FN*, 399n.)

———. 'Some Further Inquiries into the Nature and Properties of Electricity.' *Ibid.*, 93–120.

———. 'An Account of Mr. Benjamin Franklin's Treatise, lately published, intituled, *Experiments and Observations on Electricity*.' *PT*, 47 (1751–2), 202–10.

———. 'An Account of the Phaenomena of *Electricity in vacuo*.' *Ibid.*, 362–76.

Weaver, William D. *Catalogue of the Wheeler Gift of Books, Pamphlets and Periodicals in the Library of the American Institute of Electrical Engineers*. 2 vols. New York, 1909.

Weber, Heinrich. *Geschichte der gelehrten Schulen im Hochstift Bamberg von 1007–1803*. Bamberg, 1879–82. (Historischer Verein, Bamberg. *Bericht über Bestand und Wirken*, 42–4.)

Webster, Charles. 'Henry Power's Experimental Philosophy.' *Ambix*, 14 (1967), 150–78.

———. 'Henry More and Descartes: Some New Sources.' *BJHS*, 4 (1969), 359–77.

Webster, John. *Academiarum examen, or the Examination of Academies*. London, 1654. (Debus. *Science and Education*, 67–192.)

Weld, Charles Richard. *A History of the Royal Society, with Memoirs of Presidents*. 2 vols. London, 1848.

Werner, Karl. 'Die Cartesisch-Malebranche'sche Philosophie in Italien.' Akademie der Wissenschaften, Vienna. Phil.-Hist. Kl. *Sitzungsberichte*, 102 (1883), 75–141, 679–754.

———. *Franz Suarez und die Scholastik der letzten Jahrhunderte*. 2 vols. Regensburg, 1889[2].

Wesley, John. *The Journal of the Rev. John Wesley*. Nehemiah Curnock, ed. 8 vols. London, 1909–16.

Westenrieder, Lorenz von. *Geschichte der Bayerischen Akademie der Wissenschaften*. 2 vols. Munich, 1784–1807.

Westfall, Richard S. 'Unpublished Boyle Papers Relating to Scientific Method.' *Ann. Sci.*, 12 (1956), 63–73, 103–17.

———. *Force in Newton's Physics*. London and New York, 1971.

Wheler, G. 'Some Electrical Experiments, chiefly regarding the Repulsive Force of Electrical Bodies.' *PT*, 41 (1739–40), 98–117.

Whewell, William. *History of the Inductive Sciences*. 2 vols. New York, 1858[3].

Whiston, William. *Memoirs*. London, 1749.

Whiteside, Derek T. 'Newton's Early Thoughts on Planetary Motion: A Fresh Look.' *BJHS*, 2 (1964), 117–37.

Whitmore, P. J. S. *The Order of Minims in 17th-Century France*. The Hague, 1967.

Whittaker, Sir Edmund Taylor. *A History of the Theories of Aether and Electricity*. I. *The Classical Theories*. London, 1951[2].

Whyte, Lancelot Law, ed. *Roger Joseph Boscovich, S.J., F.R.S., 1711–1787*. London, 1961.

Wieleitner, E. 'Die *Elémens de géometrie* des Paters J. G. Pardies.' *Archiv für Geschichte der Naturwissenschaften und Technik*, 1 (1908–9), 436–42.

Wightman, William P. D. 'The Copley Medal and the Work of Some of the Early Recipients.' *Physis*, 3 (1961), 344–55.

Wilcke, Johan Carl. 'Fernere Untersuchung von den entgegengesetzten Elektricitäten bei der Ladung und den dazu gehörenden Theilen.' *AKSA*, 24 (1762), 213–35, 253–74.

———. 'Elektrische Versuche mit Phosphorus.' *AKSA*, 25 (1763), 207–26.

———. 'Neue Versuche vom Gefrieren des Wassers.' *AKSA*, 31 (1769), 87–108.

———. 'Von des Schnees Kälte beim Schmelzen.' *AKSA*, 34 (1772), 93–116.

———. 'Über die specifische Menge des Feuers in festen Körpern, und derselben Abmessung.' *AKSA*, 2 (1781), 48–79.

Wilkins, John, and Seth Ward. *Vindiciae academicarum, containing Some Briefe Animadversions upon Mr. Webster's Book, stiled, The Examination of Academies*. Oxford, 1654. (Debus. *Science and Education*, 193–259.)

Will, Georg Andreas. *Nürnbergisches Gelehrten-Lexikon*. 4 vols. Nuremberg and Altdorf, 1755–8.

Williams, Anna. *Miscellanies in Prose and Verse*. London, 1766.

Wilson, Arthur M. *Diderot. The Testing Years, 1713–1759*. Oxford, 1957.

Wilson, Benjamin. *An Essay towards the Explication of Electricity deduced from the Aether of Sir Isaac Newton*. London, 1746.

———. *A Treatise on Electricity*. London, 1750; 1752[2].

———. 'Experiments on the Tourmalin.' *PT*, 51 (1759), 308–39.

———. 'Dissent from the [Henley] Report.' *PT*, 68 (1778), 239–42.

———. 'New Experiments and Observations on the Nature and Use of Conductors.' *Ibid.*, 245–313.

Wilson, George. *The Life of the Honorable Henry Cavendish*. London, 1851.

Winkler, Johann Heinrich. *Gedanken von den Eigenschaften, Wirkungen und Ursachen der Electricität, nebst einer Beschreibung zwo neuer electrischen Maschinen*. Leipzig, 1744. ('Essai sur la nature, les effets et les causes de l'électricité, avec une description de deux nouvelles machines à l'électricité.' *Recueil de traltés sur l'électricité traduits de l'allemand et de l'anglois*. Paris, 1748. 1e partie. References in the text are to this translation, as the more accessible version.)

———. *Die Eigenschaften der elektrischen Materie und des elektrischen Feuers aus verschiedenen neuen Versuchen erkläret und nebst etlichen neuen Maschinen zum Elektrisiren*. Leipzig, 1745.

———. 'Concerning the Effects of Electricity upon Himself and his Wife.' *PT*, 44 (1746), 211–12.

———. *Die Stärke der electrischen Kraft des Wassers in gläsernen Gefässen, welche durch den Musschenbrökischen Versuch bekannt geworden*. Leipzig, 1746.

Winstanley, Denys Arthur. *Unreformed Cambridge. A Study of Certain Aspects of the University in the 18th Century*. Cambridge, 1935.

Winter, Eduard. *Die Registres der Berliner Akademie der Wissenschaften, 1746–1766. Dokumente für das Wirken Leonhard Eulers in Berlin*. Berlin, 1957.

Wolf, Abraham. *A History of Science, Technology and Philosophy in the 16th and 17th Centuries*. New York, 1950[2].

———. *A History of Science, Technology and Philosophy in the 18th Century*. London, 1952[2].

[Wolff, Christian.] 'Responsio ad imputationes Johannis Freindii in Transactionibus anglicis.' *Acta eruditorum* (1713), 307–14.

———. *Cosmologia generalis, methodo scientifica pertracta*. Frankfurt and Leipzig, 1731; 1737[2].

———. *Ausführliche Nachricht von seinen eigenen Schrifften die er in deutscher Sprache . . . herausgegeben*. Frankfurt, 1733[2].

———. *Allerhand nützliche Versuche, dadurch zu genauer Erkäntniss der Natur und Kunst der Weg gebähnet wird* [1721–3]. 3 vols. Halle, 1745–7[3].

———. 'Lebensbeschreibung.' Wuttke, ed. *Lebensbeschreibung*, 107–201.

———. *Briefe aus den Jahren 1719–1753. Ein Beitrag zur Geschichte der Kaiserlichen Akademie der Wissenschaften zu St. Petersburg*. St. Petersburg, 1860.

Wordsworth, Christopher. *Scholae academicae. Some Account of the Studies at the English Universities in the 18th Century*. Cambridge, 1877.

Wren, Christopher, ed. *Parentalia, or, Memoirs of the Family of the Wrens*. London, 1750.

Wuttke, Heinrich. *Christian Wolffs eigene Lebensbeschreibung*. Leipzig, 1841.

Yamazaki, Elizio. 'L'abbé Nollet et Benjamin Franklin. Une phase finale de la physique cartésienne: la théorie de la conservation de l'électricité et de l'expérience de Leyde.' *Japanese Studies in the History of Science*, 15 (1976), 37–64.

Yates, Frances A. 'The Hermetic Tradition in Renaissance Science.' C. S. Singleton, ed. *Art, Science and History in the Renaissance*. Baltimore, 1968. Pp. 255–74.

Zeleny, John. 'Observations and Experiments on Condensers with Removable Coats.' *American Journal of Physics*, 12 (1944), 329–39.

Zentgraf, Eduard. *Aus der Geschichte der Naturwissenschaften an der Universität Freiburg i. Br*. Freiburg i. Br., 1957 (*Beiträge zur Freiburger Wissenschafts- und Universitätsgeschichte*, Heft 18.)

Ziggelaar, August. *Le physicien Ignace Gaston Pardies, S. J. (1636–1673)*. Odense, 1971. (*Acta historica scientiarum et medicinalium*, XXVI.)

Zilsel, E. 'The Origins of Gilbert's Scientific Method.' *JHI*, 2 (1941), 1–32.

Zubov, Vasiliĭ Pavlovich. 'Kalorimetricheskaia formula Rikǐmana i ee predistoriia.' IIET. *Trudy*, 5 (1955), 69–93.

———. 'La formule calorimétrique et ses origines.' *Mélanges Alexandre Koyré*, I, 654–61.

Zucchi, Nicola. *Nova de machinis philosophia* . . . . Rome, 1649.

Index

Besides the usual references to texts and notes, the Index refers to all early-modern authors mentioned in the notes and to modern authors whenever their opinions are quoted or paraphrased. Otherwise sources are not indexed. A given item may appear several times in the notes, or in both notes and text, on the same page. In these cases only one reference is given. Names of institutions will be found under their locations, e.g., Paris, Académie des Sciences; cross references are provided when the name does not indicate place. Long entries are usually arranged from the general to the particular. The Index supplements the text by giving full names of persons cited and birth and death dates, when available, of the early-modern physicists.

The following abbreviations are used:

AS.	Académie des Sciences, Accademia delle Scienze. No distinction is made between an AS and a general academy, e.g., Académie des Sciences, Arts, et Belles-lettres.
AW.	Akademie der Wissenschaften or its equivalent in Dutch or Swedish.
BF.	Benjamin Franklin.
El.	Electric, Electrical.
Expt(s)	Experiment, Experiments.
Px.	Physique expérimentale, Experimental Physics.
SJ(s).	Jesuit, Jesuits.
SJC.	Jesuit College.
U(s).	University, Universities.